NFPA's Electrical References

NFPA's Electrical References

Charles R. Miller

National Fire Protection Association
Quincy, Massachusetts

Senior Product Manager: Charles Durang
Editorial-Production Services: Omegatype Typography, Inc.
Composition: Omegatype Typography, Inc.
Cover Design: Natasha Bumbeck
Manufacturing Manager: Ellen Glisker
Printer: R. R. Donnelley/Crawfordsville

 Copyright © 2004
National Fire Protection Association, Inc.
One Batterymarch Park
Quincy, Massachusetts 02169-7471

All rights reserved. No part of the material protected by this copyright notice may be reproduced or utilized in any form without acknowledgment of the copyright owner nor may it be used in any form for resale without written permission from the copyright owner.

Notice Concerning Liability: Publication of this work is for the purpose of circulating information and opinion among those concerned for fire and life safety and related subjects. While every effort has been made to achieve a work of high quality, neither the NFPA nor the authors and contributors to this work guarantee the accuracy or completeness of or assume any liability in connection with the information and opinions contained in this work. The NFPA and the authors and contributors shall in no event be liable for any personal injury, property, or other damages of any nature whatsoever, whether special, indirect, consequential, or compensatory, directly or indirectly resulting from the publication, use of or reliance upon this work.

This work is published with the understanding that the NFPA and the authors and contributors to this work are supplying information and opinion but are not attempting to render engineering or other professional services. If such services are required, the assistance of an appropriate professional should be sought.

The following are registered trademarks of the National Fire Protection Association:

National Electrical Code® and *NEC*®

Chapter 22, "First Aid Emergency Procedures," © 2004 by Coyne, Inc. Used by permission.

NFPA No.: ELREF05
ISBN: 0-87765-489-1

Library of Congress Control No.: 2004112923

Printed in the United States of America
04   05   06   07   08   5   4   3   2   1

CONTENTS

CHAPTER 1
NATIONAL ELECTRICAL CODE® ORGANIZATION   1

Tables
  Table 1.1   NEC® Brief Contents   1
  Table 1.2   NEC® Articles Arranged Alphabetically by Title   2
  Table 1.3   NEC® Articles Arranged Numerically   5
  Table 1.4   Common Numbering System for Chapter 3 Raceway and Cable Articles   9

CHAPTER 2
ELECTRICAL ABBREVIATIONS, SYMBOLS, AND FIELD TERMS   11

Tables
  Table 2.1   Electrical Abbreviations   11
  Table 2.2   Cable and Raceway Abbreviations   13
  Table 2.3   Low-Voltage Cable Abbreviations   13
  Table 2.4   International System of Units (SI)   14
  Table 2.5   Electrical Field Terms   14

CHAPTER 3
MATHEMATICAL FORMULAS AND TABLES   17

Fractions   17
Finding the Least Common Multiple   18
Multiplying Fractions   20
Dividing Fractions   21
Equations   22
Quick Reference Tables   24

Table
  Table 3.1   Squares and Square Roots   24

CONTENTS

Table 3.2	Scientific Notation (Powers of 10)	25
Table 3.3	Inch and Millimeter Equivalents of Fractions	26

CHAPTER 4
GEOMETRIC FORMULAS   29

Circle   29

Square and Rectangle   32

Triangle   32

Ellipse   35

Cube   35

Rectangular Prism   36

Pyramid   36

Cone   37

Cylinder   37

Sphere   38

Figures

Figure 4.1	Geometric terms associated with the circle	30
Figure 4.2	Square and rectangle	32
Figure 4.3	Right triangle	32
Figure 4.4	Illustration of Pythagorean triangle	33
Figure 4.5	Oblique triangle	33
Figure 4.6	Oblique triangle calculated as two right triangles	34
Figure 4.7	Equilateral triangle	34
Figure 4.8	Ellipses	35
Figure 4.9	Cube	35
Figure 4.10	Rectangular prism	36
Figure 4.11	Pyramid	36
Figure 4.12	Cone	37
Figure 4.13	Cylinder	37
Figure 4.14	Sphere	38

CHAPTER 5
CONVERSION FACTORS   39

Tables

Table 5.1	Length and Distance	39
Table 5.2	Area	41

Table 5.3	Volume	42
Table 5.4	Weight	43
Table 5.5	Pressure	44
Table 5.6	Force	46
Table 5.7	Power	46
Table 5.8	Energy	48
Table 5.9	Illuminance	49
Table 5.10	Temperature	50

CHAPTER 6
OHM'S LAW AND BASIC CIRCUITS 53

Ohm's Law 53

Circuits 58

Tables

Table 6.1	Components of Ohm's Law	53

Figures

Figure 6.1	Ohm's Law wheel	54
Figure 6.2	PIE and EIR charts	56
Figure 6.3	Using the PIE chart	56
Figure 6.4	Using the EIR chart	57
Figure 6.5	Series circuit	58
Figure 6.6	Parallel circuit (1)	60
Figure 6.7	Parallel circuit (2)	61
Figure 6.8	All resistors of equal value	62
Figure 6.9	Alternative method for calculating the total resistance	62
Figure 6.10	Series-parallel circuit (1)	63
Figure 6.11	Series-parallel circuit redrawn (1)	64
Figure 6.12	Series-parallel circuit (2)	64
Figure 6.13	Series-parallel circuit redrawn (2)	65
Figure 6.14	Series-parallel circuit with known factors (1)	65
Figure 6.15	Series-parallel circuit with known factors (2)	66

CONTENTS

vii

CONTENTS

CHAPTER 7
ELECTRICAL FORMULAS AND TABLES   67

Amperes (I)   67

Volts (E)   72

Ohms (R)   74

Resistors in Circuits   75

Inductance (L)   78

Capacitance (C)   80

Power-Factor Correction   84

Impedance (Z)   86

Reactance (X)   86

Watts (W)   88

Kilowatts   90

Volt-Amperes   91

Kilovolt-Amperes   91

Efficiency   92

Power Factor   93

Horsepower   96

Tables

Table 7.1	Three-Phase Voltage Values 72	
Table 7.2	Four-Band: ±2%, ±5%, and ±10% Tolerance 76	
Table 7.3	Five-Band: ±.1%, ±.25%, ±.5%, and ±1% Tolerance 77	
Table 7.4	Preferred Values for Resistors (E24 5%) 78	
Table 7.5	Color Codes for Molded Mica and Molded Paper Capacitors 82	
Table 7.6	Preferred Values for Capacitors 84	
Table 7.7	Power-Factor Correction 85	

Figures

Figure 7.1	Color code for four-band resistor 76	
Figure 7.2	Color code for five-band resistors 77	
Figure 7.3	Resistor 77	
Figure 7.4	Inductors connected in series 79	
Figure 7.5	Inductors connected in parallel 80	
Figure 7.6	Capacitors connected in series 81	
Figure 7.7	Capacitors connected in parallel 82	
Figure 7.8	Mica capacitors 83	

CHAPTER 8
ELECTRICAL PLAN SYMBOLS   97

Tables
- Table 8.1   Lighting Outlets   97
- Table 8.2   Switch Outlets   98
- Table 8.3   Receptacle Outlets   98
- Table 8.4   Miscellaneous Symbols   99

CHAPTER 9
LOAD CALCULATIONS   101

Dwelling Load   101

Optional One-Family Dwelling Load   108

Nondwelling Load   109

NEC Exhibits
- Exhibit 9.1   *NEC* Table 220.12, General Lighting Loads by Occupancy   102
- Exhibit 9.2   *NEC* Table 220.42, Lighting Load Demand Factors [Dwelling Units]   103
- Exhibit 9.3   *NEC* Table 220.54, Demand Factors for Household Electric Clothes Dryers   105
- Exhibit 9.4   *NEC* Table 220.55, Demand Factors and Loads for Household Electric Ranges, Wall-Mounted Ovens, Counter-Mounted Cooking Units, and Other Household Cooking Appliances over 1-3/4 kW Rating   106
- Exhibit 9.5   *NEC* Table 220.44, Demand Factors for Non-dwelling Receptacle Loads   110
- Exhibit 9.6   *NEC* Table 220.42, Lighting Load Demand Factors   112
- Exhibit 9.7   *NEC* Table 220.56, Demand Factors for Kitchen Equipment—Other Than Dwelling Unit(s)   113

CHAPTER 10
CLEARANCE AND COVER REQUIREMENTS   115

Wiring and Working Space Clearances   115

Working Space Clearances (Over 600 Volts, Nominal)   118

Wiring Space Clearances in Switchboards and Panelboards   120

Overhead Conductor Clearances   121

x

CONTENTS

Services   124

Minimum Cover Requirements   125

NEC Exhibits

Exhibit 10.1   *NEC* Table 110.26(A)(1), Working Spaces   116

Exhibit 10.2   *NEC* Table 110.31, Minimum Distance from Fence to Live Parts   118

Exhibit 10.3   *NEC* Table 110.34(A), Minimum Depth of Clear Working Space at Electrical Equipment   119

Exhibit 10.4   *NEC* Table 110.34(E), Elevation of Unguarded Live Parts Above Working Space   119

Exhibit 10.5   *NEC* Table 408.5, Clearance for Conductors Entering Bus Enclosures   120

Exhibit 10.6   *NEC* Table 408.56, Minimum Spacings Between Bare Metal Parts   120

Exhibit 10.7   *NEC* Table 490.24, Minimum Clearance of Live Parts   121

Exhibit 10.8   *NEC* Table 225.60, Clearances over Roadways, Walkways, Rail, Water, and Open Land   123

Exhibit 10.9   *NEC* Table 225.61, Clearances over Buildings and Other Structures   124

Exhibit 10.10   *NEC* Table 300.5, Minimum Cover Requirements, 0 to 600 Volts, Nominal, Burial in Millimeters (Inches)   126

Exhibit 10.11   *NEC* Table 300.50, Minimum Cover Requirements   127

CHAPTER 11
BOXES AND ENCLOSURES   129

Box Fill Calculations   129

Conduit Bodies   134

Cast Device Boxes   135

Pull and Junction Box Calculations   136

NEMA Enclosures   138

Tables

Table 11.1   Box Fill Calculations   131

Table 11.2   Conduit Bodies   134

Table 11.3   Cast Device Boxes   135

Table 11.4   NEMA Enclosure Types   138

NEC Exhibits
Exhibit 11.1   *NEC* Table 314.16(A), Metal Boxes   130
Exhibit 11.2   *NEC* Table 314.16(B), Volume Allowance Required per Conductor   130

CHAPTER 12
RACEWAYS AND CABLE TRAYS   141

Metric Designators and Trade Sizes   141

Bends   141

Supports for Intermediate Metal Conduit and Rigid Metal Conduit   142

Conductors in Flexible Metal Conduit   143

Support and Expansion Characteristics of Rigid Nonmetallic Conduit   144

Surface Nonmetallic Raceways   146

Cable Trays   148

Conduit and Tubing Fill Tables   151

Annex C Tables—Conduit and Tubing Fill for Conductors and Fixture Wires of the Same Size   165

Figures
Figure 12.1   Conductor fill table for various surface metal raceways   146

NEC Exhibits
Exhibit 12.1   *NEC* Table 300.1(C), Metric Designator and Trade Sizes   141
Exhibit 12.2   *NEC* Table 2, Radius of Conduit and Tubing Bends   142
Exhibit 12.3   *NEC* Table 344.30(B)(2), Supports for Rigid Metal Conduit   143
Exhibit 12.4   *NEC* Table 348.22, Maximum Number of Insulated Conductors in Metric Designator 12 (Trade Size 3/8) Flexible Metal Conduit*   143
Exhibit 12.5   *NEC* Table 352.30(B), Support of Rigid Nonmetallic Conduit (RNC)   144
Exhibit 12.6   *NEC* Table 352.44(A), Expansion Characteristics of PVC Rigid Nonmetallic Conduit Coefficient of

CONTENTS

xi

CONTENTS

	Thermal Expansion = 6.084×10^{-5} mm/mm/°C (3.38×10^{-5} in./in./°F) 145
Exhibit 12.7	*NEC* Table 352.44(B), Expansion Characteristics of Reinforced Thermosetting Resin Conduit (RTRC) Coefficient of Thermal Expansion = 2.7×10^{-5} mm/mm/°C (1.5×10^{-5} in./in./°F) 145
Exhibit 12.8	*NEC* Table 392.7(B), Metal Area Requirements for Cable Trays Used as Equipment Grounding Conductor 148
Exhibit 12.9	*NEC* Table 392.9, Allowable Cable Fill Area for Multiconductor Cables in Ladder, Ventilated Trough, or Solid Bottom Cable Trays for Cables Rated 2000 Volts or Less 149
Exhibit 12.10	*NEC* Table 392.9(E), Allowable Cable Fill Area for Multiconductor Cables in Ventilated Channel Cable Trays for Cables Rated 2000 Volts or Less 149
Exhibit 12.11	*NEC* Table 392.9(F), Allowable Cable Fill Area for Multiconductor Cables in Solid Channel Cable Trays for Cables Rated 2000 Volts or Less 150
Exhibit 12.12	*NEC* Table 392.10(A), Allowable Cable Fill Area for Single-Conductor Cables in Ladder or Ventilated Trough Cable Trays for Cables Rated 2000 Volts or Less 150
Exhibit 12.13	*NEC* Table 1, Percent of Cross Section of Conduit and Tubing for Conductors 151
Exhibit 12.14	*NEC* Table 4 [Abridged], Dimensions and Percent Area of Conduit and Tubing 152
Exhibit 12.15	*NEC* Table 5 [Abridged], Dimensions of Insulated Conductors and Fixture Wires 162
Exhibit 12.16	*NEC* Table C1 [Abridged], Maximum Number of Conductors or Fixture Wires in Electrical Metallic Tubing (EMT) 166
Exhibit 12.17	*NEC* Table C2 [Abridged], Maximum Number of Conductors or Fixture Wires in Electrical Nonmetallic Tubing (ENT) 167
Exhibit 12.18	*NEC* Table C3 [Abridged], Maximum Number of Conductors or Fixture Wires in Flexible Metal Conduit (FMC) 169
Exhibit 12.19	*NEC* Table C4 [Abridged], Maximum Number of Conductors or Fixture Wires in Intermediate Metal Conduit (IMC) 170

Exhibit 12.20 *NEC* Table C5 [Abridged], Maximum Number of Conductors or Fixture Wires in Liquidtight Flexible Nonmetallic Conduit (Type LFNC-B)   171

Exhibit 12.21 *NEC* Table C6 [Abridged], Maximum Number of Conductors or Fixture Wires in Liquidtight Flexible Nonmetallic Conduit (Type LFNC-A)   173

Exhibit 12.22 *NEC* Table C7 [Abridged], Maximum Number of Conductors or Fixture Wires in Liquidtight Flexible Metal Conduit (LFMC)   174

Exhibit 12.23 *NEC* Table C8 [Abridged], Maximum Number of Conductors or Fixture Wires in Rigid Metal Conduit (RMC)   175

Exhibit 12.24 *NEC* Table C9 [Abridged], Maximum Number of Conductors or Fixture Wires in Rigid PVC Conduit, Schedule 80   177

Exhibit 12.25 *NEC* Table C10 [Abridged], Maximum Number of Conductors or Fixture Wires in Rigid PVC Conduit, Schedule 40 and HDPE Conduit   178

CHAPTER 13
CONDUIT BENDING   181

Bending 90 Degree Stubs   181
Back-to-Back Bends   184
Offset Bends   185
Three-Point (or 3-Bend) Saddles   197
Segment Bends   201
Concentric Bends   205
Natural Trigonometric Functions   207

Tables

Table 13.1	90 Degree Stubs	182
Table 13.2	Distance Between Bends	187
Table 13.3	Conduit Shrink	187
Table 13.4	10 Degree Angle of Offset Bends	189
Table 13.5	22-1/2 Degree Angle of Offset Bends	189
Table 13.6	30 Degree Angle of Offset Bends	191
Table 13.7	45 Degree Angle of Offset Bends	193
Table 13.8	60 Degree Angle of Offset Bends	194
Table 13.9	Three-Point Saddle Bends	201

CONTENTS

Table 13.10   Angles per Bend and Multipliers   202

Table 13.11   Distance Between Bends for Segment Bends   204

Table 13.12   Natural Trigonometric Functions   207

Figures

Figure 13.1   Typical markings on a conduit bender   181

Figure 13.2   Take-up in 90 degree stub   182

Figure 13.3   Determining the required length   183

Figure 13.4   Aligning the bender   183

Figure 13.5   Bending the conduit to 90 degrees   183

Figure 13.6   Allowing for take-up length   183

Figure 13.7   Aligning the bender   184

Figure 13.8   Back-to-back bend   184

Figure 13.9   Marking the conduit at 50 in. and aligning with star-point symbol   185

Figure 13.10   Bending the conduit   186

Figure 13.11   Offset bend   186

Figure 13.12   Offset angles   187

Figure 13.13   The first mark at 52 in.   188

Figure 13.14   The second mark at 16 in.   188

Figure 13.15   The conduit bend   196

Figure 13.16   Three-point saddle.   197

Figure 13.17   Three-point saddle with one 45 degree bend and two 22-1/2 degree bends   197

Figure 13.18   Marking the center   199

Figure 13.19   Distance from center   199

Figure 13.20   Bending the 45 degree angle   199

Figure 13.21   Bending the first 22-1/2 degree angle   200

Figure 13.22   Aligning for and bending the second 22-1/2 degree angle   200

Figure 13.23   Installing the conduit   200

Figure 13.24   Segment bends   202

Figure 13.25   Distance between bends   203

Figure 13.26   22-in. radius at the centerline   203

Figure 13.27   Segment bends   206

CHAPTER 14
CONDUCTORS AND GROUNDING   211

Conductors   211

Grounding   230

Tables

Table 14.1	Adjusted Ampacities for 60°C (140°F) Conductors (Copper) 216
Table 14.2	Adjusted Ampacities for 75°C (167°F) Conductors (Copper) 217
Table 14.3	Adjusted Ampacities for 90°C (194°F) Conductors (Copper) 217
Table 14.4	Adjusted Ampacities for 60°C (140°F) Conductors (Aluminum or Copper-Clad Aluminum) 218
Table 14.5	Adjusted Ampacities for 75°C (167°F) Conductors (Aluminum or Copper-Clad Aluminum) 219
Table 14.6	Adjusted Ampacities for 90°C (194°F) Conductors (Aluminum or Copper-Clad Aluminum) 220

NEC Exhibits

Exhibit 14.1	*NEC* Table 300.19(A), Spacings for Conductor Supports 211
Exhibit 14.2	*NEC* Table 310.5, Minimum Size of Conductors 212
Exhibit 14.3	*NEC* Table 312.6(A), Minimum Wire-Bending Space at Terminals and Minimum Width of Wiring Gutters 212
Exhibit 14.4	*NEC* Table 312.6(B), Minimum Wire-Bending Space at Terminals 214
Exhibit 14.5	*NEC* Table 310.15(B)(2)(a), Adjustment Factors for More Than Three Current-Carrying Conductors in a Raceway or Cable 215
Exhibit 14.6	*NEC* Table 310.15(B)(6), Conductor Types and Sizes for 120/240-Volt, 3-Wire, Single-Phase Dwelling Services and Feeders. Conductor Types RHH, RHW, RHW-2, THHN, THHW, THW, THW-2, THWN, THWN-2, XHHW, XHHW-2, SE, USE, USE-2 222
Exhibit 14.7	*NEC* Table 310.16, Allowable Ampacities of Insulated Conductors Rated 0 Through 2000 Volts, 60°C Through 90°C (140°F Through 194°F), Not More Than Three Current-Carrying Conductors in Raceway, Cable, or Earth

CONTENTS

(Directly Buried), Based on Ambient Temperature of 30°C (86°F)   223

Exhibit 14.8   *NEC* Table 402.5, Allowable Ampacity for Fixture Wires   224

Exhibit 14.9   *NEC* Table 8, Conductor Properties   225

Exhibit 14.10   *NEC* Table 9, Alternating-Current Resistance and Reactance for 600-Volt Cables, 3-Phase, 60 Hz, 75°C (167°F)—Three Single Conductors in Conduit   226

Exhibit 14.11   *NEC* Table 250.66, Grounding Electrode Conductor for Alternating-Current Systems   230

Exhibit 14.12   *NEC* Table 250.122, Minimum Size Equipment Grounding Conductors for Grounding Raceway and Equipment   232

CHAPTER 15
VOLTAGE DROP   235

Determining Voltage Drop   235

Voltage Drop Formulas   235

Calculating Exact *K* Values   240

Alternative Voltage Drop Formulas Using Ohm's Law   245

Examples of Alternative Voltage Drop Formulas Using Ohm's Law   246

Tables

Table 15.1   Direct-Current Resistance and *K* Values for Uncoated Copper Conductors   240

Table 15.2   Direct-Current Resistance and *K* Values for Coated Copper Conductors   241

Table 15.3   Direct-Current Resistance and *K* Values for Aluminum Conductors   242

Table 15.4   Alternating-Current Resistance and *K* Values for Uncoated Copper Conductors at 75°C (167°F)   243

Table 15.5   Alternating-Current Resistance and *K* Values for Aluminum Conductors at 75°C (167°F)   244

CHAPTER 16
RECEPTACLES   249

Branch-Circuit Requirements   249

Receptacle and Plug Configurations   250

Figures

Figure 16.1 Configuration chart for general-purpose locking plugs and receptacles   251

Figure 16.2 Configuration chart for specific-purpose locking plugs and receptacles   252

NEC Exhibits

Exhibit 16.1 *NEC* Table 210.21(B)(2), Maximum Cord-and-Plug-Connected Load to Receptacle   249

Exhibit 16.2 *NEC* Table 210.21(B)(3), Receptacle Ratings for Various Size Circuits   249

Exhibit 16.3 *NEC* Table 210.24, Summary of Branch-Circuit Requirements   250

CHAPTER 17
SWITCHES AND LIGHTING   253

Switches   253

Lighting   267

Figures

Figure 17.1 Supply feeding the lighting outlet   254

Figure 17.2 Supply feeding the wall switch   255

Figure 17.3 Supply feeding the lighting outlet with wiring running from the lighting outlet to each switch   256

Figure 17.4 Supply feeding the lighting outlet with wiring running from the lighting outlet to one switch, and from there to the other switch   257

Figure 17.5 Supply feeding one wall switch with wiring running from the switch supplied with power to the lighting outlet, and from there to the other switch   258

Figure 17.6 Supply feeding one wall switch with wiring running from the switch supplied with power to the other switch, and then to the lighting outlet   259

Figure 17.7 Supply feeding one wall switch with wiring running from the switch supplied with power to the other switch and also to the lighting outlet   260

Figure 17.8 Supply feeding the lighting outlet with wiring running from the lighting outlet to one 3-way switch and also the 4-way

CONTENTS

	switch; wiring also runs from the 4-way switch to the other 3-way switch 261
Figure 17.9	Supply feeding the lighting outlet with wiring running from the lighting outlet to one 3-way switch...then to the 4-way switch...and then to the other 3-way switch 262
Figure 17.10	Supply feeding one 3-way wall switch with wiring running from that switch to the 4-way switch...to the lighting outlet...and then to the other 3-way switch 263
Figure 17.11	Supply feeding one 3-way wall switch with wiring running from that switch to the 4-way switch...to the other 3-way switch...and then to the lighting outlet 264
Figure 17.12	Supply feeding one 3-way wall switch with wiring running from that switch to the 4-way switch and also to the lighting outlet; wiring also runs from the 4-way switch to the other 3-way switch 265
Figure 17.13	Installing 4-way switches between two 3-way switches 266

NEC Exhibits

Exhibit 17.1	*NEC* Figure 410.8, Closet Storage Space 267

CHAPTER 18
MOTORS   269

NEC Article 430 Tables   **269**

Motor Control Abbreviations and Symbols   **278**

Wiring Diagrams   **287**

NEMA Starters   **291**

Terminal Markings and Connections   **293**

Tables

Table 18.1	Motor Control Abbreviations 278
Table 18.2	Motor Control Symbols 279
Table 18.3	Electrical Ratings for NEMA Starters 291
Table 18.4	Maximum Horsepower (Three-Phase Motors) 292

Figures

Figure 18.1	Basic diagram of two-wire control circuit 287
Figure 18.2	Wiring diagram of starter (two-wire control) 287
Figure 18.3	Control circuit only 288

Figure 18.4   Basic three-wire control circuit   288
Figure 18.5   Wiring diagram of starter (three-wire control)   289
Figure 18.6   Control circuit only   289
Figure 18.7   Multiple start and stop stations   290
Figure 18.8   Start push button with jog selector switch   290
Figure 18.9   Reversing starter   290
Figure 18.10   Reversing starter with limit switches   291
Figure 18.11   Standard numbering for a wye-connected motor   293
Figure 18.12   High-voltage connections for wye-connected motors   294
Figure 18.13   Low-voltage connections for wye-connected motors   294
Figure 18.14   Standard numbering for a delta-connected motor   295
Figure 18.15   High-voltage connections for delta-connected motors   295
Figure 18.16   Low-voltage connections for delta-connected motors   296

NEC Exhibits

Exhibit 18.1   *NEC* Table 430.7(B), Locked-Rotor Indicating Code Letters   269
Exhibit 18.2   *NEC* Table 430.37, Overload Units   270
Exhibit 18.3   *NEC* Table 430.52, Maximum Rating or Setting of Motor Branch-Circuit Short-Circuit and Ground-Fault Protective Devices   271
Exhibit 18.4   *NEC* Table 430.91, Motor Controller Enclosure Selection   272
Exhibit 18.5   *NEC* Table 430.247, Full-Load Current in Amperes, Direct-Current Motors   273
Exhibit 18.6   *NEC* Table 430.248, Full-Load Currents in Amperes, Single-Phase Alternating-Current Motors   274
Exhibit 18.7   *NEC* Table 430.249, Full-Load Current, Two-Phase Alternating-Current Motors (4-Wire)   274
Exhibit 18.8   *NEC* Table 430.250, Full-Load Current, Three-Phase Alternating-Current Motors   275
Exhibit 18.9   *NEC* Table 430.251(A), Conversion Table of Single-Phase Locked-Rotor Currents for Selection of Disconnecting Means and Controllers as Determined from Horsepower and Voltage Rating   276
Exhibit 18.10   *NEC* Table 430.251(B), Conversion Table of Polyphase Design B, C, and D Maximum Locked-Rotor Currents for

CONTENTS

Selection of Disconnecting Means and Controllers as
Determined from Horsepower and Voltage Rating and
Design Letter   277

CHAPTER 19
TRANSFORMERS   297

Transformer Ratings   297

Common Transformer Connections   300

Transformer Formulas Using Voltage, Current, and Turns   306

Formula Variations   306

Delta and Wye Transformer Formulas   309

Buck-Boost Transformers   312

Tables

Table 19.1	Single-Phase Transformers	299
Table 19.2	Three-Phase Transformers	299
Table 19.3	Boosting	313
Table 19.4	Bucking	315

Figures

Figure 19.1	Subtractive polarity	301
Figure 19.2	Additive polarity	301
Figure 19.3	Secondary windings connected in parallel	301
Figure 19.4	Secondary windings connected in series	302
Figure 19.5	Symbols representing three-phase transformer connections	302
Figure 19.6	Three-phase delta-connected transformer	302
Figure 19.7	Three-phase wye-connected transformer	303
Figure 19.8	Three single-phase transformers connected to form a delta transformer bank	303
Figure 19.9	Three single-phase transformers connected to form a wye transformer bank	303
Figure 19.10	Delta connected primary and delta connected secondary	304
Figure 19.11	Delta connected primary and wye connected secondary	304
Figure 19.12	Wye connected primary and wye connected secondary	305

Figure 19.13 Wye connected primary and delta connected secondary 305
Figure 19.14 Center-tap grounded, three-phase, four-wire delta transformer 310
Figure 19.15 Three-phase, four-wire wye transformer 311
Figure 19.16 Typical diagram of transformer terminations. 313
Figure 19.17 Diagram A1 316
Figure 19.18 Diagram B1 316
Figure 19.19 Diagram C1 316
Figure 19.20 Diagram D1 317
Figure 19.21 Diagram E 317
Figure 19.22 Diagram F 317
Figure 19.23 Diagram G 318
Figure 19.24 Diagram A2 318
Figure 19.25 Diagram B2 319
Figure 19.26 Diagram C2 319
Figure 19.27 Diagram D2 319

NEC Exhibits

Exhibit 19.1 *NEC* Table 450.3(A), Maximum Rating or Setting of Overcurrent Protection for Transformers over 600 Volts (as a Percentage of Transformer-Rated Current) 297

Exhibit 19.2 *NEC* Table 450.3(B), Maximum Rating or Setting of Overcurrent Protection for Transformers 600 Volts and Less (as a Percentage of Transformer-Rated Current) 298

CHAPTER 20
GENERAL REFERENCE 321

Tightening Torques 321
Drill and Hole Saw Sizes 323
Pulley Calculations 332
Portable Fire Extinguishers 333
Knots 336
Standard Crane Hand Signals 344
Phone and Data Jack Wiring Diagrams 353

CONTENTS

Tables

Table 20.1	Tightening Torques for Screws in Pound–Inches	321
Table 20.2	Torques in Pound–Inches for Slotted Head Screws Smaller Than No. 10, for Use with 8 AWG and Smaller Conductors 322	
Table 20.3	Torques for Recessed Allen Head Screws	323
Table 20.4	Lug-Bolting Torques for Connection of Wire Connectors to Busbars 323	
Table 20.5	Tap Drill Sizes 324	
Table 20.6	Metric Tap Drill Sizes	329
Table 20.7	Hole Saw Sizes for Conduits	332
Table 20.8	Classes of Fires 333	
Table 20.9	USOC (Universal Service Order Code) 2-Pair Configuration 353	
Table 20.10	USOC 2-Pair Configuration	354
Table 20.11	USOC 3-Pair Configuration	355
Table 20.12	USOC 4-Pair Configuration	355
Table 20.13	RJ-45 TIA/EIA 568A 4-Pair Configuration	356
Table 20.14	RJ-45 TIA/EIA 568B 4-Pair Configuration	357

Figures

Figure 20.1	Traditional labeling system for extinguishers	334
Figure 20.2	Pictograph labeling system for extinguishers	335
Figure 20.3	Figure eight knot 336	
Figure 20.4	Half hitch 336	
Figure 20.5	Two half hitches 336	
Figure 20.6	Clove hitch 337	
Figure 20.7	Timber hitch 337	
Figure 20.8	Timber hitch with half hitch	337
Figure 20.9	Pipe hitch 338	
Figure 20.10	Bowline 338	
Figure 20.11	Bowline on a bight 339	
Figure 20.12	Square knot 339	
Figure 20.13	Surgeon's knot 339	
Figure 20.14	Fisherman's knot 340	
Figure 20.15	Sheet bend 340	
Figure 20.16	Double sheet bend 340	

Figure 20.17   Single blackwell   341
Figure 20.18   Double blackwell   341
Figure 20.19   Cats-paw   342
Figure 20.20   Miller's (constrictor) knot   342
Figure 20.21   Strap knot   343
Figure 20.22   Sheepshank   343
Figure 20.23   Stop   344
Figure 20.24   Emergency stop   345
Figure 20.25   Use main hoist   345
Figure 20.26   Use whip line (auxiliary hoist)   346
Figure 20.27   Raise load (hoist)   346
Figure 20.28   Lower load   347
Figure 20.29   Raise boom   347
Figure 20.30   Lower boom   348
Figure 20.31   Extend boom (telescoping booms)   348
Figure 20.32   Retract boom (telescoping booms)   349
Figure 20.33   Swing boom   349
Figure 20.34   Raise the boom and lower the load   350
Figure 20.35   Lower the boom and raise the load   350
Figure 20.36   Travel   351
Figure 20.37   Travel (one track)   351
Figure 20.38   Travel (both tracks)   352
Figure 20.39   Move slowly   352
Figure 20.40   Dog everything   353
Figure 20.41   Front view of female connector in a 4-position/ 4-wire jack   354
Figure 20.42   Front view of female connector in a 6-position/ 4-wire jack   354
Figure 20.43   Front view of female connector in a 6-position/ 6-wire jack   355
Figure 20.44   Front view of female connector in a USOC 8-position/ 8-wire jack   356
Figure 20.45   Front view of female connector in a 568A 8-position/ 8-wire jack   356
Figure 20.46   Front view of female connector in a 568B 8-position/ 8-wire jack   357

CONTENTS

CHAPTER 21
WEBSITE ADDRESSES   359

CHAPTER 22
FIRST AID EMERGENCY PROCEDURES   365

Rescue Breathing—Adult   366

One Rescuer CPR—Adult   366

CPR—Arrival of Second Rescuer   367

Obstructed Airway—Adult   367

External Bleeding   368

Internal Bleeding   368

Traumatic Shock   369

Burns   370

Electrical Burns/Injury   371

Chemical Burns   371

Poisons   372

Heat Exhaustion   372

Heat Stroke   373

Frostbite   373

Hypothermia   374

PREFACE

NFPA's Electrical References was written to provide a quick and handy electrical reference book. Although the original intention was to write a book containing electrical reference material for use by apprentices and electricians while in the field, it has evolved into much more. Contractors, estimators, designers, engineers, inspectors, and more have benefited from the material in this book. Not only is this electrical reference book a valuable tool while on the job, it is also a valuable resource while at home or at the office. This book is not limited to any particular type of electrical occupation. Whether you work in residential, commercial, industrial, health care, or another field, the material in this electrical reference book is of great value.

A wide variety of topics are covered throughout these chapters. The first chapter focuses on the *National Electrical Code®* (*NEC®*). This chapter shows the chapter and article arrangement and the common numbering system within the raceway and cable articles. Along with the different categories of electrical abbreviations in Chapter 2 is a table containing field terms and phrases used in the electrical industry. The next three chapters focus on basic math and conversion factors. Ohm's Law is covered in Chapter 6, and Chapter 7 contains electrical formulas and examples.

Chapters 8 and 9 are a valuable resource for plan symbols and load calculations. Chapter 10 includes specific clearance and cover requirements from the *NEC*. Chapters 11 and 12 cover boxes, enclosures, raceways, and cable trays. Chapter 13 contains detailed instructions and illustrations of conduit bending. This chapter also includes tables that eliminate the need to execute calculations.

While conductors and grounding are covered in Chapter 14, voltage drop is in Chapter 15. Chapter 16 contains extracted *NEC* material from Article 210 and Article 406 that pertains to receptacles. Switches and lighting are covered in Chapter 17. This chapter includes easy-to-understand wiring diagrams for single-pole, 3-way, and 4-way switches.

Electrical reference material pertaining to motors and transformers are in Chapters 18 and 19. Chapter 20 provides general-use information that is not found elsewhere in the book. Because of computers and the Internet, Chapter 21 is a valuable reference: It contains website addresses for organizations with interests related to electrical installations. Last but not least, Chapter 22 includes first aid emergency procedures.

Some material in *NFPA's Electrical References* has been extracted from the 2005 edition of the *National Electrical Code*. Material extracted from the *NEC®* is indented and set in a different typeface for easy identification.

Although this electrical reference contains material from the *NEC*, it is not intended to be a substitute for the *NEC*.

CONTENTS

CHAPTER 1
NATIONAL ELECTRICAL CODE® ORGANIZATION

Chapter 1 contains four tables designed to help the user locate information in the *National Electrical Code*® (abbreviated *NEC*® or *Code*). The *NEC* is organized into chapters, articles, and annexes. Each chapter is divided into articles, with the exception of Article 90, Introduction, which stands alone. The articles contain enforceable code requirements, and the annexes contain explanatory, nonmandatory information.

See Table 1.1 for a brief *NEC* table of contents that contains chapter and annex titles and article numbers within each chapter. Tables 1.2 and 1.3 list the *Code* articles alphabetically and numerically.

Table 1.4 outlines the common article numbering system for the raceway and cable articles in Chapter 3. The common numbering format assists users in locating and comparing requirements that are common to each raceway and cable article. For example, Section 3xx.12 covers uses not permitted in each relevant article.

Table 1.1 *NEC*® Brief Contents

Chapter/Annex	Title	Articles
Article 90	Introduction	
Chapter 1	General	100–110
Chapter 2	Wiring and Protection	200–285
Chapter 3	Wiring Methods and Materials	300–398
Chapter 4	Equipment for General Use	400–490
Chapter 5	Special Occupancies	500–590
Chapter 6	Special Equipment	600–695
Chapter 7	Special Conditions	700–780
Chapter 8	Communications Systems	800–830
Chapter 9	Tables	
Annex A	Product Safety Standards	
Annex B	Application Information for Ampacity Calculation	
Annex C	Conduit and Tubing Fill Tables for Conductors and Fixture Wires of the Same Size	
Annex D	Examples	
Annex E	Types of Construction	
Annex F	Chapter 3 Cross-Reference Tables 2005 *NEC*—2002 *NEC*, 2002 *NEC*—1999 *NEC*	
Annex G	Administration and Enforcement	

2

NATIONAL ELECTRICAL CODE® ORGANIZATION

Table 1.2   *NEC®* Articles Arranged Alphabetically by Title

Title	Article
Agricultural Buildings	547
Air-Conditioning and Refrigerating Equipment	440
Aircraft Hangars	513
Appliances	422
Armored Cable: Type AC	320
Audio Signal Processing, Amplification, and Reproduction Equipment	640
Auxiliary Gutters	366
Branch Circuits	210
Branch-Circuit, Feeder, and Service Calculations	220
Bulk Storage Plants	515
Busways	368
Cabinets, Cutout Boxes, and Meter Socket Enclosures	312
Cable Trays	392
Cablebus	370
Capacitors	460
Carnivals, Circuses, Fairs, and Similar Events	525
Cellular Concrete Floor Raceways	372
Cellular Metal Floor Raceways	374
Circuits and Equipment Operating at Less Than 50 Volts	720
Class 1, Class 2, and Class 3 Remote-Control, Signaling, and Power-Limited Circuits	725
Control Panels	409
Class I Locations	501
Class I, Zone 0, 1, and 2 Locations	505
Class II Locations	502
Class III Locations	503
Closed-Loop and Programmed Power Distribution	780
Commercial Garages, Repair and Storage	511
Communications Circuits	800
Community Antenna Television and Radio Distribution Systems	820
Concealed Knob-and-Tube Wiring	394
Conductors for General Wiring	310
Cranes and Hoists	610
Definitions	100
Electric Signs and Outline Lighting	600
Electric Vehicle Charging System	625
Electric Welders	630
Electrical Metallic Tubing: Type EMT	358
Electrical Nonmetallic Tubing: Type ENT	362
Electrically Driven or Controlled Irrigation Machines	675
Electrolytic Cells	668
Electroplating	669
Elevators, Dumbwaiters, Escalators, Moving Walks, Wheelchair Lifts, and Stairway Chair Lifts	620

Table continues below.

Table 1.2 NEC® Articles Arranged Alphabetically by Title *(continued)*

Title	Article
Emergency Systems	700
Equipment, Over 600 Volts, Nominal	490
Feeders	215
Fire Alarm Systems	760
Fire Pumps	695
Fixed Electric Heating Equipment for Pipelines and Vessels	427
Fixed Electric Space-Heating Equipment	424
Fixed Outdoor Electric Deicing and Snow-Melting Equipment	426
Fixture Wires	402
Flat Cable Assemblies: Type FC	322
Flat Conductor Cable: Type FCC	324
Flexible Cords and Cables	400
Flexible Metal Conduit: Type FMC	348
Flexible Metallic Tubing: Type FMT	360
Floating Buildings	553
Fuel Cell Systems	692
Generators	445
Grounding and Bonding	250
Hazardous (Classified) Locations—Specific	510
Hazardous (Classified) Locations, Classes I, II, and III, Divisions 1 and 2	500
Health Care Facilities	517
Induction and Dielectric Heating Equipment	665
Industrial Machinery	670
Information Technology Equipment	645
Instrumentation Tray Cable: Type ITC	727
Integrated Electrical Systems	685
Integrated Gas Spacer Cable: Type IGS	326
Interconnected Electric Power Production Sources	705
Intermediate Metal Conduit: Type IMC	342
Intrinsically Safe Systems	504
Introduction	90
Legally Required Standby Systems	701
Lighting Systems Operating at 30 Volts or Less	411
Liquidtight Flexible Metal Conduit: Type LFMC	350
Liquidtight Flexible Nonmetallic Conduit: Type LFNC	356
Luminaires (Lighting Fixtures),Lampholders, and Lamps	410
Manufactured Buildings	545
Manufactured Wiring Systems	604
Marinas and Boatyards	555
Medium Voltage Cable: Type MV	328
Messenger Supported Wiring	396
Metal Wireways	376
Metal-Clad Cable: Type MC	330
Mineral-Insulated, Metal-Sheathed Cable: Type MI	332
Mobile Homes, Manufactured Homes, and Mobile Home Parks	550
Motion Picture and Television Studios and Similar Locations	530

Continued

NATIONAL ELECTRICAL CODE® ORGANIZATION

NATIONAL ELECTRICAL CODE® ORGANIZATION

Table 1.2 NEC® Articles Arranged Alphabetically by Title (continued)

Title	Article
Motion Picture Projection Rooms	540
Motor Fuel Dispensing Facilities	514
Motors, Motor Circuits, and Controllers	430
Multioutlet Assembly	380
Natural and Artificially Made Bodies of Water	682
Network-Powered Broadband Communications Systems	830
Nonmetallic Extensions	382
Nonmetallic Underground Conduit with Conductors: Type NUCC	354
Nonmetallic Wireways	378
Nonmetallic-Sheathed Cable: Types NM, NMC, and NMS	334
Office Furnishings (Consisting of Lighting Accessories and Wired Partitions)	605
Open Wiring on Insulators	398
Optical Fiber Cables and Raceways	770
Optional Standby Systems	702
Outlet, Device, Pull, and Junction Boxes; Conduit Bodies; Fittings; and Handhole Enclosures	314
Outside Branch Circuits and Feeders	225
Overcurrent Protection	240
Park Trailers	552
Phase Converters	455
Pipe Organs	650
Assembly Occupancies	518
Power and Control Tray Cable: Type TC	336
Radio and Television Equipment	810
Receptacles, Cord Connectors, and Attachment Plugs (Caps)	406
Recreational Vehicles and Recreational Vehicle Parks	551
Requirements for Electrical Installations	110
Resistors and Reactors	470
Rigid Metal Conduit: Type RMC	344
Rigid Nonmetallic Conduit: Type RNC	352
Sensitive Electronic Equipment	647
Service-Entrance Cable: Types SE and USE	338
Services	230
Solar Photovoltaic Systems	690
Spray Application, Dipping, and Coating Processes	516
Storage Batteries	480
Strut-Type Channel Raceway	384
Surface Metal Raceways	386
Surface Nonmetallic Raceways	388
Surge Arresters	280
Swimming Pools, Fountains, and Similar Installations	680
Switchboards and Panelboards	408

Table continues below.

Table 1.2 *NEC*® Articles Arranged Alphabetically by Title *(continued)*

Title	Article
Switches	404
Temporary Installations	590
Theaters, Audience Areas of Motion Picture and Television Studios, Performance Areas, and Similar Locations	520
Transformers and Transformer Vaults (Including Secondary Ties)	450
Transient Voltage Surge Suppressors: TVSSs	285
Underfloor Raceways	390
Underground Feeder and Branch-Circuit Cable: Type UF	340
Use and Identification of Grounded Conductors	200
Wiring Methods	300
X-Ray Equipment	660
Zone 20, 21, and 22 Locations for Flammable Dusts, Fibers and Flyings	506

Table 1.3 *NEC*® Articles Arranged Numerically

Article	Title
90	Introduction
100	Definitions
110	Requirements for Electrical Installation
200	Use and Identification of Grounded Conductors
210	Branch Circuits
215	Feeders
220	Branch-Circuit, Feeder, and Service Calculations
225	Outside Branch Circuits and Feeders
230	Services
240	Overcurrent Protection
250	Grounding and Bonding
280	Surge Arresters
285	Transient Voltage Surge Suppressors: TVSSs
300	Wiring Methods
310	Conductors for General Wiring
312	Cabinets, Cutout Boxes, and Meter Socket Enclosures
314	Outlet, Device, Pull, and Junction Boxes; Conduit Bodies; Fittings; and Handhole Enclosures
320	Armored Cable: Type AC
322	Flat Cable Assemblies: Type FC
324	Flat Conductor Cable: Type FCC
326	Integrated Gas Spacer Cable: Type IGS
328	Medium Voltage Cable: Type MV
330	Metal-Clad Cable: Type MC
332	Mineral-Insulated, Metal-Sheathed Cable: Type MI
334	Nonmetallic-Sheathed Cable: Types NM, NMC, and NMS

Continued

NATIONAL ELECTRICAL CODE® ORGANIZATION

NATIONAL ELECTRICAL CODE® ORGANIZATION

Table 1.3 NEC® Articles Arranged Numerically (continued)

Article	Title
336	Power and Control Tray Cable: Type TC
338	Service-Entrance Cable: Types SE and USE
340	Underground Feeder and Branch-Circuit Cable: Type UF
342	Intermediate Metal Conduit: Type IMC
344	Rigid Metal Conduit: Type RMC
348	Flexible Metal Conduit: Type FMC
350	Liquidtight Flexible Metal Conduit: Type LFMC
352	Rigid Nonmetallic Conduit: Type RNC
354	Nonmetallic Underground Conduit with Conductors: Type NUCC
356	Liquidtight Flexible Nonmetallic Conduit: Type LFNC
358	Electrical Metallic Tubing: Type EMT
360	Flexible Metallic Tubing: Type FMT
362	Electrical Nonmetallic Tubing: Type ENT
366	Auxiliary Gutters
368	Busways
370	Cablebus
372	Cellular Concrete Floor Raceways
374	Cellular Metal Floor Raceways
376	Metal Wireways
378	Nonmetallic Wireways
380	Multioutlet Assembly
382	Nonmetallic Extensions
384	Strut-Type Channel Raceway
386	Surface Metal Raceways
388	Surface Nonmetallic Raceways
390	Underfloor Raceways
392	Cable Trays
394	Concealed Knob-and-Tube Wiring
396	Messenger Supported Wiring
398	Open Wiring on Insulators
400	Flexible Cords and Cables
402	Fixture Wires
404	Switches
406	Receptacles, Cord Connectors, and Attachment Plugs (Caps)
408	Switchboards and Panelboards
409	Control Panels
410	Luminaires (Lighting Fixtures), Lampholders, and Lamps
411	Lighting Systems Operating at 30 Volts or Less
422	Appliances
424	Fixed Electric Space-Heating Equipment
426	Fixed Outdoor Electric Deicing and Snow-Melting Equipment
427	Fixed Electric Heating Equipment for Pipelines and Vessels
430	Motors, Motor Circuits, and Controllers
440	Air-Conditioning and Refrigerating Equipment
445	Generators

Table continues below.

Table 1.3 NEC® Articles Arranged Numerically *(continued)*

Article	Title
450	Transformers and Transformer Vaults (Including Secondary Ties)
455	Phase Converters
460	Capacitors
470	Resistors and Reactors
480	Storage Batteries
490	Equipment, Over 600 Volts, Nominal
500	Hazardous (Classified) Locations, Classes I, II, and III, Divisions 1 and 2
501	Class I Locations
502	Class II Locations
503	Class III Locations
504	Intrinsically Safe Systems
505	Class I, Zone 0, 1, and 2 Locations
506	Zone 20, 21 and 22 Locations for Flammable Dust, Fibers and Flyings
510	Hazardous (Classified) Locations—Specific
511	Commercial Garages, Repair and Storage
513	Aircraft Hangars
514	Motor Fuel Dispensing Facilities
515	Bulk Storage Plants
516	Spray Application, Dipping, and Coating Processes
517	Health Care Facilities
518	Assembly Occupancies
520	Theaters, Audience Areas of Motion Picture and Television Studios, Performance Areas, and Similar Locations
525	Carnivals, Circuses, Fairs, and Similar Events
530	Motion Picture and Television Studios and Similar Locations
540	Motion Picture Projection Rooms
545	Manufactured Buildings
547	Agricultural Buildings
550	Mobile Homes, Manufactured Homes, and Mobile Home Parks
551	Recreational Vehicles and Recreational Vehicle Parks
552	Park Trailers
553	Floating Buildings
555	Marinas and Boatyards
590	Temporary Installations
600	Electric Signs and Outline Lighting
604	Manufactured Wiring Systems
605	Office Furnishings (Consisting of Lighting Accessories and Wired Partitions)
610	Cranes and Hoists
620	Elevators, Dumbwaiters, Escalators, Moving Walks, Wheelchair Lifts, and Stairway Chair Lifts
625	Electric Vehicle Charging System
630	Electric Welders
640	Audio Signal Processing, Amplification, and Reproduction Equipment
645	Information Technology Equipment

Continued

NATIONAL ELECTRICAL CODE® ORGANIZATION

8

NATIONAL ELECTRICAL CODE® ORGANIZATION

Table 1.3 NEC® Articles Arranged Numerically (continued)

Article	Title
647	Sensitive Electronic Equipment
650	Pipe Organs
660	X-Ray Equipment
665	Induction and Dielectric Heating Equipment
668	Electrolytic Cells
669	Electroplating
670	Industrial Machinery
675	Electrically Driven or Controlled Irrigation Machines
680	Swimming Pools, Fountains, and Similar Installations
682	Natural and Artificially Made Bodies of Water
685	Integrated Electrical Systems
690	Solar Photovoltaic Systems
692	Fuel Cell Systems
695	Fire Pumps
700	Emergency Systems
701	Legally Required Standby Systems
702	Optional Standby Systems
705	Interconnected Electric Power Production Sources
720	Circuits and Equipment Operating at Less Than 50 Volts
725	Class 1, Class 2, and Class 3 Remote-Control, Signaling, and Power-Limited Circuits
727	Instrumentation Tray Cable: Type ITC
760	Fire Alarm Systems
770	Optical Fiber Cables and Raceways
780	Closed-Loop and Programmed Power Distribution
800	Communications Circuits
810	Radio and Television Equipment
820	Community Antenna Television and Radio Distribution Systems
830	Network-Powered Broadband Communications Systems

Table 1.4 Common Numbering System for Chapter 3 Raceway and Cable Articles

I. **General**

3xx.1	Scope
3xx.2	Definitions
3xx.3	Other Articles
3xx.4	Listing Requirements

II. **Installation**

3xx.10	Uses Permitted
3xx.12	Uses Not Permitted
3xx.14	Dissimilar Metals
3xx.16	Temperature Limits
3xx.20	Size
3xx.22	Number of Conductors
3xx.24	Bends-How Made
3xx.26	Bends-Number in One Run
3xx.28	Trimming (or Reaming and Threading)
3xx.30	Securing and Supporting
3xx.40	Boxes and Fittings
3xx.42	Couplings and Connectors
3xx.44	Expansion Fittings
3xx.46	Bushings
3xx.48	Joints
3xx.50	Conductor Terminations
3xx.56	Splices and Taps
3xx.60	Grounding

III. **Construction Specifications**

3xx.100	Construction
3xx.104	Conductors
3xx.108	Equipment Grounding
3xx.110	Corrosion Protection
3xx.112	Insulation
3xx.116	Sheath or Jacket
3xx.120	Marking
3xx.130	Standard Lengths
3xx.140	Conductors and Cable
3xx.150	Conductor Fill

NATIONAL ELECTRICAL CODE® ORGANIZATION

CHAPTER 2
ELECTRICAL ABBREVIATIONS, SYMBOLS, AND FIELD TERMS

Chapter 2 provides abbreviations and/or symbols for common electrical terms (see Table 2.1), cable and raceway abbreviations (see Tables 2.2 and 2.3), symbols used in the International System of Units (SI) (see Table 2.4), and descriptions of electrical field terms used in the trade (see Table 2.5).

Table 2.1 Electrical Abbreviations

Abbreviation/ Symbol	Term	Abbreviation/ Symbol	Term
Δ	delta	C	Celsius; centigrade; capacitance
Ω	ohm		
>	is greater than (5 > 4)	CAP	capacitor
<	is less than (4 < 5)	CATV	community antenna television
≥	is greater than or equal to	CB	circuit breaker
≤	is less than or equal to	CCTV	closed-circuit television
±	plus or minus	CCW	counterclockwise
Ø	phase	CKT	circuit
1Ø	single phase	CMIL	circular mil
3Ø	three phase	CONT	continuous
3Ø 3W	three phase, three wire	CPS	cycles per second
3Ø 4W	three phase, four wire	CR	control relay
1P	one pole	CSA	Canadian Standards Association
2P	two pole		
3P	three pole	CT	current transformer
2P2W	two pole, two wire	CU	copper
3P3W	three pole, three wire	CW	clockwise
3P4W	three pole, four wire	dc	direct current
A	ampere; amp; ammeter	DISC	disconnect
ac	alternating current	DP	double pole
AHJ	authority having jurisdiction	DPDT	double pole, double throw
A-hr	amp-hour	DPST	double pole, single throw
AIC	ampere interrupting capacity	DVM	digital voltmeter
AL	aluminum	E	voltage
AL-CU	aluminum and copper	EMF	electromotive force; voltage
AM	ammeter	f	frequency
AMP	ampere	F	Fahrenheit; farad
ARM	armature	FLC	full-load current
AWG	American Wire Gage	FREQ	frequency
BLDG	building	ft^2	square feet

Continued

ELECTRICAL ABBREVIATIONS, SYMBOLS, AND FIELD TERMS

Table 2.1   Electrical Abbreviations *(continued)*

Abbreviation/ Symbol	Term	Abbreviation/ Symbol	Term
G	conductance	NPT	National (American) Standard Pipe Taper
GEN	generator		
GFCI	ground-fault circuit interrupter	NRTL	Nationally Recognized Testing Laboratories
GFI	ground-fault interrupter		
GFPE	ground-fault protection of equipment	OCPD	overcurrent protective device
		P	pole; power
GND	ground	PE	professional engineer
GRC	galvanized rigid conduit	PF	power factor
H	transformer, primary side; henry	PLFA	power-limited fire alarm circuit
		PNL	panelboard
HID	high intensity discharge	PRI	primary
HP	horsepower	PT	potential transformer
HV	high voltage	PVC	polyvinyl chloride
HVAC	heating, ventilating, and air conditioning	R	resistance, radius
		REC	receptacle
Hz	hertz (cycle) per second	RF	radio frequency
I	current	RFI	radio frequency interference
IC	interrupting capacity	RMS	root-mean-square
ID	inside diameter	RPM	revolutions per minute
IG	isolated ground	SE	service entrance
JB	junction box	SI	International System of Units
K/O	knockout	SN	solid neutral
kCMIL	one thousand circular mils	SP	single pole
kVA	kilovolt-amperes	SPDT	single pole, double throw
kVAr	kilovolt-amperes reactive	Specs	specifications
kW	kilowatts	SPST	single pole, single throw
L	line; load; inductance	SW	switch
LED	light-emitting diode	TEMP	temperature, temporary
LRC	locked rotor current	UL	Underwriters Laboratories
LS	limited smoke	V	volt
LV	low voltage	VA	volt-ampere
m^2	square meter	VAC	volts alternating current
MCB	main circuit breaker	VAr	volt-amperes reactive
MCC	motor control center	VD	voltage drop
MCM	one thousand circular mils	VDC	volts direct current
MDP	main distribution panel	VOM	volt-ohm-multimeter
MISC	miscellaneous	W	watt
MLO	main lugs only	W-hr	watt-hour
MTW	machine tool wire	WM	wattmeter
NEC®	*National Electrical Code®*	WP	weatherproof
NEMA	National Electrical Manufacturers Association	X	transformer, secondary side
		XFMR	transformer
NFPA	National Fire Protection Association	Y	wye
		Z	impedance
NPLFA	non-power-limited fire alarm circuit		

Table 2.2 Cable and Raceway Abbreviations

Abbreviation	Cable or Raceway	Abbreviation	Cable or Raceway
AC Cable	armored cable	NMC Cable	nonmetallic-sheathed cable w/ corrosion resistant nonmetallic jacket
EMT	electrical metallic tubing		
ENT	electrical nonmetallic tubing	NMS Cable	nonmetallic-sheathed cable w/ signaling, data, and communications conductors plus power or control conductors
FC Assemblies	flat cable assemblies		
FCC Cable	flat conductor cable		
FMC	flexible metal conduit		
FMT	flexible metal tubing		
IGS Cable	integrated gas spacer cable	NUCC	nonmetallic underground conduit with conductors
IMC	intermediate metal conduit		
LFMC	liquidtight metal conduit	RMC	rigid metal conduit
LFNC	liquidtight flexible nonmetallic conduit	RNC	rigid nonmetallic conduit
		SE Cable	service-entrance cable
MC Cable	metal-clad cable	TC Cable	power and control tray cable
MI Cable	mineral-insulated, metal-sheathed cable	UF Cable	underground feeder and branch-circuit cable
MV Cable	medium voltage cable	USE Cable	service-entrance cable (underground)
NM Cable	nonmetallic-sheathed cable		

Table 2.3 Low-Voltage Cable Abbreviations

Abbreviation	Cable	Abbreviation	Cable
CATV	coaxial general-purpose cable	CMUC	undercarpet communications wire and cable
CATVP	coaxial plenum cable		
CATVR	coaxial riser cable	CMX	communications cable, limited use
CATVX	coaxial cable, limited use		
CL2	Class 2 cable	FPL	power-limited fire alarm cable
CL2P	Class 2 plenum cable	FPLP	power-limited fire alarm plenum cable
CL2R	Class 2 riser cable		
CL2X	Class 2 cable, limited use	FPLR	power-limited fire alarm riser cable
CL3	Class 3 cable		
CL3P	Class 3 plenum cable	NPLF	non-power-limited fire alarm cable
CL3R	Class 3 riser cable		
CL3X	Class 3 cable, limited use	NPLFP	non-power-limited fire alarm cable (environmental air spaces)
CM	communications general-purpose cable		
CMG	communications general-purpose cable	NPLFR	non-power-limited fire alarm riser cable
CMP	communications plenum cable	PLTC	power-limited tray cable
CMR	communications riser cable		

ELECTRICAL ABBREVIATIONS, SYMBOLS, AND FIELD TERMS

ELECTRICAL ABBREVIATIONS, SYMBOLS, AND FIELD TERMS

Table 2.4   International System of Units (SI)

Symbol	Prefix	Multiplying Factor	Power of 10
T	tera	1 000 000 000 000	10^{12}
G	giga	1 000 000 000	10^{9}
M	mega	1 000 000	10^{6}
k	kilo	1000	10^{3}
h	hekto	100	10^{2}
da	deka	10	10^{1}
Base unit		1	10^{0}
d	deci	0.1	10^{-1}
c	centi	0.01	10^{-2}
m	milli	0.001	10^{-3}
μ	micro	0.000 001	10^{-6}
n	nano	0.000 000 001	10^{-9}
p	pico	0.000 000 000 001	10^{-12}

Table 2.5   Electrical Field Terms

Field Term	Description
11B	4-11/16-in. square metal box
1900 box	4-in. square metal box
4–11 box	4-11/16-in. square metal box
4-S blank	blank cover for 4-in. square box
4-S box	4-in. square metal box
4-square box	4-in. square metal box
5-S blank	blank cover for 4-11/16-in. square box
5-S box	4-11/16-in. square metal box
Acorn	ground rod clamp (teardrop)
Banger	cable fault locater
Bat wing	conduit hanger from flange, rod, or support wire
Battleships	box holders for cut-in box
Bell box	weatherproof junction box
Bell wire	thermostat wire
Blues	lineman's pliers
Box holdits	box holders for cut-in box
Brick box	3-1/2-in. metal octagon box
Bug	split bolt connector
Burndy	split bolt connector
Bus duct	busway
BX	armored cable: Type AC
Can	panelboard or control center cabinet (enclosure)
Channel locks	tongue and groove pliers
Cheater	ground adapter; 3-prong adapter

Table continues below.

Table 2.5 Electrical Field Terms *(continued)*

Field Term	Description
Chicago bender	mechanical conduit bender
Chinese finger	wire mesh pulling basket for conductors
Condulet	conduit body
Crescent wrench	adjustable wrench
Cut-in	device box with ears
Dikes	diagonal cutting pliers
Donut	reducing washer
Earthquake clip	lay-in (troffer) light fixture support clip
Erickson	3-piece coupling
Extension ring	box extension
F-clips (F-strap)	box holders for cut-in box
Flex	flexible metal conduit
Flower pot	concrete pipe sleeve
Frog eye	2-head emergency light
FS box	weatherproof junction box
Gang box	mobile storage box or cabinet; gangable metal boxes
Gem box	1-gang metal device box with ears
Goof plate	large or jumbo cover plate
Greenfield	flexible metal conduit
Ground tail	green ground wire (pigtail) with screw
Gutter	wireway
Handy box	4-in. × 2-1/8-in. metal device box
Heavy-wall	rigid metal conduit
Hickey	rigid conduit hand bender
Hi-hat	can for recessed light
Hold eases	box holders for cut-in box
Hot box	PVC heater
Hot dog	wire marker dispenser (red) with 10 flip-top slots
Hurricane clip	lay-in (troffer) light fixture support clip
Ice skates	box holders for cut-in box
Jake	pulling elbow
J-box	junction box
Jiffy clip	one-hole conduit strap
Kearney	split bolt connector
Kellums grip	wire mesh pulling basket for conductors
Kindorf	strut-type channel raceway
Kindorf straps	strut-type channel straps
Klines	lineman's pliers
LB	conduit body
Load center	panelboard
Loaded hot dog	wire marker dispenser (red) full of wire markers
Lube	wire pulling lubricant
Madison clips	box holders for cut-in box
Megger	conductor insulation tester
Minnies	minerallac strap
Monkey wrench	adjustable pipe wrench

Continued

ELECTRICAL ABBREVIATIONS, SYMBOLS, AND FIELD TERMS

ELECTRICAL ABBREVIATIONS, SYMBOLS, AND FIELD·TERMS

Table 2.5 Electrical Field Terms *(continued)*

Field Term	Description
Mud ring	plaster ring
Myers hub	threadless wp hub for threaded conduit
Nail-on	single-gang plastic nail-on device box
O bushing	insulating (anti-short) bushing
Piggyback breaker	twin breaker
Plugmold	multioutlet assembly
P-ring	plaster ring
Push-penny	knockout closure
Quad breaker	twin double pole breaker
R E	reducing bushing
R E bushing	reducing bushing
Red devil	insulating (anti-short) bushing
Red head	insulating (anti-short) bushing
Romex	nonmetallic-sheathed cable
Sealtight	liquidtight flexible metal conduit
Simplex receptacle	single receptacle
Smurf tube (pipe)	electrical nonmetallic tubing (ENT)
Snake	fish tape
Snap cover	weatherproof cover
Soap	wire pulling lubricant
Stand off strap	minerallac strap
Stillson wrench	adjustable pipe wrench
Stinger	green ground wire (pigtail) with screw
Strip	fluorescent light fixture
Sub-panel	panelboard; remote panelboard
T-11	weatherproof junction box
Teardrop	ground rod clamp
Thin-wall	electrical metallic tubing (EMT)
Thumper	cable fault locater
Tick tracer	non-contact voltage detector
Troffer	2 × 4 lay-in light fixture
Trough	wireway
Tub	panelboard or control center cabinet (enclosure)
Tugger	electric wire/cable puller
Unistrut	strut-type channel raceway
Utility box	4-in. × 2-1/8-in. metal device box
Wiggy	solenoid-type voltage tester
Wiremold	surface metal and nonmetallic raceway
Yellow 77	wire pulling lubricant
Zip cord	2-wire lamp cord
Zip ties	cable ties

CHAPTER 3
MATHEMATICAL FORMULAS AND TABLES

Chapter 3 covers mathematical formulas and tables that are relevant to electricians. It begins with a description of the fraction and its components, including common mathematical functions, such as addition, subtraction, multiplication, and division. A definition and description of equations follows, and the chapter concludes with three useful tables: Table 3.1 provides squares and square roots for whole numbers 1 through 100. The numerical values and names for scientific notation (powers of 10) are located in Table 3.2. Table 3.3 provides decimal and millimeter equivalents of fractions.

FRACTIONS

A fraction can be expressed as a decimal (0.5), or with a numerator and denominator (1/2). A fraction is a quantity less than a whole number. Fractions are also defined as any quantity expressed by a numerator and denominator.

Numerators and denominators can be expressed above and below a line, such as $\frac{a}{b}$ or to the left and right of a slanted line, such as *a/b*.

The top number (or left number) is the dividend, or numerator.
The bottom number (or right number) is the divisor, or denominator.

$$\frac{\text{Numerator}}{\text{Denominator}} = \text{Decimal Equivalent}$$

In the fraction $\frac{1}{2}$, 1 is the numerator and 2 is the denominator.

To find the decimal equivalent of a proper fraction, divide the numerator by the denominator.

For example, to find the decimal equivalent of $\frac{1}{2}$, divide 1 by 2 (1 ÷ 2 = 0.5).

There are four main types of fractions: (1) proper fraction, (2) improper fraction, (3) mixed number, and (4) complex fraction.

A proper fraction is one with the numerator (top number) smaller than the denominator (bottom number). The denominator must not be zero.

Proper fraction: $\frac{4}{5}$

(the numerator is smaller than the denominator)

In an improper fraction, the numerator is equal to or greater than the denominator.

MATHEMATICAL FORMULAS AND TABLES

Improper fraction: $\dfrac{5}{4}$

(the numerator is larger than the denominator)

A mixed number contains a whole number and (added to) a proper fraction.

Mixed number: $1\dfrac{1}{4}$

(a whole number and a proper fraction)

A complex (or compound) fraction contains one or more fractions within a fraction. A complex fraction is a fraction that is made up of fractions.

Complex fraction: $\dfrac{1}{\dfrac{1}{10} + \dfrac{1}{20} + \dfrac{1}{30}}$

FINDING THE LEAST COMMON MULTIPLE

One function in working with fractions is to find the least common multiple (or lowest common denominator). The least common multiple is the smallest whole number that each denominator (bottom number) will divide into without a remainder. One method for finding the least common multiple is to list the multiples of each number (multiply by 2, 3, 4, etc.), and then select the smallest common number in each list.

Example

Find the least common multiple of 3, 5, 6, and 15.

Step 1: List the multiples of each number.

Multiples of **3** are 6, 9, 12, 15, 18, 21, 24, 27, **30,** 33, 36
Multiples of **5** are 10, 15, 20, 25, **30,** 35, 40
Multiples of **6** are 12, 18, 24, **30,** 36, 42
Multiples of **15** are 15, **30,** 45, 60

Step 2: Select the smallest common number in the collection of numbers. Since 30 is the smallest common number in each list, the least common multiple of 3, 5, 6, and 15 is 30.

Adding and Subtracting Fractions

The first step in adding and subtracting fractions is to find the least common multiple in the collection of fractions. Next, using the least common multiple, change each fraction into an equivalent fraction. Fractions are equivalent if they can be expressed so that they indicate the same quotient (proportion). An equivalent fraction can be found by multiplying (or dividing) the numerators and denominators by the same nonzero whole number. One equivalent of 4/5 is 8/10 because both the 4 and 5 have been multiplied by 2. One equivalent of 5/10 is 1/2 because both the 5 and 10 have been divided by 5. After changing each fraction into an equivalent with the lowest common multiple, add or subtract the numerators (top numbers). Finally, reduce the fraction to lowest terms (lowest form). If the result is an improper fraction, change to a mixed number.

Example: Adding Fractions

Step 1: Find the least common multiple.

$$\frac{1}{2} + \frac{4}{5}$$

The least common multiple of the two bottom numbers is 10.

Step 2: Using the least common multiple, change the fractions into equivalent fractions.

$$\frac{5}{10} + \frac{8}{10}$$

Since each fraction has the same denominator, it can be rewritten as follows.

$$\frac{5}{10} + \frac{8}{10} = \frac{5+8}{10}$$

Step 3: Add the numerators and simplify the resulting fraction.

$$\frac{5+8}{10} = \frac{13}{10} = 1\frac{3}{10}$$

Example: Subtracting Fractions

Step 1: Find the least common multiple.

$$\frac{7}{8} - \frac{2}{3}$$

The least common multiple of the two denominators is 24.

Step 2: Using the least common multiple, change the fractions into equivalent fractions.

$$\frac{21}{24} - \frac{16}{24}$$

MATHEMATICAL FORMULAS AND TABLES

MATHEMATICAL FORMULAS AND TABLES

Step 3: Rewrite the fraction.

$$\frac{21}{24} - \frac{16}{24} = \frac{21 - 16}{24}$$

Step 4: Subtract the numerators and reduce to lowest terms.

$$\frac{21 - 16}{24} = \frac{5}{24}$$

(This fraction is already at lowest terms.)

Example: Adding and Subtracting Fractions

Step 1: Find the least common multiple.

$$\frac{15}{16} - \frac{19}{32} + \frac{5}{8} - \frac{1}{4} - \frac{7}{32}$$

The least common multiple in this collection is 32.

Step 2: Using the least common multiple, change the fractions into equivalent fractions.

$$\frac{30}{32} - \frac{19}{32} + \frac{20}{32} - \frac{8}{32} - \frac{7}{32}$$

Step 3: Rewrite the fraction.

$$\frac{30}{32} - \frac{19}{32} + \frac{20}{32} - \frac{8}{32} - \frac{7}{32} = \frac{30 - 19 + 20 - 8 - 7}{32}$$

Step 4: Add and subtract the numerators and simplify.

$$\frac{30 - 19 + 20 - 8 - 7}{32} = \frac{16}{32} = \frac{1}{2}$$

MULTIPLYING FRACTIONS

The first step in multiplying fractions is to multiply the numerators (top numbers). The product (result) is the numerator in the answer. The numbers on the top line remain on the top line. Next, multiply the denominators (bottom numbers). The product is the denominator in the answer. The numbers on the bottom line remain on the bottom line. Finally, reduce the fraction to lowest terms. If the result is an improper fraction, change to a mixed number.

Example: Multiplying Fractions

Step 1: Multiply the numerators.

$$\frac{3}{4} \times \frac{2}{3} = \frac{3 \times 2}{4} = \frac{6}{}$$

Next, multiply the denominators.

$$\frac{3}{4} \times \frac{2}{3} = \frac{}{4 \times 3} = \frac{}{12}$$

Step 2: Reduce the fraction to lowest terms.

$$\frac{3 \times 2}{4 \times 3} = \frac{6}{12} = \frac{1}{2}$$

DIVIDING FRACTIONS

The process of dividing fractions actually involves multiplying. To divide fractions, multiply the reciprocal of the divisor by the dividend. The dividend (first fraction) will remain the same. The divisor (second fraction) must be changed in order to perform the calculation. The first step in dividing fractions is to replace the divisor (second fraction) with its reciprocal. To get the reciprocal, swap the numbers on the top line with the numbers on the bottom line. Thus, the numerator becomes the denominator and the denominator becomes the numerator. For example, the reciprocal of 1/2 is 2/1. During the same step, replace the division sign with a multiplication sign. Next, multiply the fractions. Finally, reduce the fraction to lowest terms.

Example 1: Dividing Fractions

Step 1: Replace the divisor (second fraction) with its reciprocal. Also, replace the division sign with a multiplication sign.

$$\frac{9}{32} \div \frac{3}{8} = \frac{9}{32} \times \frac{8}{3}$$

Step 2: Multiply the numerators and denominators.

$$\frac{9}{32} \times \frac{8}{3} = \frac{9 \times 8}{32 \times 3} \times \frac{72}{96}$$

Step 3: Reduce the fraction to lowest terms.

$$\frac{72}{96} = \frac{3}{4}$$

(Both numbers can be divided by 24 without remainder.)

Example 2: Dividing Fractions

During the second step in Example 1, prior to the multiplication, the fraction could have been simplified. The numerator of the first fraction and the denominator of the second fraction can be divided by a common factor. Likewise, the denominator of

MATHEMATICAL FORMULAS AND TABLES

the first fraction and the numerator of the second fraction can be divided by a common factor. Both 9 and 3 can be divided evenly by 3. Both 32 and 8 can be divided evenly by 8.

Step 1: Divide 9 and 3 by the highest whole number that will divide into each without remainder. Both can be divided evenly by 3.

$$\frac{\overset{3}{\cancel{9}}}{} \times \frac{}{\cancel{3}_1} = \frac{3}{} \times \frac{}{1}$$

Step 2: Divide 32 and 8 by the highest whole number that will divide into each without remainder. Both can be divided evenly by 8.

$$\frac{}{_4\cancel{32}} \times \frac{\overset{1}{\cancel{8}}}{} = \frac{}{4} \times \frac{1}{}$$

Step 3: Finish solving the problem.

$$\frac{\overset{3}{\cancel{9}}}{_4\cancel{32}} \times \frac{\overset{1}{\cancel{8}}}{\cancel{3}_1} = \frac{3}{4} \times \frac{1}{1} = \frac{3 \times 1}{4 \times 1} = \frac{3}{4}$$

By reducing the fraction before multiplication, the result is already at lowest terms. To find the decimal equivalent of a proper fraction, divide the numerator (top number) by the denominator (bottom number).

The decimal equivalent of 3/4 is 0.75 because $3 \div 4 = 0.75$.

EQUATIONS

The term *equation* is defined as an expression of equality between two quantities as shown by the equal mark (=). Every equation has two parts, the left-hand side (LHS) and the right-hand side (RHS). In a true equation, the number shown on the left-hand side of the equal mark is the same as the number shown on the right-hand side. For example, $4 + 6 = 2 \times 5$. The sum on the left-hand side (10) is equal to the product on the right-hand side (10).

If the right-hand side is equal to the left-hand side, the same amount can be added to both sides and the results will remain equal. If $a = b$, then $a + c = b + c$. The same rule holds true for subtracting, multiplying, and dividing.

$a + c = b + c$ Add the same number or quantity to both sides of an equation.

$a - c = b - c$ Subtract the same number or quantity from both sides of an equation.

$a \times c = b \times c$ Multiply the same number or quantity by both sides of an equation.

$a \div c = b \div c$ Divide the same number or quantity into both sides of an equation.

For example, in the equation 4 + 6 = 2 × 5, both sides can be divided by the number 10.

$$\frac{4+6}{10} = \frac{2 \times 5}{10}$$

The result on the LHS is equal to the result on the RHS: 1 = 1

An equation can be a procedure or formula that is used to find an unknown factor. In the electrical industry, equations are used daily. With certain known factors, equations can be used to find an unknown factor. For example, the formula $E = I \times IR$ is used to find voltage (E) when the known factors are current (I) and resistance (R). This formula can also be written $E = IR$ (when letters are side by side they must be multiplied). Letters, or variables, used in formulas must be replaced by actual numbers to solve for unknowns. Formulas are usually shown with one letter on the left-hand side and two or more letters and/or numbers on the right. For example, the formula $I = \frac{watts}{E \times 1.732 \times Eff \times PF}$ is used to find current (I) when watts, voltage (three-phase) (E), efficiency percent (Eff), and power factor percent (PF) are known.

Formulas can be rearranged or transposed. Although this guide contains the variations of each formula, at times it may be necessary to transpose an equation. The formula shown above, for example, can be transposed or rearranged to find watts if voltage (three-phase) (E), current (I), efficiency percent (Eff), and power factor percent (PF) are known. This rearranged formula would be written as follows:

$$watts = E \times 1.732 \times I \times Eff \times PF$$

Transposing Equations

Letters, or variables, can be moved from one side of the equal sign (=) to the other by changing the sign of each term moved. This procedure is called transposing. Since the same number can be added to both sides of an equation, $X - 5 = 10$ can be expressed as $X - 5 + 5 = 10 + 5$. The result is $X = 15$. Alternatively, a number or term can be moved from one side to the other by changing the sign. A negative number can be moved to the other side of the equal sign by changing it to a positive number. In the previous example, the −5 can be moved to the other side of the equation by changing it to a +5. For example, $X - 5 = 10$ becomes $X = 10 + 5$. The result again is $X = 15$.

A positive number can be moved to the other side of the equal sign by changing it to a negative number. For example, $X + 5 = 10$ becomes $X = 10 - 5$. The result is $X = 5$.

A multiplier can be moved to the other side of the equal sign by changing it to a divisor. A divisor can be moved to the other side by changing it to a multiplier.

The formula to find voltage when current and resistance are known is $E = I \times R$. To find current when voltage and resistance are known, transpose this equation by moving the R to the other side. Since the R on the RHS is a multiplier, move it to the LHS as a divisor.

$$\frac{E}{R} = I \text{ or } I = \frac{E}{R}$$

MATHEMATICAL FORMULAS AND TABLES

MATHEMATICAL FORMULAS AND TABLES

QUICK REFERENCE TABLES

The following tables are especially helpful when a calculator is not handy. Table 3.1 is a quick reference for finding squares and square roots for whole numbers 1 through 100. Table 3.2 gives the names and numerical values for scientific notation (powers of 10). See the beginning of Chapter 5 for an explanation of scientific notation. The last table, Table 3.3, provides inch and millimeter equivalents of fractions of an inch. Because of this table's arrangement, it is helpful in several ways. It is easy to convert a fraction of an inch (down to sixty-fourths) to its decimal equivalency, or to its millimeter equivalency. Finding the closest fraction of an inch to a decimal or millimeter size is also possible.

Table 3.1 Squares and Square Roots

n	n^2	$\sqrt{n}$	n	n^2	$\sqrt{n}$	n	n^2	$\sqrt{n}$	n	n^2	$\sqrt{n}$
1	1	1.0	26	676	5.099	51	2,601	7.1414	76	5,776	8.7178
2	4	1.4142	27	729	5.1962	52	2,704	7.2111	77	5,929	8.775
3	9	1.7321	28	784	5.2915	53	2,809	7.2801	78	6,084	8.8318
4	16	2.0	29	841	5.3852	54	2,916	7.3485	79	6,241	8.8882
5	25	2.2361	30	900	5.4772	55	3,025	7.4162	80	6,400	8.9443
6	36	2.4495	31	961	5.5678	56	3,136	7.4833	81	6,561	9.0
7	49	2.6458	32	1,024	5.6569	57	3,249	7.5498	82	6,724	9.0554
8	64	2.8284	33	1,089	5.7446	58	3,364	7.6158	83	6,889	9.1104
9	81	3.0	34	1,156	5.831	59	3,481	7.6811	84	7,056	9.1652
10	100	3.1623	35	1,225	5.9161	60	3,600	7.746	85	7,225	9.2195
11	121	3.3166	36	1,296	6.0	61	3,721	7.8102	86	7,396	9.2736
12	144	3.4641	37	1,369	6.0828	62	3,844	7.874	87	7,569	9.3274
13	169	3.6056	38	1,444	6.1644	63	3,969	7.9373	88	7,744	9.3808
14	196	3.7417	39	1,521	6.245	64	4,096	8.0	89	7,921	9.434
15	225	3.873	40	1,600	6.3246	65	4,225	8.0623	90	8,100	9.4868
16	256	4.0	41	1,681	6.4031	66	4,356	8.124	91	8,281	9.5394
17	289	4.1231	42	1,764	6.4807	67	4,489	8.1854	92	8,464	9.5917
18	324	4.2426	43	1,849	6.5574	68	4,624	8.2462	93	8,649	9.6437
19	361	4.3589	44	1,936	6.6332	69	4,761	8.3066	94	8,836	9.6954
20	400	4.4721	45	2,025	6.7082	70	4,900	8.3666	95	9,025	9.7468
21	441	4.5826	46	2,116	6.7823	71	5,041	8.4261	96	9,216	9.798
22	484	4.6904	47	2,209	6.8557	72	5,184	8.4853	97	9,409	9.8489
23	529	4.7958	48	2,304	6.9282	73	5,329	8.544	98	9,604	9.8995
24	576	4.899	49	2,401	7.0	74	5,476	8.6023	99	9,801	9.9499
25	625	5.0	50	2,500	7.0711	75	5,625	8.6603	100	10,000	10.0

Table 3.2 Scientific Notation (Powers of 10)

Power of 10	Numerical Value	Name
10^{18}	1 000 000 000 000 000 000	quintillion
10^{15}	1 000 000 000 000 000	quadrillion
10^{12}	1 000 000 000 000	trillion
10^{9}	1 000 000 000	billion
10^{8}	100 000 000	one-hundred million
10^{7}	10 000 000	ten million
10^{6}	1 000 000	million
10^{5}	100 000	one-hundred thousand
10^{4}	10 000	ten thousand
10^{3}	1 000	thousand
10^{2}	100	hundred
10^{1}	10	ten
10^{0}	1	one
10^{-1}	0.1	one-tenth
10^{-2}	0.01	one-hundredth
10^{-3}	0.001	one-thousandth
10^{-4}	0.000 1	one-ten thousandth
10^{-5}	0.000 01	one-hundred thousandth
10^{-6}	0.000 001	one-millionth
10^{-7}	0.000 000 1	one-ten millionth
10^{-8}	0.000 000 01	one-hundred millionth
10^{-9}	0.000 000 001	one-billionth
10^{-12}	0.000 000 000 001	one-trillionth
10^{-15}	0.000 000 000 000 001	one-quadrillionth
10^{-18}	0.000 000 000 000 000 001	one-quintillionth

MATHEMATICAL FORMULAS AND TABLES

26

MATHEMATICAL FORMULAS AND TABLES

Table 3.3   Inch and Millimeter Equivalents of Fractions

Halves	4ths	8ths	16ths	32nds	64ths	Decimals of an Inch	Millimeters
...	...	...	...	...	1/64	0.015625	0.3969
...	...	...	...	1/32	...	0.03125	0.7938
...	...	...	...	...	3/64	0.046875	1.1906
...	...	...	1/16	...	...	0.0625	1.5875
...	...	...	...	...	5/64	0.078125	1.9844
...	...	...	...	3/32	...	0.09375	2.3813
...	...	...	...	...	7/64	0.109375	2.7781
...	...	1/8	...	...	...	0.125	3.1750
...	...	...	...	...	9/64	0.140625	3.5719
...	...	...	...	5/32	...	0.15625	3.9688
...	...	...	...	...	11/64	0.171875	4.3656
...	...	...	3/16	...	...	0.1875	4.7625
...	...	...	...	...	13/64	0.203125	5.1594
...	...	...	...	7/32	...	0.21875	5.5563
...	...	...	...	...	15/64	0.234375	5.9531
...	1/4	...	...	...	...	0.25	6.3500
...	...	...	...	...	17/64	0.265625	6.7469
...	...	...	...	9/32	...	0.28125	7.1438
...	...	...	...	...	19/64	0.296875	7.5406
...	...	...	5/16	...	...	0.3125	7.9375
...	...	...	...	...	21/64	0.328125	8.3344
...	...	...	...	11/32	...	0.34375	8.7313
...	...	...	...	...	23/64	0.359375	9.1281
...	...	3/8	...	...	...	0.375	9.5250
...	...	...	...	...	25/64	0.390625	9.9219
...	...	...	...	13/32	...	0.40625	10.3188
...	...	...	...	...	27/64	0.421875	10.7156
...	...	...	7/16	...	...	0.4375	11.1125
...	...	...	...	...	29/64	0.453125	11.5094
...	...	...	...	15/32	...	0.46875	11.9063
...	...	...	...	...	31/64	0.484375	12.3031
1/2	...	...	...	...	...	0.5	12.7000
...	...	...	...	...	33/64	0.515625	13.0969
...	...	...	...	17/32	...	0.53125	13.4938
...	...	...	...	...	35/64	0.546875	13.8906
...	...	...	9/16	...	...	0.5625	14.2875
...	...	...	...	...	37/64	0.578125	14.6844
...	...	...	...	19/32	...	0.59375	15.0813
...	...	...	...	...	39/64	0.609375	15.4781
...	...	5/8	...	...	...	0.625	15.8750
...	...	...	...	...	41/64	0.640625	16.2719
...	...	...	...	21/32	...	0.65625	16.6688
...	...	...	...	...	43/64	0.671875	17.0656

Table continues below.

Table 3.3 Inch and Millimeter Equivalents of Fractions (*continued*)

Halves	4^{ths}	8^{ths}	16^{ths}	32^{nds}	64^{ths}	Decimals of an Inch	Millimeters
...	...	...	11/16	...	...	0.6875	17.4625
...	...	...	...	...	45/64	0.703125	17.8594
...	...	...	...	23/32	...	0.71875	18.2563
...	...	...	...	...	47/64	0.734375	18.6531
...	3/4	...	...	...	...	0.75	19.0500
...	...	...	...	...	49/64	0.765625	19.4469
...	...	...	...	25/32	...	0.78125	19.8438
...	...	...	...	...	51/64	0.796875	20.2406
...	...	...	13/16	...	...	0.8125	20.6375
...	...	...	...	...	53/64	0.828125	21.0344
...	...	...	...	27/32	...	0.84375	21.4313
...	...	...	...	...	55/64	0.859375	21.8281
...	...	7/8	...	...	...	0.875	22.2250
...	...	...	...	...	57/64	0.890625	22.6219
...	...	...	...	29/32	...	0.90625	23.0188
...	...	...	...	...	59/64	0.921875	23.4156
...	...	...	15/16	...	...	0.9375	23.8125
...	...	...	...	...	61/64	0.953125	24.2094
...	...	...	...	31/32	...	0.96875	24.6063
...	...	...	...	...	63/64	0.984375	25.0031
...	...	...	...	...	...	1 inch	25.4 mm

MATHEMATICAL FORMULAS AND TABLES

CHAPTER 4
GEOMETRIC FORMULAS

Chapter 4 provides a convenient reference for geometric formulas. These formulas are useful for calculating area, surface area, perimeter, length, and volume. The geometric shapes include circles, squares, rectangles, triangles, ellipses, cubes, rectangular prisms, pyramids, cones, cylinders, and spheres.

One component that is common to many of the following formulas is pi (π), which is the ratio of the circumference to the diameter of a circle. Pi is always a constant; $\pi = 3.1416$ (3.14159265359).

CIRCLE

A circle is a closed curve in a plane with every point equal in distance from a fixed point called the center. The following terms are associated with a circle and are illustrated in Figure 4.1:

Circumference—the distance around the circle

Tangent—a line that touches only one point of a circle and is perpendicular to the radius at that point

Diameter—a straight line passing through the center of the circle (twice the radius)

Radius—a straight line from the center of the circle to the curve (half the diameter)

Secant—a straight line passing through a circle

Chord—a line segment with endpoints on the circle

Arc—the portion of the circumference intercepted (cut off) by a chord

Sector—a wedge within the circle (like a slice of pie)

pi (π)—the circumference divided by the diameter (the ratio of the circumference to the diameter of a circle is always a constant); $\pi = 3.1416$

Circumference of a Circle

Either of the following formulas can be used to determine the circumference of a circle.

Circumference = $2\pi R$ or ($2 \times 3.1416 \times$ radius) or ($6.2832 \times$ radius)

Circumference = πD or ($3.1416 \times$ diameter)

30

GEOMETRIC FORMULAS

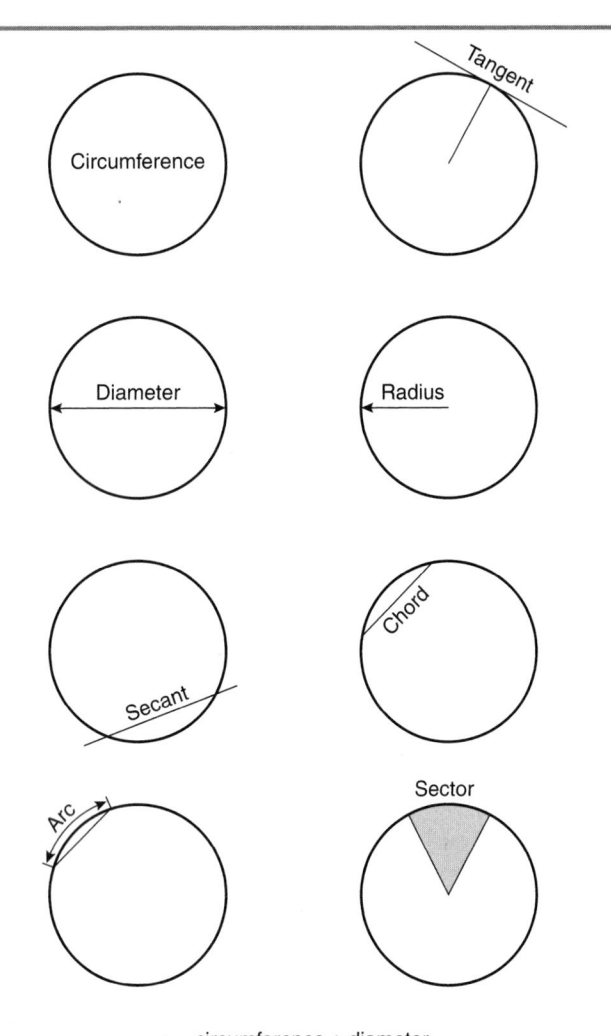

π = circumference ÷ diameter

Figure 4.1   Geometric terms associated with the circle.

Degrees of a Circle

360 Degrees = circumference
360 Degrees = $2\pi R$ or $(2 \times 3.1416 \times radius)$ or $(6.2832 \times radius)$
1 Degree = circumference ÷ 360
1 Degree = radius × 0.01745
1 Degree = diameter × 0.00873

Area of a Circle

Area = πR^2 or $(3.1416 \times radius \times radius)$

Area = $(\pi D^2)/4$ or $\left(\dfrac{3.1416 \times diameter \times diameter}{4}\right)$ or $(diameter \times diameter \times 0.7854)$

Area = $circumference^2 \times 0.0796$ or $(circumference \times circumference \times 0.0796)$

Area = half of circumference × half of diameter or $\left(\dfrac{circumference}{2} \times \dfrac{diameter}{2}\right)$

Radius of a Circle

Radius = half the length of the diameter or $\left(\dfrac{diameter}{2}\right)$
Radius = circumference × 0.15916
Radius = $\sqrt{area} \times 0.56419$ or (square root of the area × 0.56419)

Diameter of a Circle

Diameter = circumference × 0.31831
Diameter = $\sqrt{area} \times 1.12838$ or (square root of the area × 1.12838)

Length of a Circular Arc

Arc = degrees × $(\pi/180) \times R$ or (degrees × 0.01745 × radius)

Area of a Circle Sector

Sector = half of arc's length × radius or $\left(\dfrac{length\ of\ arc \times radius}{2}\right)$

Sector = $(degrees/360) \times \pi \times r^2$ or $\left(\dfrac{degrees}{360} \times 3.1416 \times radius \times radius\right)$

GEOMETRIC FORMULAS

SQUARE AND RECTANGLE

Figure 4.2 shows a square and a rectangle.

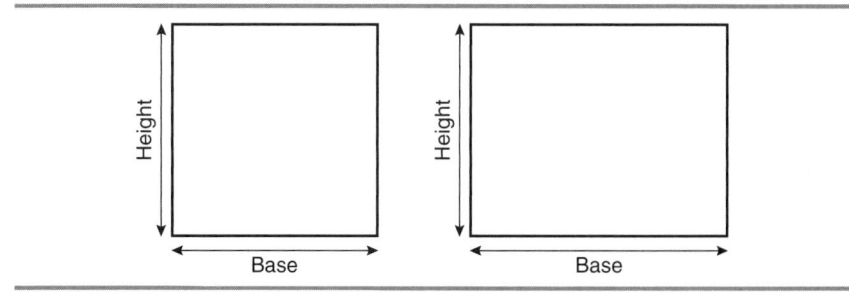

Figure 4.2   Square and rectangle.

Area = base × height

TRIANGLE

A triangle is a three-sided polygon. The sum of the angles of a triangle is 180 degrees. Triangles are classified as acute, obtuse, and right triangles. All the angles on an acute triangle are less than 90 degrees. An obtuse triangle has one angle greater than 90 degrees. A right triangle (see Figures 4.3 and 4.4) has one 90 degree angle. An oblique triangle (see Figures 4.5 and 4.6) is any triangle not having a right (90 degree) angle. An oblique triangle can be an acute or obtuse triangle.

A triangle having all three sides of equal length is an equilateral triangle (see Figure 4.7). Therefore, each angle measures 60 degrees. A triangle with two sides of equal length is an isosceles triangle. Two angles of an isosceles triangle will be the same. A triangle with each side having a different length is a scalene triangle.

Right Triangle

Figure 4.3 shows a right triangle.

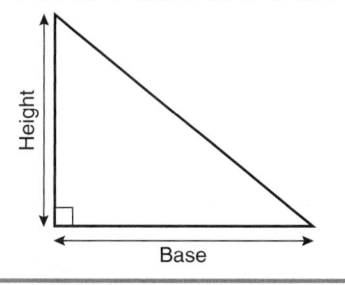

Figure 4.3   Right triangle.

Area = base × height × 0.5

The two sides that form the right (90 degree) angle are called the legs. The side opposite the right angle is called the hypotenuse. The Pythagorean Theorem states that the square of the hypotenuse of a right triangle is equal to the sum of the squares of the two legs.

Pythagorean Theorem (see Figure 4.4):

$A^2 + B^2 = C^2$

$A = \sqrt{C^2 - B^2}$

$B = \sqrt{C^2 - A^2}$

$C = \sqrt{A^2 + B^2}$

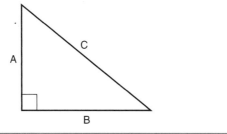

Figure 4.4   Illustration of Pythagorean triangle.

Oblique Triangle

Figure 4.5 shows an oblique triangle.

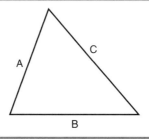

Figure 4.5   Oblique triangle.

GEOMETRIC FORMULAS

GEOMETRIC FORMULAS

The area of an oblique triangle can be determined as follows:

Area = $\sqrt{S \times (S - A) \times (S - B) \times (S - C)}$

Where S = (A + B + C) × 0.5

The area of an oblique triangle can also be calculated as two right triangles (see Figure 4.6).

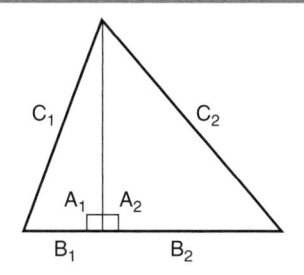

Figure 4.6   Oblique triangle calculated as two right triangles.

Equilateral Triangle

Figure 4.7 shows an equilateral triangle.

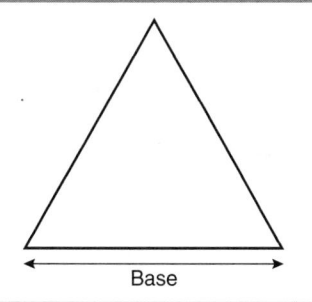

Figure 4.7   Equilateral triangle.

Height (Altitude) = base × $\sqrt{3}$ × 0.5 or (base × 0.866)

Area = base × height × 0.5

Area = base^2 × $\sqrt{3}$ × 0.25 or (base × base × 0.433)

ELLIPSE

Figure 4.8 shows ellipses.

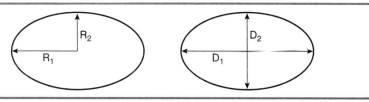

Figure 4.8 Ellipses.

Area = $\pi R_1 R_2$ or (3.1416 × radius$_1$ × radius$_2$)
Area = π × diameter$_1$ × diameter$_2$ × 0.25 or (diameter$_1$ × diameter$_2$ × 0.7854)

CUBE

Figure 4.9 shows a cube.

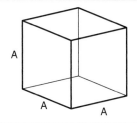

Figure 4.9 Cube.

Surface Area = $6 \times A^2$ or (6 × A × A)
Volume = A^3 or (A × A × A)

GEOMETRIC FORMULAS

36

GEOMETRIC FORMULAS

RECTANGULAR PRISM

Figure 4.10 shows a rectangular prism.

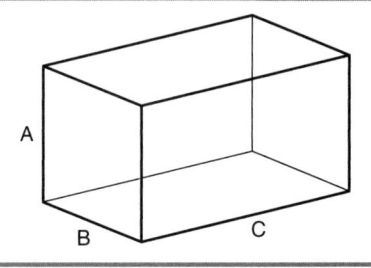

Figure 4.10   Rectangular prism.

Surface Area = (2 × A × B) + (2 × B × C) + (2 × A × C)

Volume = A × B × C

PYRAMID

Figure 4.11 shows a pyramid.

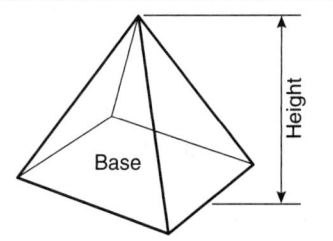

Figure 4.11   Pyramid.

Volume = (area of base) × (height of pyramid) × (1/3)

CONE

Figure 4.12 shows a cone.

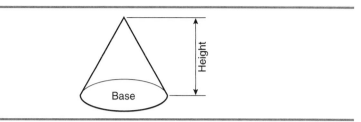

Figure 4.12 Cone.

Volume = (area of base) × (height of pyramid) × (1/3)

CYLINDER

Figure 4.13 shows a cylinder.

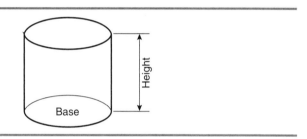

Figure 4.13 Cylinder.

Volume = (area of base) × (height)

38

GEOMETRIC FORMULAS

SPHERE

Figure 4.14 shows a sphere.

Figure 4.14   Sphere.

Surface Area = $4 \times \pi \times R^2$ or (12.5664 × radius × radius)

Surface Area = $\pi \times D^2$ or (3.1416 × diameter × diameter)

Volume = $4/3 \times \pi \times R^3$ or (4.1888 × radius × radius × radius)

Volume = $0.5236 \times D^3$ or (0.5236 × diameter × diameter × diameter)

CHAPTER 5
CONVERSION FACTORS

Chapter 5 contains ten tables that may be used by electrical system designers, electrical engineers, and on occasion, electricians for making a wide variety of conversions between SI and U.S. customary units of measurement.

The conversion factors are divided into ten tables (see Tables 5.1 through 5.10). Except for the last table (temperature), all tables include scientific notation (powers of 10). With the use of scientific notation, the conversion factors have manageable numbers. The conversion factors consist of five digits, and most are followed by the power of 10. The exponent of 10 is the raised number to the right of 10. An exponent is the number of places the decimal must be moved. Where the exponent of 10 is a positive number, move the decimal point to the right. Where the exponent is a negative number, move the decimal point to the left. For example, for a positive exponent (move decimal point to the right): $1.9735 \times 10^5 = 197{,}350$. For a negative exponent (move decimal point to the left): $1.5783 \times 10^{-5} = 0.000015783$.

Table 5.1 Length and Distance

Convert From	Conversion Factor		To
Centimeter	× 3.9370×10^{-1}	=	____ Inch
Centimeter	× 1.0000×10^{-2}	=	____ Meter
Centimeter	× 1.0000×10^{1}	=	____ Millimeter
Chain	× 6.6000×10^{1}	=	____ Foot
Chain	× 1.0000×10^{-1}	=	____ Furlong
Chain	× 4.0000	=	____ Rod
Chain	× 2.2000×10^{1}	=	____ Yard
Cubit	× 1.5000	=	____ Foot
Fathom	× 6.0000	=	____ Foot
Foot	× 3.0480×10^{1}	=	____ Centimeter
Foot	× 1.5152×10^{-2}	=	____ Chain
Foot	× 6.6667×10^{-1}	=	____ Cubit
Foot	× 1.6667×10^{-1}	=	____ Fathom
Foot	× 1.5152×10^{-3}	=	____ Furlong
Foot	× 1.2000×10^{1}	=	____ Inch
Foot	× 3.0480×10^{-1}	=	____ Meter
Foot	× 6.0606×10^{-2}	=	____ Rod
Foot	× 3.3333×10^{-1}	=	____ Yard
Furlong	× 1.0000×10^{1}	=	____ Chain
Furlong	× 6.6000×10^{2}	=	____ Foot
Furlong	× 4.0000×10^{1}	=	____ Rod
Furlong	× 2.2000×10^{2}	=	____ Yard
Inch	× 2.5400	=	____ Centimeter
Inch	× 8.3333×10^{-2}	=	____ Foot

Continued

40

CONVERSION FACTORS

Table 5.1   Length and Distance *(continued)*

Convert From	Conversion Factor		To
Inch	×	1.0000×10^3 =	____ Mil
Inch	×	2.5400×10^1 =	____ Millimeter
Inch	×	2.7778×10^{-2} =	____ Yard
Kilometer	×	3.2808×10^3 =	____ Foot
Kilometer	×	1.0000×10^3 =	____ Meter
Kilometer	×	6.2137×10^{-1} =	____ Mile
Kilometer	×	1.0936×10^3 =	____ Yard
Knot	×	1.0000 =	____ Miles (nautical) per hour
Knot	×	1.1508 =	____ Miles per hour
League	×	3.0000 =	____ Mile
Meter	×	1.0000×10^2 =	____ Centimeter
Meter	×	3.2808 =	____ Foot
Meter	×	3.9370×10^1 =	____ Inch
Meter	×	1.0000×10^{-3} =	____ Kilometer
Meter	×	6.2137×10^{-4} =	____ Mile
Meter	×	1.0000×10^3 =	____ Millimeter
Meter	×	1.9884×10^{-1} =	____ Rod
Meter	×	1.0936 =	____ Yard
Mil	×	1.0000×10^{-3} =	____ Inch
Mil	×	2.5400×10^{-2} =	____ Millimeter
Mile	×	8.0000×10^1 =	____ Chain
Mile	×	5.2800×10^3 =	____ Foot
Mile	×	8.0000 =	____ Furlong
Mile	×	1.6093 =	____ Kilometer
Mile	×	3.3333×10^{-1} =	____ League
Mile	×	3.2000×10^2 =	____ Rod
Mile	×	1.7600×10^3 =	____ Yard
Mile (nautical)	×	1.1508 =	____ Mile (statute)
Mile (statute)	×	8.6898×10^{-1} =	____ Mile (nautical)
Miles (nautical) per hour	×	1.0000 =	____ Knot
Miles per hour	×	8.6898×10^{-1} =	____ Knot
Millimeter	×	1.0000×10^{-1} =	____ Centimeter
Millimeter	×	3.9370×10^{-2} =	____ Inch
Nautical league	×	3.0000 =	____ Nautical mile
Nautical mile	×	3.3333×10^{-1} =	____ Nautical league
Rod	×	2.5000×10^{-1} =	____ Chain
Rod	×	2.7500 =	____ Fathom
Rod	×	2.5000×10^{-2} =	____ Furlong
Rod	×	5.5000 =	____ Yard
Yard	×	3.0000 =	____ Foot
Yard	×	3.6000×10^1 =	____ Inch
Yard	×	9.1440×10^{-1} =	____ Meter

Note: Unless otherwise noted, Mile = mile (statute)
Note: Thousandth of an inch = mil

Table 5.2 Area

Convert From	Conversion Factor		To
Acre	× 4.0000	=	___ Rood
Acre	× 1.0000 × 10^1	=	___ Square chain
Acre	× 4.3560 × 10^4	=	___ Square foot
Acre	× 1.6000 × 10^2	=	___ Square rod
Circular inch	× 7.8540 × 10^{-1}	=	___ Square inch
Circular mil	× 7.8540 × 10^{-1}	=	___ Square mil
Hectare	× 2.4711	=	___ Acre
Rood	× 4.0000 × 10^1	=	___ Square rod
Square centimeter	× 1.5500 × 10^{-1}	=	___ Square inch
Square chain	× 1.0000 × 10^{-1}	=	___ Acre
Square foot	× 1.4400 × 10^2	=	___ Square inch
Square foot	× 1.1111 × 10^{-1}	=	___ Square yard
Square inch	× 1.2732	=	___ Circular inch
Square inch	× 6.4516	=	___ Square centimeter
Square kilometer	× 2.4711 × 10^2	=	___ Acre
Square kilometer	× 3.8610 × 10^{-1}	=	___ Square mile
Square meter	× 1.0764 × 10^1	=	___ Square foot
Square meter	× 1.1960	=	___ Square yard
Square mil	× 1.2732	=	___ Circular mil
Square mile	× 6.4000 × 10^2	=	___ Acre
Square mile	× 2.5900	=	___ Square kilometer
Square mile	× 7.5512 × 10^{-1}	=	___ Square nautical mile
Square millimeter	× 1.9735 × 10^3	=	___ Circular mil
Square millimeter	× 1.5500 × 10^{-3}	=	___ Square inch
Square millimeter	× 1.5500 × 10^3	=	___ Square mil
Square nautical mile	× 1.3243	=	___ Square mile
Square rod	× 3.0250 × 10^1	=	___ Square yard
Square yard	× 9.0000	=	___ Square foot
Square yard	× 8.3613 × 10^{-1}	=	___ Square meter

CONVERSION FACTORS

42

CONVERSION FACTORS

Table 5.3   Volume

Convert From		Conversion Factor		To
Bushel	×	1.2445	= ___	Cubic foot
Bushel	×	4.0000	= ___	Peck
Cord	×	1.2800×10^2	= ___	Cubic foot
Cubic centimeter	×	6.1024×10^{-2}	= ___	Cubic inch
Cubic foot	×	8.0356×10^{-1}	= ___	Bushel
Cubic foot	×	1.7280×10^3	= ___	Cubic inch
Cubic foot	×	3.7037×10^{-2}	= ___	Cubic yard
Cubic foot	×	2.8316×10^1	= ___	Liter
Cubic foot	×	7.4805	= ___	US gallon
Cubic inch	×	1.6387×10^1	= ___	Cubic centimeter
Cubic inch	×	5.5411×10^{-1}	= ___	US fluid ounce
Cubic meter	×	3.5315×10^1	= ___	Cubic foot
Cubic meter	×	1.3080	= ___	Cubic yard
Cubic meter	×	1.0000×10^3	= ___	Liter
Cubic meter	×	1.0000	= ___	Stere
Cubic yard	×	2.7000×10^1	= ___	Cubic foot
Cubic yard	×	7.6455×10^{-1}	= ___	Cubic meter
Deciliter	×	1.0000×10^{-1}	= ___	Liter
Dekaliter	×	1.0000×10^1	= ___	Liter
Liter	×	1.0000×10^3	= ___	Cubic centimeter
Liter	×	6.1025×10^1	= ___	Cubic inch
Liter	×	1.7598	= ___	UK pint
Liter	×	1.0567	= ___	US quart
Milliliter	×	1.0000×10^{-1}	= ___	Centiliter
Peck	×	2.5000×10^{-1}	= ___	Bushel
Stere	×	1.0000	= ___	Cubic meter
UK fluid ounce	×	9.6076×10^{-1}	= ___	US fluid ounce
UK gallon	×	4.5460	= ___	Liter
UK gallon	×	8.0000	= ___	UK pint
UK gallon	×	1.2009	= ___	US gallon
UK pint	×	1.2009	= ___	US pint
US fluid ounce	×	1.8047	= ___	Cubic inch
US fluid ounce	×	1.0408	= ___	UK fluid ounce
US fluid ounce	×	6.2500×10^{-2}	= ___	US pint
US gallon	×	3.2000×10^1	= ___	Gill (US)
US gallon	×	3.7853	= ___	Liter
US gallon	×	8.3267×10^{-1}	= ___	UK gallon
US gallon	×	1.2800×10^2	= ___	US fluid ounce
US gallon	×	8.5937×10^{-1}	= ___	US gallon (dry)
US gallon	×	8.0000	= ___	US pint
US gallon	×	4.0000	= ___	US quart
US gallon (dry)	×	1.1636	= ___	US gallon
US pint	×	2.8875×10^1	= ___	Cubic inch
US pint	×	8.3267×10^{-1}	= ___	UK pint
US pint	×	1.6000×10^1	= ___	US fluid ounce

Table continues below.

Table 5.3 Volume *(continued)*

Convert From	Conversion Factor		To
US pint	× 8.8594 × 10⁻¹	=	___ US pint (dry)
US pint	× 5.0000 × 10⁻¹	=	___ US quart
US pint (dry)	× 1.1636	=	___ US pint
US quart	× 5.7500 × 10¹	=	___ Cubic inch
US quart	× 8.0000 × 10¹	=	___ Gill (US)
US quart	× 9.4635 × 10⁻¹	=	___ Liter
US quart	× 8.3267 × 10⁻¹	=	___ UK quart
US quart	× 3.2000 × 10¹	=	___ US fluid ounce
US quart	× 2.5000 × 10⁻¹	=	___ US gallon
US quart	× 2.0000	=	___ US pint
US quart	× 8.5937 × 10⁻¹	=	___ US quart (dry)
US quart (dry)	× 1.1636	=	___ US quart

Note: Unless otherwise noted, US pint = US pint (liquid)
Note: Unless otherwise noted, US quart = US quart (liquid)
Note: Unless otherwise noted, US gallon = US gallon (liquid)

Table 5.4 Weight

Convert From	Conversion Factor		To
Carat	× 2.0000 × 10⁻¹	=	___ Gram
Dram	× 1.7719	=	___ Gram
Grain	× 6.4799 × 10⁻²	=	___ Gram
Grain	× 6.4799 × 10¹	=	___ Milligram
Grain	× 4.1667 × 10⁻²	=	___ Pennyweight
Gram	× 5.0000	=	___ Carat
Gram	× 5.6437 × 10⁻¹	=	___ Dram
Gram	× 1.5432 × 10¹	=	___ Grain
Gram	× 3.5274 × 10⁻²	=	___ Ounce
Kilogram	× 3.5274 × 10¹	=	___ Ounce
Kilogram	× 2.2046	=	___ Pound
Kilogram	× 1.5747 × 10⁻¹	=	___ Stone
Metric ton (tonne)	× 2.2046 × 10³	=	___ Pound
Metric ton (tonne)	× 9.8421 × 10⁻¹	=	___ Ton (long)
Metric ton (tonne)	× 1.1023	=	___ Ton (short)
Ounce	× 1.6000 × 10¹	=	___ Dram
Ounce	× 4.3750 × 10²	=	___ Grain
Ounce	× 2.8350 × 10¹	=	___ Gram
Ounce	× 1.8229 × 10¹	=	___ Pennyweight
Ounce	× 6.2500 × 10⁻²	=	___ Pound
Pennyweight	× 2.4000 × 10¹	=	___ Grain
Pennyweight	× 5.4857 × 10⁻²	=	___ Ounce
Pound	× 2.5600 × 10²	=	___ Dram

Continued

CONVERSION FACTORS

44

CONVERSION FACTORS

Table 5.4   Weight (continued)

Convert From	Conversion Factor			To
Pound	×	7.0000×10^3	= ___	Grain
Pound	×	4.5359×10^2	= ___	Gram
Pound	×	1.6000×10^1	= ___	Ounce
Pound	×	3.1081×10^{-2}	= ___	Slug
Slug	×	1.4594×10^1	= ___	Kilogram
Slug	×	3.2174×10^1	= ___	Pound
Stone	×	6.3503	= ___	Kilogram
Stone	×	1.4000×10^1	= ___	Pound
Ton (long)	×	1.0160	= ___	Metric ton (tonne)
Ton (long)	×	2.2400×10^3	= ___	Pound
Ton (long)	×	1.1200	= ___	Ton (short)
Ton (short)	×	9.0718×10^{-1}	= ___	Metric ton (tonne)
Ton (short)	×	2.0000×10^3	= ___	Pound
Ton (short)	×	6.2162×10^1	= ___	Slug
Ton (short)	×	8.9286×10^{-1}	= ___	Ton (long)

Note: Unless otherwise noted, Dram = Dram (avoirdupois)
Note: Unless otherwise noted, Ounce = Ounce (avoirdupois)
Note: Unless otherwise noted, Pound = Pound (avoirdupois)

Table 5.5   Pressure

Convert From	Math Function			To
Atmosphere	×	1.0133	= ___	Bar
Atmosphere	×	7.6000×10^1	= ___	Centimeter of mercury (0°C)
Atmosphere	×	3.3899×10^1	= ___	Foot of water (4°C)
Atmosphere	×	2.9921×10^1	= ___	Inch of mercury (0°C)
Atmosphere	×	4.0678×10^2	= ___	Inch of water (4°C)
Atmosphere	×	1.4696×10^1	= ___	Pound-force/Square inch
Atmosphere	×	8.5027	= ___	Ton-force/Square yard
Bar	×	9.8692×10^{-1}	= ___	Atmosphere
Dyne/Square centimeter	×	1.0000×10^{-6}	= ___	Bar
Foot of seawater	×	9.7419×10^{-1}	= ___	Foot of water (4°C)
Foot of water (4°C)	×	2.9500×10^{-2}	= ___	Atmosphere
Foot of water (4°C)	×	1.0265	= ___	Foot of seawater
Foot of water (4°C)	×	8.8267×10^{-1}	= ___	Inch of mercury (0°C)
Foot of water (4°C)	×	1.2000×10^1	= ___	Inch of water (4°C)
Foot of water (4°C)	×	3.0480×10^2	= ___	Kilogram-force/Square meter
Foot of water (4°C)	×	6.2428×10^1	= ___	Pound-force/Square foot
Inch of mercury (0°C)	×	3.3421×10^{-2}	= ___	Atmosphere
Inch of mercury (0°C)	×	1.1329	= ___	Foot of water (4°C)
Inch of mercury (0°C)	×	1.3595×10^1	= ___	Inch of water (4°C)
Inch of mercury (0°C)	×	3.4532×10^2	= ___	Kilogram-force/Square meter

Table continues below.

Table 5.5 Pressure *(continued)*

Convert From	Math Function	To
Inch of mercury (0°C)	× 2.5400×10^1	___ Millimeter of mercury (0°C)
Inch of mercury (0°C)	× 4.9115×10^{-1}	___ Pound-force/Square inch
Inch of water (4°C)	× 7.3556×10^{-2} =	___ Inch of mercury (0°C)
Inch of water (4°C) Square	× 2.5400×10^1 =	___ Kilogram-force/Square meter
Inch of water (4°C)	× 1.8683 =	___ Millimeter of mercury (0°C)
Inch of water (4°C)	× 5.2023 =	___ Pound-force/Square foot
Kilogram-force/Square meter	× 9.8067 =	___ Pascal (Newton/Square meter)
Kilogram-force/Square meter	× 6.5898 =	___ Poundal/Square foot
Kilogram-force/Square meter	× 2.0482×10^{-1} =	___ Pound-force/Square foot
Millibar	× 1.0000×10^2 =	___ Pascal (Newton/Square meter)
Millimeter of mercury (0°C)	× 5.3524×10^{-1} =	___ Inch of water (4°C)
Millimeter of mercury (0°C)	× 1.3595×10^1 =	___ Kilogram-force/Square meter
Millimeter of mercury (0°C)	× 2.7845 =	___ Pound-force/Square foot
Ounce-force/Square inch	× 1.7300 =	___ Inch of water (4°C)
Pascal (Newton/Square meter)	× 1.0000×10^1 =	___ Dyne/Square centimeter
Pascal (Newton/Square meter)	× 1.0197×10^{-1} =	___ Kilogram-force/Square meter
Pascal (Newton/Square meter)	× 2.0885×10^{-2} =	___ Pound-force/Square foot
Poundal/Square foot	× 1.4882 =	___ Pascal (Newton/Square meter)
Poundal/Square foot	× 3.1081×10^{-2} =	___ Pound-force/Square foot
Poundal/Square inch	× 8.6032×10^{-1} =	___ Inch of water (4°C)
Poundal/Square inch	× 4.4757 =	___ Pound-force/Square foot
Pound-force/Square foot	× 1.9222×10^{-1} =	___ Inch of water (4°C)
Pound-force/Square foot	× 4.8824 =	___ Kilogram-force/Square meter
Pound-force/Square foot	× 4.7880×10^1 =	___ Pascal (Newton/Square meter)
Pound-force/Square foot	× 3.2174×10^1 =	___ Poundal/Square foot
Pound-force/Square inch	× 6.8046×10^{-2} =	___ Atmosphere
Pound-force/Square inch	× 2.3067 =	___ Foot of water (4°C)
Pound-force/Square inch	× 2.0360 =	___ Inch of mercury (0°C)
Pound-force/Square inch	× 2.7680×10^1 =	___ Inch of water (4°C)
Pound-force/Square inch	× 5.1715×10^1 =	___ Millimeter of mercury (0°C)
Pound-force/Square inch	× 1.6000×10^1 =	___ Ounce-force/Square inch
Pound-force/Square inch	× 3.2174×10^1 =	___ Poundal/Square inch
Ton-force/Square foot	× 1.0585 =	___ Atmosphere
Ton-force/Square foot	× 2.2400×10^3 =	___ Pound-force/Square foot
Ton-force/Square inch	× 1.4400×10^2 =	___ Ton-force/Square foot
Ton-force/Square yard	× 1.1761×10^{-1} =	___ Atmosphere
Ton-force/Square yard	× 3.9868 =	___ Foot of water (4°C)
Ton-force/Square yard	× 3.5190 =	___ Inch of mercury (0°C)
Ton-force/Square yard	× 4.7842×10^1 =	___ Inch of water (4°C)
Ton-force/Square yard	× 1.7284 =	___ Pound-force/Square inch

Note: Kilogram-force/Square meter = millimeter of water
Note: Pound-force/Square inch = PSI

CONVERSION FACTORS

46

CONVERSION FACTORS

Table 5.6 Force

Convert From	Conversion Factor		To
Dyne	×	1.0197×10^{-3}	= ___ Gram-force
Dyne	×	1.0000×10^{-5}	= ___ Newton
Gram-force	×	9.8067×10^{2}	= ___ Dyne
Gram-force	×	2.2046×10^{-3}	= ___ Pound-force
Kilogram-force (kilopond)	×	9.8067	= ___ Newton
Kilogram-force (kilopond)	×	2.2046	= ___ Pound-force
Newton	×	1.0000×10^{5}	= ___ Dyne
Newton	×	3.5969	= ___ Ounce-force
Newton	×	2.2481×10^{-1}	= ___ Pound-force
Ounce-force	×	2.8350×10^{1}	= ___ Gram-force
Ounce-force	×	2.7801×10^{-1}	= ___ Newton
Poundal	×	1.4098×10^{1}	= ___ Gram-force
Poundal	×	3.1081×10^{-2}	= ___ Pound-force
Pound-force	×	4.5359×10^{2}	= ___ Gram-force
Pound-force	×	4.5359×10^{-1}	= ___ Kilogram-force (kilopond)
Pound-force	×	4.4482	= ___ Newton
Pound-force	×	3.2174×10^{1}	= ___ Poundal
Ton-force	×	9.9640×10^{3}	= ___ Newton
Ton-force	×	2.2400×10^{3}	= ___ Pound-force

Table 5.7 Power

Convert From	Conversion Factor		To
Btu/Hour	×	2.9307×10^{-4}	= ___ Kilowatt
Btu/Hour	×	2.9307×10^{-1}	= ___ Watt
Btu/Minute	×	4.1999	= ___ Calorie/Second
Btu/Minute	×	1.2969×10^{1}	= ___ Foot-pound/Second
Btu/Minute	×	2.3581×10^{-2}	= ___ Horsepower
Btu/Minute	×	1.7931	= ___ Kilogram-meter/Second
Btu/Minute	×	1.7584×10^{1}	= ___ Watt
Btu/Second	×	1.0550×10^{3}	= ___ Watt
Calorie/Second	×	2.3810×10^{-1}	= ___ Btu/Minute
Erg/Second	×	1.0197×10^{-3}	= ___ Gram-centimeter/Second
Erg/Second	×	1.0000×10^{-7}	= ___ Watt
Foot-pound/Minute	×	3.0303×10^{-5}	= ___ Horsepower
Foot-pound/Minute	×	2.2597×10^{-5}	= ___ Kilowatt
Foot-pound/Minute	×	2.2597×10^{-2}	= ___ Watt
Foot-pound/Second	×	7.7104×10^{-2}	= ___ Btu/Minute
Foot-pound/Second	×	3.2174×10^{1}	= ___ Foot-poundal/Second
Foot-pound/Second	×	1.8182×10^{-3}	= ___ Horsepower
Foot-pound/Second	×	1.1658	= ___ Kilocalorie/Hour

Table continues below.

Table 5.7 Power *(continued)*

Convert From	Conversion Factor		To
Foot-pound/Second	× 1.3825×10^{-1}	= ___	Kilogram-meter/Second
Foot-pound/Second	× 1.3558×10^{-3}	= ___	Kilowatt
Foot-pound/Second	× 1.3558	= ___	Watt
Foot-poundal/Second	× 3.1081×10^{-2}	= ___	Foot-pound/Second
Gram-centimeter/Second	× 9.8067×10^{2}	= ___	Erg/Second
Gram-centimeter/Second	× 9.8067×10^{-5}	= ___	Watt
Horsepower	× 2.5444×10^{3}	= ___	Btu/Hour
Horsepower	× 4.2407×10^{1}	= ___	Btu/Minute
Horsepower	× 3.3000×10^{4}	= ___	Foot-pound/Minute
Horsepower	× 5.5000×10^{2}	= ___	Foot-pound/Second
Horsepower	× 1.0139	= ___	Horsepower (metric)
Horsepower	× 1.0686×10^{1}	= ___	Kilocalorie/Minute
Horsepower	× 7.6040×10^{1}	= ___	Kilogram-meter/Second
Horsepower	× 7.4570×10^{2}	= ___	Watt
Horsepower (metric)	× 4.1827×10^{1}	= ___	Btu/Minute
Horsepower (metric)	× 32550	= ___	Foot-pound/Minute
Horsepower (metric)	× 9.8632×10^{-1}	= ___	Horsepower
Horsepower (metric)	× 7.5000×10^{1}	= ___	Kilogram-meter/Second
Horsepower (metric)	× 7.3550×10^{2}	= ___	Watt
Kilocalorie/Day	× 4.8458×10^{-2}	= ___	Watt
Kilocalorie/Hour	× 8.5778×10^{-1}	= ___	Foot-pound/Second
Kilocalorie/Hour	× 1.1630	= ___	Watt
Kilocalorie/Minute	× 5.1467×10^{1}	= ___	Foot-pound/Second
Kilocalorie/Minute	× 7.1156	= ___	Kilogram-meter/Second
Kilocalorie/Minute	× 6.9780×10^{1}	= ___	Watt
Kilocalorie/Second	× 2.3810×10^{2}	= ___	Btu/Minute
Kilocalorie/Second	× 6.0000×10^{1}	= ___	Kilocalorie/Minute
Kilocalorie/Second	× 4.1868	= ___	Kilowatt
Kilogram-meter/Second	× 5.5769×10^{-1}	= ___	Btu/Minute
Kilogram-meter/Second	× 7.2330	= ___	Foot-pound/Second
Kilogram-meter/Second	× 9.8067	= ___	Watt
Kilowatt	× 3.4121×10^{3}	= ___	Btu/Hour
Kilowatt	× 5.6869×10^{1}	= ___	Btu/Minute
Kilowatt	× 4.4254×10^{4}	= ___	Foot-pound/Minute
Kilowatt	× 7.3756×10^{2}	= ___	Foot-pound/Second
Kilowatt	× 1.3410	= ___	Horsepower
Kilowatt	× 1.3596	= ___	Horsepower (metric)
Kilowatt	× 1.0197×10^{2}	= ___	Kilogram-meter/Second
Watt	× 3.4121	= ___	Btu/Hour
Watt	× 5.6869×10^{-2}	= ___	Btu/Minute
Watt	× 1.0000×10^{7}	= ___	Erg/Second
Watt	× 4.4254×10^{1}	= ___	Foot-pound/Minute
Watt	× 1.0197×10^{4}	= ___	Gram-centimeter/Second
Watt	× 1.3410×10^{-3}	= ___	Horsepower
Watt	× 2.0636×10^{1}	= ___	Kilocalorie/Day
Watt	× 6.1182	= ___	Kilogram-meter/Minute

CONVERSION FACTORS

48

CONVERSION FACTORS

Table 5.8   Energy

Convert From	Conversion Factor		To
British thermal unit (Btu)	×	5.5556×10^{-1}	= ___ Celsius heat unit
Btu	×	2.5200×10^{2}	= ___ Calorie
Btu	×	7.7817×10^{2}	= ___ Foot-pound
Btu	×	3.9301×10^{-4}	= ___ Horsepower-hour
Btu	×	1.0551×10^{3}	= ___ Joule
Btu	×	1.0759×10^{2}	= ___ Kilogram-meter
Btu	×	2.9307×10^{-4}	= ___ Kilowatt-hour
Btu	×	2.9307×10^{-1}	= ___ Watt-hour
Calorie	×	3.9683×10^{-3}	= ___ Btu
Calorie	×	3.0880	= ___ Foot-pound
Calorie	×	4.1868	= ___ Joule
Calorie	×	4.2693×10^{-1}	= ___ Kilogram-meter
Calorie	×	1.1630×10^{-3}	= ___ Watt-hour
Celsius heat unit	×	1.8000	= ___ Btu
Dyne-centimeter	×	1.0000	= ___ Erg
Erg	×	1.0000	= ___ Dyne-centimeter
Erg	×	7.3756×10^{-8}	= ___ Foot-pound
Erg	×	1.0197×10^{-3}	= ___ Gram-centimeter
Erg	×	2.7778×10^{-14}	= ___ Kilowatt-hour
Foot-pound	×	1.2851×10^{-3}	= ___ Btu
Foot-pound	×	3.2383×10^{-1}	= ___ Calorie
Foot-pound	×	1.3558×10^{7}	= ___ Erg
Foot-pound	×	3.2174×10^{1}	= ___ Foot-poundal
Foot-pound	×	5.0505×10^{-7}	= ___ Horsepower-hour
Foot-pound	×	1.3558	= ___ Joule
Foot-pound	×	1.3825×10^{-1}	= ___ Kilogram-meter
Foot-pound	×	3.7662×10^{-7}	= ___ Kilowatt-hour
Foot-pound	×	3.7662×10^{-4}	= ___ Watt-hour
Foot-poundal	×	3.1081×10^{-2}	= ___ Foot-pound
Gram-centimeter	×	9.8067×10^{2}	= ___ Erg
Horsepower-hour	×	2.5444×10^{3}	= ___ Btu
Horsepower-hour	×	1.9800×10^{6}	= ___ Foot-pound
Horsepower-hour	×	6.4119×10^{2}	= ___ Kilocalorie
Horsepower-hour	×	7.4570×10^{-1}	= ___ Kilowatt-hour
Horsepower-hour	×	7.4570×10^{2}	= ___ Watt-hour
Horsepower-hour (metric)	×	2.5096×10^{3}	= ___ Btu
Horsepower-hour (metric)	×	7.3550×10^{-1}	= ___ Kilowatt-hour
Joule	×	9.4782×10^{-4}	= ___ Btu
Joule	×	2.3885×10^{-1}	= ___ Calorie
Joule	×	7.3756×10^{-1}	= ___ Foot-pound
Joule	×	1.0197×10^{-1}	= ___ Kilogram-meter
Joule	×	2.7778×10^{-7}	= ___ Kilowatt-hour
Joule	×	2.7778×10^{-4}	= ___ Watt-hour
Joule	×	1.0000	= ___ Watt-second
Kilocalorie	×	3.9683	= ___ Btu

Table continues below.

Table 5.8 Energy *(continued)*

Convert From	Conversion Factor		To
Kilocalorie	× 4.1868×10^3	=	___ Joule
Kilocalorie	× 1.1630	=	___ Watt-hour
Kilogram-meter	× 2.3423	=	___ Calorie
Kilogram-meter	× 7.2330	=	___ Foot-pound
Kilogram-meter	× 9.8067	=	___ Joule
Kilogram-meter	× 2.7241×10^{-3}	=	___ Watt-hour
Kilowatt-hour	× 3.4121×10^3	=	___ Btu
Kilowatt-hour	× 3.6000×10^{13}	=	___ Erg
Kilowatt-hour	× 2.6552×10^6	=	___ Foot-pound
Kilowatt-hour	× 1.3410	=	___ Horsepower-hour
Kilowatt-hour	× 1.3597	=	___ Horsepower-hour (metric)
Kilowatt-hour	× 3.6000×10^6	=	___ Joule
Kilowatt-hour	× 8.5985×10^2	=	___ Kilocalorie
Therm	× 1.0000×10^5	=	___ Btu
Watt-hour	× 3.4121	=	___ Btu
Watt-hour	× 8.5985×10^2	=	___ Calorie
Watt-hour	× 2.6552×10^3	=	___ Foot-pound
Watt-hour	× 1.3410×10^{-3}	=	___ Horsepower-hour
Watt-hour	× 3.6000×10^3	=	___ Joule
Watt-hour	× 3.6710×10^2	=	___ Kilogram-meter
Watt-second	× 1.0000	=	___ Joule

Table 5.9 Illuminance

Convert From	Conversion Factor		To
Candle power (sphere)	× 1.2566×10^1	=	___ Lumen
Foot-candle	× 1.0000	=	___ Lumen/Square foot
Foot-candle	× 1.0764×10^1	=	___ Lumen/Square meter
Foot-candle	× 1.0764×10^1	=	___ Lux
Lumen	× 1.0000	=	___ Candela steradian
Lumen	× 7.9577×10^{-2}	=	___ Candle power (sphere)
Lumen	× 1.4960×10^{-3}	=	___ Watt
Lumen/Square centimeter	× 9.2903×10^2	=	___ Foot-candle
Lumen/Square centimeter	× 9.2903×10^2	=	___ Lumen/Square foot
Lumen/Square centimeter	× 1.0000	=	___ Phot
Lumen/Square foot	× 1.0000	=	___ Foot-candle
Lumen/Square foot	× 1.0764×10^1	=	___ Lumen/Square meter
Lumen/Square foot	× 1.0764×10^1	=	___ Lux
Lumen/Square meter	× 9.2903×10^{-2}	=	___ Foot-candle
Lumen/Square meter	× 9.2903×10^{-2}	=	___ Lumen/Square foot

Continued

CONVERSION FACTORS

CONVERSION FACTORS

Table 5.9 Illuminance *(continued)*

Convert From	Conversion Factor			To
Lumen/Square meter	×	1.0000	= ___	Lux
Lux	×	9.2903×10^{-2}	= ___	Foot-candle
Lux	×	9.2903×10^{-2}	= ___	Lumen/Square foot
Lux	×	1.0000	= ___	Lumen/Square meter
Lux	×	1.0000×10^{-4}	= ___	Phot
Phot	×	9.2903×10^{2}	= ___	Foot-candle
Phot	×	9.2903×10^{2}	= ___	Lumen/Square foot
Phot	×	1.0000×10^{4}	= ___	Lumen/Square meter

FAHRENHEIT

The conversion temperatures for Fahrenheit and Celsius have been calculated according to the following formulas and are shown in Table 5.10.

Fahrenheit to Celsius

Temperature Celsius = 5/9 × (Temperature Fahrenheit − 32)
or: °C = 0.55556 × (°F − 32)

Celsius to Fahrenheit

Temperature Fahrenheit = (9/5 × Temperature Celsius) + 32
or: °F = (1.8 × °C) + 32

Table 5.10 Temperature

Celsius	Fahrenheit	Celsius	Fahrenheit	Celsius	Fahrenheit
−30	−22.0	−16	3.2	−2	28.4
−29	−20.2	−15	5.0	−1	30.2
−28	−18.4	−14	6.8	0	32.0
−27	−16.6	−13	8.6	1	33.8
−26	−14.8	−12	10.4	2	35.6
−25	−13.0	−11	12.2	3	37.4
−24	−11.2	−10	14.0	4	39.2
−23	−9.4	−9	15.8	5	41.0
−22	−7.6	−8	17.6	6	42.8
−21	−5.8	−7	19.4	7	44.6
−20	−4.0	−6	21.2	8	46.4
−19	−2.2	−5	23.0	9	48.2
−18	−0.4	−4	24.8	10	50.0
−17	1.4	−3	26.6	11	51.8

Table continues below.

Table 5.10 Temperature *(continued)*

Celsius	Fahrenheit	Celsius	Fahrenheit	Celsius	Fahrenheit
12	53.6	59	138.2	106	222.8
13	55.4	60	140.0	107	224.6
14	57.2	61	141.8	108	226.4
15	59.0	62	143.6	109	228.2
16	60.8	63	145.4	110	230.0
17	62.6	64	147.2	111	231.8
18	64.4	65	149.0	112	233.6
19	66.2	66	150.8	113	235.4
20	68.0	67	152.6	114	237.2
21	69.8	68	154.4	115	239.0
22	71.6	69	156.2	116	240.8
23	73.4	70	158.0	117	242.6
24	75.2	71	159.8	118	244.4
25	77.0	72	161.6	119	246.2
26	78.8	73	163.4	120	248.0
27	80.6	74	165.2	121	249.8
28	82.4	75	167.0	122	251.6
29	84.2	76	168.8	123	253.4
30	86.0	77	170.6	124	255.2
31	87.8	78	172.4	125	257.0
32	89.6	79	174.2	126	258.8
33	91.4	80	176.0	127	260.6
34	93.2	81	177.8	128	262.4
35	95.0	82	179.6	129	264.2
36	96.8	83	181.4	130	266.0
37	98.6	84	183.2	131	267.8
38	100.4	85	185.0	132	269.6
39	102.2	86	186.8	133	271.4
40	104.0	87	188.6	134	273.2
41	105.8	88	190.4	135	275.0
42	107.6	89	192.2	136	276.8
43	109.4	90	194.0	137	278.6
44	111.2	91	195.8	138	280.4
45	113.0	92	197.6	139	282.2
46	114.8	93	199.4	140	284.0
47	116.6	94	201.2	141	285.8
48	118.4	95	203.0	142	287.6
49	120.2	96	204.8	143	289.4
50	122.0	97	206.6	144	291.2
51	123.8	98	208.4	145	293.0
52	125.6	99	210.2	146	294.8
53	127.4	100	212.0	147	296.6
54	129.2	101	213.8	148	298.4
55	131.0	102	215.6	149	300.2
56	132.8	103	217.4		
57	134.6	104	219.2		
58	136.4	105	221.0		

CONVERSION FACTORS

CHAPTER 6
OHM'S LAW AND BASIC CIRCUITS

Chapter 6 covers Ohm's Law, which is the relationship between voltage (E), current (I), and resistance (R) and is used to find an unknown factor (or quantity) when two or more other quantities are known. The chapter also describes various types of circuits, series circuits, parallel circuits, and combination circuits.

OHM'S LAW

Ohm's Law states that the current in a circuit is equal to the voltage of the circuit, divided by the resistance of the circuit. The formulas for Ohm's Law are as follows:

$$I = \frac{E}{R} \quad R = \frac{E}{I} \quad E = I \times R$$

The terms that make up Ohm's Law are defined in the following paragraphs; see also Table 6.1.

Voltage: The pressure required to force one ampere through a resistance of one ohm. Voltage is the force that causes current to flow. The force (or pressure) necessary to cause electrons to flow through a conductor is known as electromotive force (EMF) or voltage. The basic unit of measurement in electricity is the volt.

Resistance: The opposition which a conductor or material offers to the flow of current. Resistance depends on four factors: (1) material, (2) length, (3) temperature, and (4) cross-sectional area. Resistance is measured in ohms (Ω).

Current: The electric current that will flow through one ohm under a pressure of one volt. Current is the flow of electrons along a conductor. Current is measured in amperes (amps).

Power: The rate at which electrical energy is delivered and consumed. Power is measured in watts. The *National Electrical Code*® expresses power as VA or volt-amps.

Table 6.1 Components of Ohm's Law

Quantity	Unit	Symbol	Quantity	Unit	Symbol
Voltage	volt	E or EMF	Current	ampere	I or A
Resistance	ohm	R or Ω	Power	watt	W, P, or VA

54

OHM'S LAW AND BASIC CIRCUITS

Ohm's Law Wheel

Figure 6.1 can be used to find a formula for any value of the terms discussed in Ohm's Law. The formulas are expressed with a unity (100%) power factor. When two factors (or quantities) are known, a third factor may be calculated from a formula derived from the Ohm's Law wheel. First, find the letter of the unknown factor inside the center circle. Next, select one of the three formulas containing the two known factors. Finally, replace the letters with the known factors and solve the problem.

Example

What is the current draw of an electric heater rated 1200 watts at 120 volts?

Solution

Current is measured in amps so the following formula is selected to solve the problem.

$$\text{Amps} = \frac{\text{Watts}}{\text{Volts}} \qquad I = \frac{W}{E} \qquad I = \frac{1200}{120} = 10 \text{ amps}$$

An application of each formula from the Ohm's Law wheel is shown on the following page. Two of the following factors (ohms, amps, volts, and watts) are employed in each calculation. (The formulas in these examples run in a clockwise pattern around the Ohm's Law wheel.)

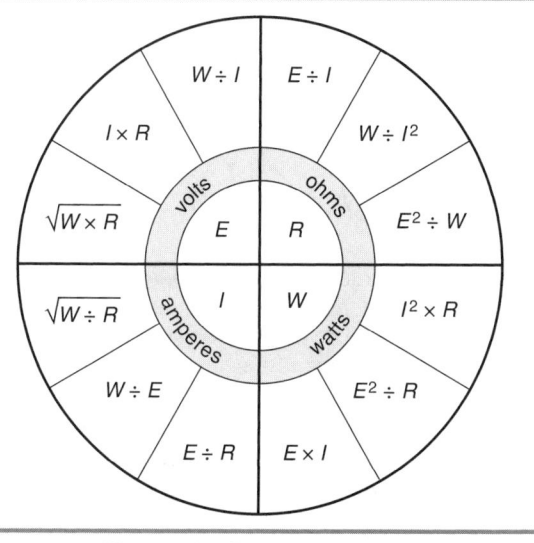

Figure 6.1   Ohm's Law wheel.

Factors

Ohms: 12
Amps: 10
Volts: 120
Watts: 1200

Formulas

Ohms = $\dfrac{\text{Volts}}{\text{Amps}}$ $R = \dfrac{E}{I}$ $R = \dfrac{120}{10} = 12$ ohms

Ohms = $\dfrac{\text{Watts}}{\text{Amps}^2}$ $R = \dfrac{W}{I^2}$ $R = \dfrac{1200}{10 \times 10} = \dfrac{1200}{100} = 12$ ohms

Ohms = $\dfrac{\text{Volts}^2}{\text{Watts}}$ $R = \dfrac{E^2}{W}$ $R = \dfrac{120 \times 120}{1200} = \dfrac{14,400}{1200} = 12$ ohms

Watts = Amps^2 × Ohms $W = I^2 \times R$ $W = 10 \times 10 \times 12 = 1200$ watts

Watts = $\dfrac{\text{Volts}^2}{\text{Ohms}}$ $W = \dfrac{E^2}{R}$ $W = \dfrac{120 \times 120}{12} = \dfrac{14,400}{12} = 1200$ watts

Watts = Volts × Amps $W = E \times I$ $W = 120 \times 10 = 1200$ watts

Amps = $\dfrac{\text{Volts}}{\text{Ohms}}$ $I = \dfrac{E}{R}$ $I = \dfrac{120}{12} = 10$ amps

Amps = $\dfrac{\text{Watts}}{\text{Volts}}$ $I = \dfrac{W}{E}$ $I = \dfrac{1200}{120} = 10$ amps

Amps = $\sqrt{\text{Watts} \div \text{Ohms}}$ $I = \sqrt{W \div R}$

$I = \sqrt{1200 \div 12} = \sqrt{100} = 10$ amps

Volts = $\sqrt{\text{Watts} \times \text{Ohms}}$ $E = \sqrt{W \times R}$

$E = \sqrt{1200 \times 12} = \sqrt{14,400} = 120$ volts

Volts = Amps × Ohms $E = I \times R$ $E = 10 \times 12 = 120$ volts

Volts = $\dfrac{\text{Watts}}{\text{Amps}}$ $E = \dfrac{W}{I}$ $E = \dfrac{1200}{10} = 120$ volts

Alternative Wheels

The "PIE" and "EIR" charts shown in Figure 6.2 may be easier to remember than the Ohm's Law wheel. By using one or a combination of both wheels, most of the unknown factors can be calculated from the two known factors. To find the formula, simply cover the unknown factor, as shown in Figure 6.3, and multiply the known factors if the letters are side by side. Divide the top known factor by the bottom

56

OHM'S LAW AND BASIC CIRCUITS

known factor if one letter is above and one below. Note that the "W" has been replaced by "P" to help in memorizing it as the "PIE" chart.

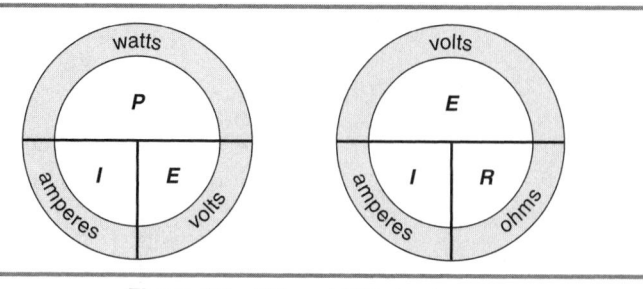

Figure 6.2   PIE and EIR charts.

Example 1

How much power does a 120-volt circuit that draws 10 amperes consume?

Solution

Power is measured in watts so *P* is the unknown that is covered in the first PIE chart shown on the left in Figure 6.3. Since the remaining two letters (*I* and *E*) are side by side, the two known factors must be multiplied. The product of 10 amps multiplied by 120 volts is 1200 watts.

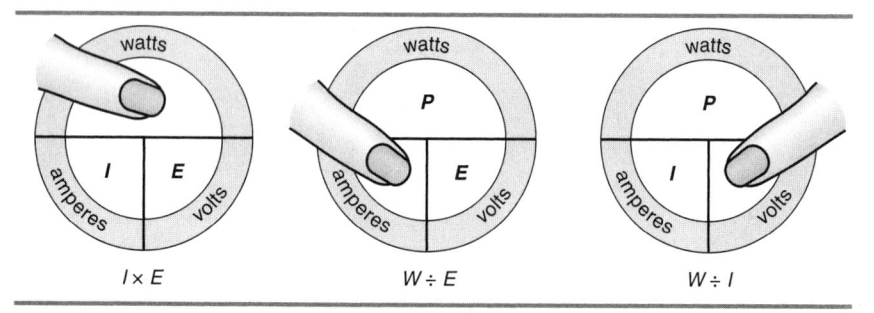

Figure 6.3   Using the PIE chart.

Watts = Amps × Volts $W = I \times E$ $W = 10 \times 120 = 1200$ watts

Example 2

If the current is 10 amperes and the resistance is 12 ohms, what is the voltage?

Solution

Voltage (*E*) is the unknown that is covered in the first EIR chart shown on the left in Figure 6.4. The two letters *I* and *R* are side by side and therefore the two known factors must be multiplied. The product of 10 amps multiplied by 12 ohms is 120 volts.

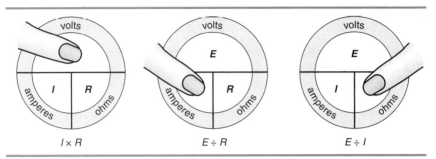

Figure 6.4 Using the EIR chart.

Volts = Amps × Ohms $E = I \times R$ $E = 10 \times 12 = 120$ volts

Each of the smaller wheels shown in Figure 6.4 provides three formulas. By using a combination of both wheels, additional unknown factors can be calculated (see Example 3).

Example 3

What is the resistance of an electric heater rated 1200 watts at 120 volts?

Solution

Step 1: Find the current draw of the heater. Select the PIE chart in Figure 6.2 and solve for current because watts and volts are the known factors. To find *I*, divide volts into watts. The result of dividing 1200 watts by 120 volts is 10 amps.

$$\text{Amps} = \frac{\text{Watts}}{\text{Volts}} \qquad I = \frac{W}{E} \qquad I = \frac{1200}{120} = 10 \text{ amps}$$

Step 2: Find the resistance by dividing the voltage by the amperage. Select the EIR chart in Figure 6.2 and now solve for resistance because volts and amps are known factors. To find *R*, divide amps into volts. The result of dividing 120 watts by 10 amps is 12 ohms.

$$\text{Ohms} = \frac{\text{Volts}}{\text{Amps}} \qquad R = \frac{E}{I} \qquad R = \frac{120}{10} = 12 \text{ ohms}$$

58

OHM'S LAW AND BASIC CIRCUITS

CIRCUITS

An electrical circuit is a closed path through which current flows or travels. Formulas for series, parallel, and combination circuits are covered in the remainder of this chapter.

Series Circuit

In a series circuit, current flows from a voltage source through every part of the electrical device in a single path before returning to the voltage source.

Properties of a Series Circuit

- Resistance is additive.

  Total Resistance $R_T = R_1 + R_2 + R_3$

- Current remains the same throughout the circuit.

  Total Current $I_T = I_1 = I_2 = I_3$

- Voltage is additive.

  Total Voltage $E_T = E_1 + E_2 + E_3$

- Power is additive.

  Total Power $W_T = W_1 + W_2 + W_3$

Note: In a series circuit, if more resistors are added, the total resistance will increase.

Example of a Series Circuit

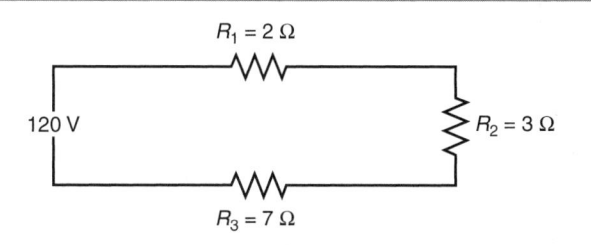

Figure 6.5   Series circuit.

Using the above-mentioned properties of a series circuit and the Ohm's Law wheel, the following problems can be solved.

1. What is the total resistance of the circuit shown in Figure 6.5?

   $R_T = R_1 + R_2 + R_3$ $R_T = 2 + 3 + 7 = 12$ ohms

2. What is the total current?

   Amps = $\dfrac{\text{Volts}}{\text{Ohms}}$ $I = \dfrac{E}{R}$ $I = \dfrac{120}{12} = 10$ amps

3. What is the voltage drop across R_1?

   Volts = Amps × Ohms $E = I \times R$ $E = 10 \times 2 = 20$ volts

4. What is the voltage drop across R_2?

   Volts = Amps × Ohms $E = I \times R$ $E = 10 \times 3 = 30$ volts

5. What is the voltage drop across R_3?

   Volts = Amps × Ohms $E = I \times R$ $E = 10 \times 7 = 70$ volts

6. What is the total voltage?

   $E_T = E_1 + E_2 + E_3$ $E_T = 20 + 30 + 70 = 120$ volts

7. How much power is being consumed by R_1?

   Watts = Volts × Amps $W = E \times I$ $W = 20 \times 10 = 200$ watts

8. How much power is being consumed by R_2?

   Watts = Volts × Amps $W = E \times I$ $W = 30 \times 10 = 300$ watts

9. How much power is being consumed by R_3?

   Watts = Volts × Amps $W = E \times I$ $W = 70 \times 10 = 700$ watts

10. What is the total power being consumed by the above circuit?

    $W_T = W_1 + W_2 + W_3$ $W_T = 200 + 300 + 700 = 1200$ watts

    or

    Watts = Volts × Amps $W = E \times I$ $W = 120 \times 10 = 1200$ watts

Parallel Circuit

In a parallel circuit, current is divided between the parallel paths. The amount of current in each parallel path depends on the resistance. The total current is the sum of the currents from each resistor. The same voltage from the voltage source flows through each resistor equally.

Properties of a Parallel Circuit

- The total resistance is less than any one resistor.

  Total Resistance (Reciprocal Formula) $R_T = \dfrac{1}{\dfrac{1}{R_1} + \dfrac{1}{R_2} + \dfrac{1}{R_3}}$

OHM'S LAW AND BASIC CIRCUITS

Total Resistance (Product/Sum Formula) $\quad R_T = \dfrac{R_1 \times R_2}{R_1 + R_2}$

- Current is additive.

  Total Current $\quad I_T = I_1 + I_2 + I_3$

- Voltage remains the same throughout the circuit. Voltage drop is the same as the voltage source in all resistors.

  Total Voltage $\quad E_T = E_1 = E_2 = E_3$

- Power is additive.

  Total Power $\quad W_T = W_1 + W_2 + W_3$

Note: In a parallel circuit, if more resistors are added, the total resistance will decrease.

Examples of Reciprocal Formula Calculations

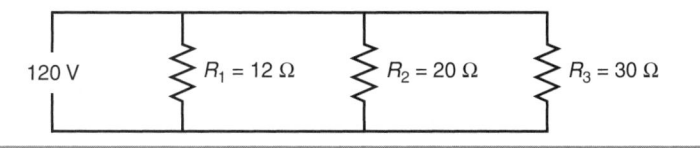

120 V $\qquad R_1 = 12\ \Omega \qquad R_2 = 20\ \Omega \qquad R_3 = 30\ \Omega$

Figure 6.6   Parallel circuit (1).

Using the above-mentioned properties of a parallel circuit and the Ohm's Law wheel, the following problems can be solved.

1. What is the total resistance in Figure 6.6?

$$R_T = \cfrac{1}{\dfrac{1}{R_2} + \dfrac{1}{R_2} + \dfrac{1}{R_3}}$$

$$R_T = \cfrac{1}{\dfrac{1}{12} + \dfrac{1}{20} + \dfrac{1}{30}} = \frac{1}{0.0833 + 0.05 + 0.0333} = \frac{1}{0.1666} = 6 \text{ ohms}$$

Note: In a parallel circuit, the total resistance is always less than the smallest resistor.

2. What is the current through R_1?

$$\text{Amps} = \frac{\text{Volts}}{\text{Ohms}} \qquad I = \frac{E}{R} \qquad I = \frac{120}{12} = 10 \text{ amps}$$

3. What is the current through R_2?

   Amps = $\dfrac{\text{Volts}}{\text{Ohms}}$ $I = \dfrac{E}{R}$ $I = \dfrac{120}{20} = 6$ amps

4. What is the current through R_3?

   Amps = $\dfrac{\text{Volts}}{\text{Ohms}}$ $I = \dfrac{E}{R}$ $I = \dfrac{120}{30} = 4$ amps

5. What is the total current?

   $I_T = I_1 + I_2 + I_3$ $I_T = 10 + 6 + 4 = 20$ amps

   The voltage drop across each resistor is 120 volts.

6. How much power is being consumed by R_1?

   Watts = Volts × Amps $W = E \times I$ $W = 120 \times 10 = 1200$ watts

7. How much power is being consumed by R_2?

   Watts = Volts × Amps $W = E \times I$ $W = 120 \times 6 = 720$ watts

8. How much power is being consumed by R_3?

   Watts = Volts × Amps $W = E \times I$ $W = 120 \times 4 = 480$ watts

9. What is the total power being consumed by the above circuit?

   $W_T = W_1 + W_2 + W_3$ $W_T = 1200 + 720 + 480 = 2400$ watts

   or

   Watts = Volts × Amps $W = E \times I$ $W = 120 \times 20 = 2400$ watts

Examples of Product/Sum Formula Calculations

The product/sum formula is an alternative to the reciprocal formula. The formula's name is derived from the top two resistors being multiplied (product), and the bottom two being added (sum). Unlike the reciprocal formula, only two resistors at a time can be calculated.

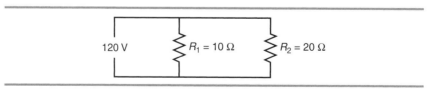

Figure 6.7   Parallel circuit (2).

OHM'S LAW AND BASIC CIRCUITS

62

OHM'S LAW AND BASIC CIRCUITS

Example 1

What is the total resistance in the drawing shown in Figure 6.7?

$$R_T = \frac{R_1 \times R_2}{R_1 + R_2} \qquad R_T = \frac{10 \times 20}{10 + 20} = \frac{200}{30} = 6.6667 \text{ ohms}$$

Example of Parallel Circuit with all Resistors of Equal Value

In a parallel circuit, if all resistors are of equal value, the calculation can be shortened. Divide the resistance of one resistor by the number of resistors. Each resistor in the parallel circuit must be of the exact same value, as shown in Figure 6.8.

Example 1

What is the total resistance in the drawing shown in Figure 6.8?

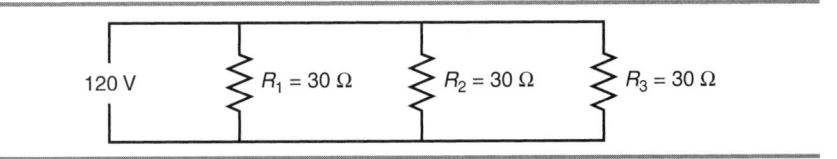

Figure 6.8   All resistors of equal value.

$$R_T = \frac{\text{Resistance of One Resistor}}{\text{Number of Resistors}} \qquad R_T = \frac{R}{N} \qquad R_T = \frac{30}{3} = 10 \text{ ohms}$$

Alternative Method for Calculating Resistance

There is an alternative method for finding the total resistance in a parallel circuit. Instead of calculating the total resistance first, calculate the total current in the circuit. Then, use Ohm's Law to find the total resistance.

Example 1

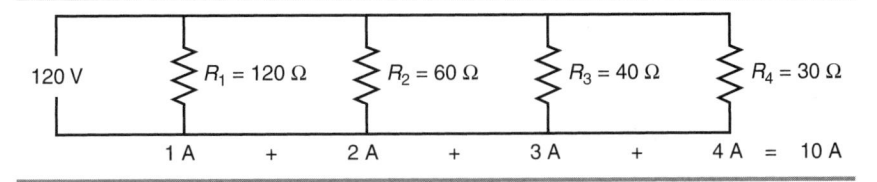

Figure 6.9   Alternative method for calculating the total resistance.

1. What is the total resistance in the drawing shown in Figure 6.9?

   To find the total resistance, first find the total current.

2. What is the current through R_1?

$$\text{Amps} = \frac{\text{Volts}}{\text{Ohms}} \qquad I = \frac{E}{R} \qquad I = \frac{120}{120} = 1 \text{ amp}$$

3. What is the current through R_2?

$$\text{Amps} = \frac{\text{Volts}}{\text{Ohms}} \qquad I = \frac{E}{R} \qquad I = \frac{120}{60} = 2 \text{ amps}$$

4. What is the current through R_3?

$$\text{Amps} = \frac{\text{Volts}}{\text{Ohms}} \qquad I = \frac{E}{R} \qquad I = \frac{120}{40} = 3 \text{ amps}$$

5. What is the current through R_4?

$$\text{Amps} = \frac{\text{Volts}}{\text{Ohms}} \qquad I = \frac{E}{R} \qquad I = \frac{120}{30} = 4 \text{ amps}$$

6. What is the total current?

$$I_T = I_1 + I_2 + I_3 + I_4 \qquad I_T = 1 + 2 + 3 + 4 = 10 \text{ amps}$$

Now, since the total current and voltage are known, the total resistance can be calculated from the Ohm's Law formula.

$$\text{Ohms} = \frac{\text{Volts}}{\text{Amps}} \qquad R = \frac{E}{I} \qquad R = \frac{120}{10} = 12 \text{ ohms}$$

Combination Circuits

A series-parallel circuit is a combination of at least one series circuit and at least one parallel circuit. To find the total resistance of a series-parallel circuit, convert the circuit to a series circuit.

Example 1

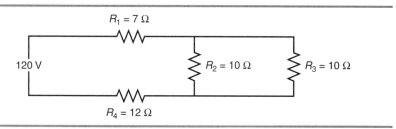

Figure 6.10 Series-parallel circuit (1).

What is the total resistance in the drawing shown in Figure 6.10?

OHM'S LAW AND BASIC CIRCUITS

64

OHM'S LAW AND BASIC CIRCUITS

Solution

Step 1: Find the combined resistance of R_2 and R_3.

Since both resistors are of equal value, divide the resistance of one resistor by the number of resistors.

$$R_T = \frac{\text{Resistance of One Resistor}}{\text{Number of Resistors}} \qquad R_T = \frac{R}{N} \qquad R_T = \frac{10}{2} = 5 \text{ ohms}$$

Step 2: Redraw the circuit and solve as a series circuit, as shown in Figure 6.11.

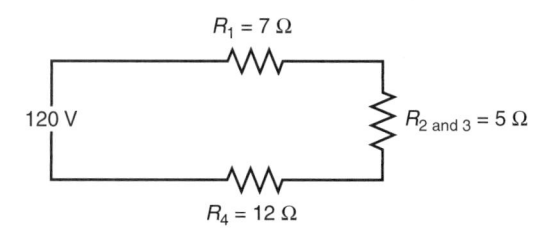

Figure 6.11   Series-parallel circuit redrawn (1).

$R_T = 7 + 5 + 12 = 24$ ohms

Example 2

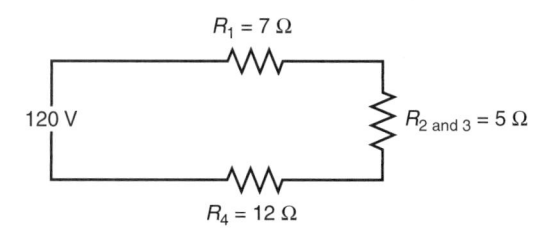

Figure 6.12   Series-parallel circuit (2).

Find the missing values for the circuit in Figure 6.12.

Step 1: Find the resistance of the two series circuits and redraw the circuit, as shown in Figure 6.13.

$R_{1\ and\ 2} = 10 + 50 = 60$ ohms

$R_{3\ and\ 4} = 20 + 100 = 120$ ohms

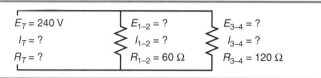

Figure 6.13   Series-parallel circuit redrawn (2).

Step 2: Find the total resistance.

$$R_T = \frac{1}{\frac{1}{60} + \frac{1}{120}} = \frac{1}{0.0167 + 0.0083} = \frac{1}{0.025} = 40 \text{ ohms}$$

Step 3: Find the total current.

$$\text{Amps} = \frac{\text{Volts}}{\text{Ohms}} \quad I = \frac{E}{R} \quad I = \frac{240}{40} = 6 \text{ amps}$$

Since voltage remains the same through a parallel circuit, each of the two combined resistors has a potential of 240 volts.

Step 4: Find the current through R_{1-2}.

$$\text{Amps} = \frac{\text{Volts}}{\text{Ohms}} \quad I = \frac{E}{R} \quad I = \frac{240}{60} = 4 \text{ amps}$$

Step 5: Find the current through R_{3-4}.

$$\text{Amps} = \frac{\text{Volts}}{\text{Ohms}} \quad I = \frac{E}{R} \quad I = \frac{240}{120} = 2 \text{ amps}$$

Step 6: Insert all the known factors into the second drawing (see Figure 6.14).

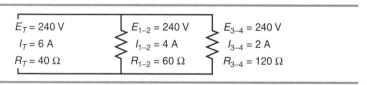

Figure 6.14   Series-parallel circuit with known factors (1).

66

OHM'S LAW AND BASIC CIRCUITS

Since current remains the same in a series circuit, the current in both I_1 and I_2 is 4 amps. The current in I_3 and I_4 is 2 amps.

Step 7: Find the voltage drop in each resistor (see Figure 6.15).

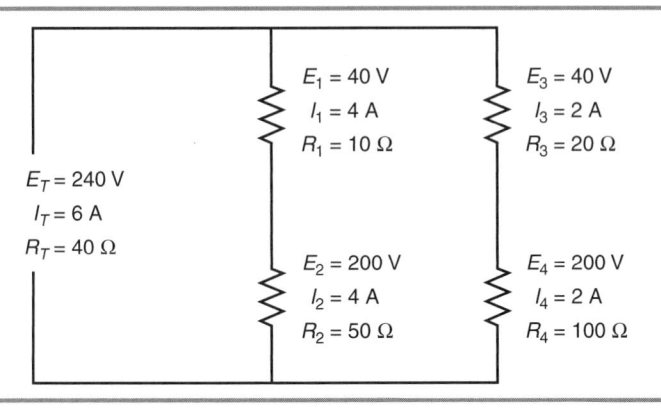

Figure 6.15   Series-parallel circuit with known factors (2).

Voltage drop across R_1:
Volts = Amps × Ohms $E = I \times R$ $E = 4 \times 10 = 40$ volts

Voltage drop across R_2:
Volts = Amps × Ohms $E = I \times R$ $E = 4 \times 50 = 200$ volts

Voltage drop across R_3:
Volts = Amps × Ohms $E = I \times R$ $E = 2 \times 20 = 40$ volts

Voltage drop across R_4:
Volts = Amps × Ohms $E = I \times R$ $E = 2 \times 100 = 200$ volts

CHAPTER 7
ELECTRICAL FORMULAS AND TABLES

Chapter 7 covers electrical formulas and relevant tables that help the electrician solve for certain unknown factors or quantities with at least two known factors. The layout and arrangement of formulas in this chapter makes it easy to find and select the right electrical formula. The Table of Contents helps in quickly locating the unknown factor. After turning to the page containing the unknown factor formulas, find the formula that includes all the known factors. Each formula in this chapter includes an example showing how to use the formula.

AMPERES (I)

Current is the flow of electrons along a conductor. Current is measured in amperes (amps). One ampere flows through one ohm under a pressure of one volt. Because the amount of current is needed to size conductors and overcurrent protection, formulas solving for amperes are among the most useful.

1. To find amperes when *volts* and *ohms* are known:

    Amps = $\dfrac{\text{Volts}}{\text{Ohms}}$ $\left[I = \dfrac{E}{R} \right]$

    Example

    What is the current draw of a circuit with a source voltage of 120 volts and a resistance of 10 ohms?

    $I = \dfrac{120}{10} = 12$ amps

2. To find amperes when *watts* and *ohms* are known:

    Amps = $\sqrt{\text{Watts} \div \text{Ohms}}$ $[I = \sqrt{W \div R}]$

    Example

    What is the current draw of an electric heater rated 1200 watts with a resistance of 12 ohms?

    $I = \sqrt{1200 \div 12} = \sqrt{100} = 10$ amps

3. To find amperes in a *single-phase* circuit when *watts* and *volts* are known:

    Amps = $\dfrac{\text{Watts}}{\text{Volts}}$ $\left[I = \dfrac{W}{E} \right]$

ELECTRICAL FORMULAS AND TABLES

ELECTRICAL FORMULAS AND TABLES

Example

What is the current draw of an electric heater rated 1200 watts at 120 volts?

$$I = \frac{1200}{120} = 10 \text{ amps}$$

4. To find amperes in a *three-phase* circuit when *watts* and *volts* are known:

$$\text{Amps} = \frac{\text{Watts}}{\text{Volts} \times 1.732} \qquad \left[I = \frac{W}{E \times 1.732} \right]$$

Example

What is the current draw of an electric heater rated 3600 watts at 208 volts, three-phase?

$$I = \frac{3600}{208 \times 1.732} = 10 \text{ amps}$$

[Note: For two-phase (4-wire) voltage, use 2 instead of 1.732.]

5. To find amperes in a *single-phase* circuit when *watts, volts,* and *power factor* are known:

$$\text{Amps} = \frac{\text{Watts}}{\text{Volts} \times \text{Power Factor}} \qquad \left[I = \frac{W}{E \times PF} \right]$$

Example

What is the current of a 2400-watt load operating at 80% power factor, supplied by a 240-volt, single-phase source?

$$I = \frac{2400}{240 \times 0.80} = 12.5 \text{ amps}$$

6. To find amperes in a *three-phase* circuit when *watts, volts,* and *power factor* are known:

$$\text{Amps} = \frac{\text{Watts}}{\text{Volts} \times 1.732 \times \text{Power Factor}} \qquad \left[I = \frac{W}{E \times 1.732 \times PF} \right]$$

Example

What is the current of a 7200-watt load operating at 80% power factor, supplied by a 208-volt, three-phase source?

$$I = \frac{7200}{208 \times 1.732 \times 0.80} = 25 \text{ amps}$$

[Note: For two-phase (4-wire) voltage, use 2 instead of 1.732.]

7. To find amperes in a *single-phase* circuit when *kilowatts* and *volts* are known:

$$\text{Amps} = \frac{\text{Kilowatts} \times 1000}{\text{Volts}} \qquad \left[I = \frac{kW \times 1000}{E} \right]$$

Example

What is the current draw of an electric heater rated 1.2 kilowatts at 120 volts?

$$I = \frac{1.2 \times 1000}{120} = 10 \text{ amps}$$

8. To find amperes in a *three-phase* circuit when *kilowatts* and *volts* are known:

$$\text{Amps} = \frac{\text{Kilowatts} \times 1000}{\text{Volts} \times 1.732} \qquad \left[I = \frac{kW \times 1000}{E \times 1.732} \right]$$

Example

What is the current draw of an electric heater rated 3.6 kilowatts at 208 volts, three-phase?

$$I = \frac{3.6 \times 1000}{208 \times 1.732} = 10 \text{ amps}$$

[Note: For two-phase (4-wire) voltage, use 2 instead of 1.732.]

9. To find amperes in a *single-phase* circuit when *kilowatts, volts,* and *power factor* are known:

$$\text{Amps} = \frac{\text{Kilowatts} \times 1000}{\text{Volts} \times \text{Power Factor}} \qquad \left[I = \frac{kW \times 1000}{E \times PF} \right]$$

Example

What is the current of a 2.4-kilowatt load operating at 80% power factor, supplied by a 240-volt, single-phase source?

$$I = \frac{2.4 \times 1000}{240 \times 0.80} = 12.5 \text{ amps}$$

10. To find amperes in a *three-phase* circuit when *kilowatts, volts,* and *power factor* are known:

$$\text{Amps} = \frac{\text{Kilowatts} \times 1000}{\text{Volts} \times 1.732 \times \text{Power Factor}} \qquad \left[I = \frac{kW \times 1000}{E \times 1.732 \times PF} \right]$$

Example

What is the current of a 7.2-kilowatt load operating at 80% power factor, supplied by a 208-volt, three-phase source?

$$I = \frac{7.2 \times 1000}{208 \times 1.732 \times 0.80} = 25 \text{ amps}$$

[Note: For two-phase (4-wire) voltage, use 2 instead of 1.732.]

ELECTRICAL FORMULAS AND TABLES

ELECTRICAL FORMULAS AND TABLES

11. To find amperes in a *single-phase* circuit when *kilovolt-amps* and *volts* are known:

$$\text{Amps} = \frac{\text{Kilovolt-amps} \times 1000}{\text{Volts}} \qquad \left[I = \frac{kVA \times 1000}{E} \right]$$

Example

What is the current draw of an electric load rated 2.4 kVA at 240 volts, single-phase?

$$I = \frac{2.4 \times 1000}{240} = 10 \text{ amps}$$

12. To find amperes in a *three-phase* circuit when *kilovolt-amps* and *volts* are known

$$\text{Amps} = \frac{\text{Kilovolt-amps} \times 1000}{\text{Volts} \times 1.732} \qquad \left[I = \frac{kVA \times 1000}{E \times 1.732} \right]$$

Example

What is the current draw of an electric load rated 3.6 kVA at 208 volts, three-phase?

$$I = \frac{3.6 \times 1000}{208 \times 1.732} = 10 \text{ amps}$$

[Note: For two-phase (4-wire) voltage, use 2 instead of 1.732.]

13. To find amperes in a *single-phase* circuit when *watts, volts, efficiency,* and *power factor* are known:

$$\text{Amps} = \frac{\text{Watts}}{\text{Volts} \times \text{Efficiency} \times \text{Power Factor}} \qquad \left[I = \frac{W}{E \times \textit{Eff.} \times PF} \right]$$

Example

What is the current of a 2500-watt load supplied by a 240 volt, single-phase source? The efficiency is 70% and the power factor is 80%.

$$I = \frac{2500}{240 \times 0.70 \times 0.80} = 18.6 \text{ amps}$$

14. To find amperes in a *three-phase* circuit when *watts, volts, efficiency,* and *power factor* are known:

$$\text{Amps} = \frac{\text{Watts}}{\text{Volts} \times 1.732 \times \text{Efficiency} \times \text{Power Factor}}$$

$$\left[I = \frac{W}{E \times 1.732 \times \textit{Eff.} \times PF} \right]$$

Example

What is the current of a 7200-watt load supplied by a 208-volt, three-phase source? The efficiency is 75% and the power factor is 80%.

$$I = \frac{7200}{208 \times 1.732 \times 0.75 \times 0.80} = 33.3 \text{ amps}$$

[Note: For two-phase (4-wire) voltage, use 2 instead of 1.732.]

15. To find amperes in a *dc* circuit when *horsepower, voltage,* and *efficiency* are known:

$$\text{Amps} = \frac{\text{Horsepower} \times 746}{\text{Volts} \times \text{Efficiency}} \quad \left[I = \frac{HP \times 746}{E \times \textit{Eff.}} \right]$$

Example

What is the current draw of a 2-hp, 120-volt, dc motor operating at 73% efficiency?

$$I = \frac{2 \times 746}{120 \times 0.73} = 17 \text{ amps}$$

16. To find amperes in a *single-phase* circuit when horsepower, voltage, efficiency, and *power factor* are known:

$$\text{Amps} = \frac{\text{Horsepower} \times 746}{\text{Volts} \times \text{Efficiency} \times \text{Power Factor}} \quad \left[I = \frac{HP \times 746}{E \times \textit{Eff.} \times PF} \right]$$

Example

What is the current draw of a 5-hp, 208-volt, single-phase motor? The motor's efficiency is 70%, and the power factor is 80%.

$$I = \frac{5 \times 746}{208 \times 0.70 \times 0.80} = 32 \text{ amps}$$

17. To find amperes in a *three-phase* circuit when *horsepower, voltage, efficiency,* and *power factor* are known:

$$\text{Amps} = \frac{\text{Horsepower} \times 746}{\text{Volts} \times 1.732 \times \text{Efficiency} \times \text{Power Factor}}$$

$$\left[I = \frac{HP \times 746}{E \times 1.732 \times \textit{Eff.} \times PF} \right]$$

Example

What is the current draw of a 5-hp, 208-volt, three-phase motor? The motor's efficiency is 85%, and the power factor is 90%.

$$I = \frac{5 \times 746}{208 \times 1.732 \times 0.85 \times 0.90} = 13.5 \text{ amps}$$

[Note: For two-phase (4-wire) voltage, use 2 instead of 1.732.]

ELECTRICAL FORMULAS AND TABLES

ELECTRICAL FORMULAS AND TABLES

VOLTS (E)

Voltage is the force that causes current to flow. The basic unit of measurement in electricity is the *volt*. The pressure required to force one ampere through a resistance of one ohm is one volt. The force (or pressure) necessary to cause electrons to flow through a conductor is known as electromotive force (EMF), or voltage.

1. To find volts when *amps* and *ohms* are known:

   Volts = Amps × Ohms $[E = I \times R]$

   Example

   What is the voltage of a circuit with 10 amps and 12 ohms?

   $E = 10 \times 12 = 120$ volts

2. To find volts when *watts* and *amps* are known:

   $\text{Volts} = \dfrac{\text{Watts}}{\text{Amps}}$ $\left[E = \dfrac{W}{I} \right]$

   Example

   What is the voltage of a circuit with 1200 watts and 10 amps?

   $E = \dfrac{1200}{10} = 120$ volts

3. To find volts when *watts* and *ohms* are known:

   $\text{Volts} = \sqrt{\text{Watts} \times \text{Ohms}}$ $E = \sqrt{W \times R}$

   Example

   What is the voltage of a circuit with 1200 watts and 12 ohms?

   $E = \sqrt{1200 \times 12} = 120$ volts

   Table 7.1 is a quick reference table showing the voltage values for common three-phase voltages.

Table 7.1   Three-Phase Voltage Values

200 × 1.732 = 346	240 × 1.732 = 416	550 × 1.732 = 953
208 × 1.732 = 360	440 × 1.732 = 762	575 × 1.732 = 996
220 × 1.732 = 381	460 × 1.732 = 797	600 × 1.732 = 1039
230 × 1.732 = 398	480 × 1.732 = 831	

4. To find volts in a *three-phase* circuit when *watts* and *amps* are known:

   $E = \dfrac{W}{I \times 1.732}$

   [Note: For two-phase (4-wire) voltage, use 2 instead of 1.732.]

5. To find volts in a *single-phase* circuit when *watts, amperes,* and *power factor* are known:

$$E = \frac{W}{I \times PF}$$

6. To find volts in a *three-phase* circuit when *watts, amperes,* and *power factor* are known:

$$E = \frac{W}{I \times 1.732 \times PF}$$

[Note: For two-phase (4-wire) voltage, use 2 instead of 1.732.]

7. To find volts in a *single-phase* circuit when *kilowatts* and *amperes* are known:

$$E = \frac{kW \times 1000}{I}$$

8. To find volts in a *three-phase* circuit when *kilowatts* and *amperes* are known:

$$E = \frac{kW \times 1000}{I \times 1.732}$$

[Note: For two-phase (4-wire) voltage, use 2 instead of 1.732.]

9. To find volts in a *single-phase* circuit when *kilowatts, amperes,* and *power factor* are known:

$$E = \frac{kW \times 1000}{I \times PF}$$

10. To find volts in a *three-phase* circuit when *kilowatts, amperes,* and *power factor* are known:

$$E = \frac{kW \times 1000}{I \times 1.732 \times PF}$$

[Note: For two-phase (4-wire) voltage, use 2 instead of 1.732.]

11. To find volts in a *single-phase* circuit when *kilovolt-amps* and *amperes* are known:

$$E = \frac{kVA \times 1000}{I}$$

12. To find volts in a *three-phase* circuit when *kilovolt-amps* and *amperes* are known:

$$E = \frac{kVA \times 1000}{I \times 1.732}$$

[Note: For two-phase (4-wire) voltage, use 2 instead of 1.732.]

ELECTRICAL FORMULAS AND TABLES

74

ELECTRICAL FORMULAS AND TABLES

13. To find volts in a *single-phase* circuit when *watts, amperes, efficiency,* and *power factor* are known:

$$E = \frac{W}{I \times \textit{Eff.} \times PF}$$

14. To find volts in a *three-phase* circuit when *watts, amperes, efficiency,* and *power factor* are known:

$$E = \frac{W}{I \times 1.732 \times \textit{Eff.} \times PF}$$

[Note: For two-phase (4-wire) voltage, use 2 instead of 1.732.]

15. To find effective or root-mean-square (RMS) value when peak value is known:

RMS = Peak × 0.707

16. To find effective or root-mean-square (RMS) value when average value is known:

RMS = Average × 1.11

17. To find peak value when effective or root-mean-square (RMS) value is known:

Peak = RMS × 1.414

18. To find peak value when average value is known:

Peak = Average × 1.57

19. To find average value when peak value is known:

Average = Peak × 0.637

20. To find average value when effective or root-mean-square (RMS) value is known:

Average = RMS × 0.9

OHMS (R)

Resistance is the opposition which a conductor or material offers to the flow of current. Resistance is measured in ohms. Resistance depends on four factors: (1) material, (2) length, (3) temperature, and (4) cross-sectional area.

1. To find ohms when *volts* and *amps* are known:

$$\text{Ohms} = \frac{\text{Volts}}{\text{Amps}} \qquad \left[R = \frac{E}{I} \right]$$

Example

What is the resistance in ohms of a circuit with 120 volts and 10 amps?

$$R = \frac{120}{10} = 12 \text{ ohms}$$

2. To find ohms when *watts* and *amps* are known:

$$\text{Ohms} = \frac{\text{Watts}}{\text{Amps}^2} \qquad \left[R = \frac{W}{I^2} \right]$$

Example

What is the resistance in ohms of a circuit with 1200 watts and 10 amps?

$$R = \frac{1200}{10^2} = 12 \text{ ohms}$$

3. To find ohms when *volts* and *watts* are known:

$$\text{Ohms} = \frac{\text{Volts}^2}{\text{Watts}} \qquad \left[R = \frac{E^2}{W} \right]$$

Example

What is the resistance in ohms of a circuit with 120 volts and 1200 watts?

$$R = \frac{120^2}{1200} = 12 \text{ ohms}$$

RESISTORS IN CIRCUITS

Calculating the total resistance of a circuit may be necessary before resistance can be used in a formula. Following are the formulas for finding the total resistance in series and parallel circuits. (See Chapter 6 for examples and detailed explanations of resistors in circuits.)

Resistors Connected in Series

Resistance is additive. In a series circuit, if more resistors are added, the total resistance will increase.

$$R_T = R_1 + R_2 + R_3$$

Resistors Connected in Parallel

For resistors connected in parallel, the total resistance is less than any one resistor.

Reciprocal Formula $\left[R_T = \dfrac{1}{\dfrac{1}{R_1} + \dfrac{1}{R_2} + \dfrac{1}{R_3}} \right]$

ELECTRICAL FORMULAS AND TABLES

76

ELECTRICAL FORMULAS AND TABLES

Product/Sum Formula $\left[R_T = \dfrac{R_1 \times R_2}{R_1 + R_2} \right]$

Color Codes for Resistors

While some resistors are marked with their actual value, smaller resistors are often marked with colored bands. The 4-band and 5-band systems show resistance values and tolerances. Each color band represents a digit, multiplier, or tolerance. See Figure 7.1 and Table 7.2 for 4-band resistors, and Figure 7.2 and Table 7.3 for 5-band resistors.

Example

What is the value of the resistor shown in Figure 7.3?

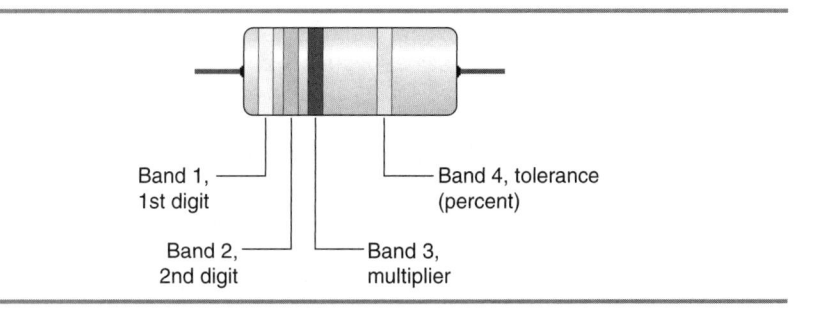

Band 1, 1st digit

Band 4, tolerance (percent)

Band 2, 2nd digit

Band 3, multiplier

Figure 7.1   Color code for four-band resistor.

Table 7.2   Four-Band: ±2%, ±5%, and ±10% Tolerance

Color	1st Band (1st Digit)	2nd Band (2nd Digit)	3rd Band (Multiplier)	4th Band (Tolerance)
No color				± 20%
Silver			$\times 10^{-2}$	± 10%
Gold			$\times 10^{-1}$	± 5%
Black	0	0	$\times 10^{0}$	
Brown	1	1	$\times 10^{1}$	
Red	2	2	$\times 10^{2}$	± 2%
Orange	3	3	$\times 10^{3}$	
Yellow	4	4	$\times 10^{4}$	
Green	5	5	$\times 10^{5}$	
Blue	6	6	$\times 10^{6}$	
Violet	7	7	$\times 10^{7}$	
Gray	8	8	$\times 10^{8}$	
White	9	9	$\times 10^{9}$	

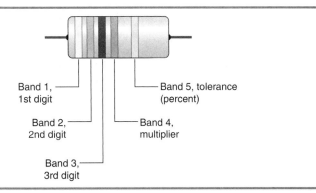

Figure 7.2   Color code for five-band resistors.

Table 7.3   Five-Band: ±.1%, ±.25%, ±.5%, and ±1% Tolerance

Color	1st Band (1st Digit)	2nd Band (2nd Digit)	3rd Band (3rd Digit)	4th Band (Multiplier)	5th Band (Tolerance)
No color					
Silver				$\times 10^{-2}$	
Gold				$\times 10^{-1}$	
Black	0	0	0	$\times 10^{0}$	
Brown	1	1	1	$\times 10^{1}$	± 1%
Red	2	2	2	$\times 10^{2}$	
Orange	3	3	3	$\times 10^{3}$	
Yellow	4	4	4	$\times 10^{4}$	
Green	5	5	5	$\times 10^{5}$	± 0.50%
Blue	6	6	6	$\times 10^{6}$	± 0.25%
Violet	7	7	7	$\times 10^{7}$	± 0.10%
Gray	8	8	8	$\times 10^{8}$	
White	9	9	9	$\times 10^{9}$	

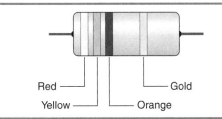

Red ———
Yellow ———
——— Gold
——— Orange

Figure 7.3   Resistor.

ELECTRICAL FORMULAS AND TABLES

ELECTRICAL FORMULAS AND TABLES

Solution

1st band, red = 2 (first number)

2nd band, yellow = 4 (second number)

3rd band, orange = 10^3 (or add three zeros after the second digit)

4th band, gold = ± 5%

Value of resistor = 24,000 ohms (or 24 kΩ) with a tolerance of ± 5%

Table 7.4   Preferred Values for Resistors (E24 5%)

Ohms	Ohms	Ohms	Ohms	Ohms	Ohms	Ohms	Ohms
1.0	10	100	1.0k	10k	100k	1.0M	10M
1.1	11	110	1.1k	11k	110k	1.1M	11M
1.2	12	120	1.2k	12k	120k	1.2M	12M
1.3	13	130	1.3k	13k	130k	1.3M	13M
1.5	15	150	1.5k	15k	150k	1.5M	15M
1.6	16	160	1.6k	16k	160k	1.6M	16M
1.8	18	180	1.8k	18k	180k	1.8M	18M
2.0	20	200	2.0k	20k	200k	2.0M	20M
2.2	22	220	2.2k	22k	220k	2.2M	22M
2.4	24	240	2.4k	24k	240k	2.4M	
2.7	27	270	2.7k	27k	270k	2.7M	
3.0	30	300	3.0k	30k	300k	3.0M	
3.3	33	330	3.3k	33k	330k	3.3M	
3.6	36	360	3.6k	36k	360k	3.6M	
3.9	39	390	3.9k	39k	390k	3.9M	
4.3	43	430	4.3k	43k	430k	4.3M	
4.7	47	470	4.7k	47k	470k	4.7M	
5.1	51	510	5.1k	51k	510k	5.1M	
5.6	56	560	5.6k	56k	560k	5.6M	
6.2	62	620	6.2k	62k	620k	6.2M	
6.8	68	680	6.8k	68k	680k	6.8M	
7.5	75	750	7.5k	75k	750k	7.5M	
8.2	82	820	8.2k	82k	820k	8.2M	
9.1	91	910	9.1k	91k	910k	9.1M	

Note: k = kilohms = 1000 ohms
Note: M = megohms = 1,000,000 ohms

INDUCTANCE (L)

Resistance is the only thing that opposes the flow of current in dc circuits. In ac circuits, resistance, inductance, and capacitance oppose the flow of current. An inductor opposes a change in current. An inductor causes current to lag voltage in a circuit. The basic unit of measurement in inductance is the henry. The symbol for henry is H. Inductance is represented by L.

Inductors Connected in Series

When inductors are connected in series, they are added together. In a series circuit, if more inductors are added, the total inductance will increase, as shown in the following formula. The formula is comparable to the formula for resistors in series.

Total Inductance $[L_T = L_1 + L_2 + L_3]$

Example

What is the total inductance in the drawing shown in Figure 7.4?

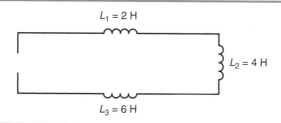

Figure 7.4   Inductors connected in series.

Solution

$L_T = L_1 + L_2 + L_3$

$L_T = 2 + 4 + 6 = 12$ henrys

A similar formula can be used to find the total inductive reactance for inductors connected in series. Inductive reactance is represented by X_L.

Inductive Reactance $[X_{LT} = X_{L1} + X_{L2} + X_{L3}]$

Inductors Connected in Parallel

For inductors connected in parallel the total inductance is less than any one inductor. The formulas are comparable to the formulas for resistors in parallel.

Total Inductance (Reciprocal Formula) $\left[L_T = \dfrac{1}{\dfrac{1}{L_1} + \dfrac{1}{L_2} + \dfrac{1}{L_3}} \right]$

Total Inductance (Product/Sum Formula) $\left[L_T = \dfrac{L_1 \times L_2}{L_1 + L_2} \right]$

ELECTRICAL FORMULAS AND TABLES

80

ELECTRICAL FORMULAS AND TABLES

Example

What is the total inductance in the drawing shown in Figure 7.5?

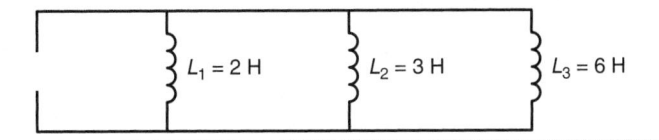

Figure 7.5   Inductors connected in parallel.

Solution

$$L_T = \frac{1}{\dfrac{1}{L_1} + \dfrac{1}{L_2} + \dfrac{1}{L_3}}$$

$$L_T = \frac{1}{\dfrac{1}{2} + \dfrac{1}{3} + \dfrac{1}{6}} = \frac{1}{0.5 + 0.3333 + 0.1667} = \frac{1}{1.0} = 1 \text{ henry}$$

Note: The total inductance of parallel inductors is always less than the smallest inductor.

Similar formulas can be used to find the total inductive reactance for inductors connected in parallel. Inductive reactance is represented by the letters X_L.

Inductive Reactance (Reciprocal Formula) $X_{LT} = \dfrac{1}{\dfrac{1}{X_{L1}} + \dfrac{1}{X_{L2}} + \dfrac{1}{X_{L3}}}$

Inductive Reactance (Product/Sum Formula) $\left[X_{LT} = \dfrac{X_{L1} \times X_{L2}}{X_{L1} + X_{L2}} \right]$

CAPACITANCE (C)

Capacitance is the storage of electric energy in a capacitor. Capacitors are devices that oppose voltage change. The devices are made up of two conductive surfaces separated by an insulating material. Capacitors store and then release energy by means of an electrostatic field. A capacitor causes current to lead voltage in a circuit. The basic unit of measurement in capacitance is the farad. The symbol for farad is F. Capacitance is represented by C.

Capacitors Connected in Series

For capacitors connected in series, the total capacitance is less than any one capacitor. The formulas are comparable to the formulas for resistors in parallel.

Total Capacitance (Reciprocal Formula) $\left[C_T = \dfrac{1}{\dfrac{1}{C_1} + \dfrac{1}{C_2} + \dfrac{1}{C_3}} \right]$

Total Capacitance (Product/Sum Formula) $\left[C_T = \dfrac{C_1 \times C_2}{C_1 + C_2} \right]$

Example

What is the total capacitance in Figure 7.6?

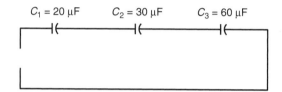

Figure 7.6 Capacitors connected in series.

Solution

$$C_T = \dfrac{1}{\dfrac{1}{C_1} + \dfrac{1}{C_2} + \dfrac{1}{C_3}}$$

$$C_T = \dfrac{1}{\dfrac{1}{20} + \dfrac{1}{30} + \dfrac{1}{60}} = \dfrac{1}{0.05 + 0.0333 + 0.0167} = \dfrac{1}{0.10} = 10 \text{ microfarads}$$

Note: The total capacitance of series capacitors is always less than the smallest capacitor.

Capacitors Connected in Parallel

Capacitance is additive. To find the total capacitance of capacitors connected in parallel, add the capacitors. The formula is comparable to the formula for resistors in series.

Total Capacitance $[C_T = C_1 + C_2 + C_3]$

ELECTRICAL FORMULAS AND TABLES

82

ELECTRICAL FORMULAS AND TABLES

Example

What is the total capacitance in Figure 7.7?

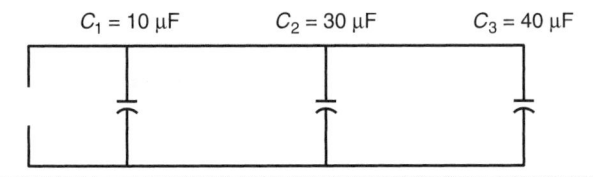

Figure 7.7   Capacitors connected in parallel.

Solution

$C_T = C_1 + C_2 + C_3$ $C_T = 10 + 30 + 40 = 80$ microfarads

Color Codes for Capacitors

All capacitors are marked in some way with their capacitance value. Tolerance and voltage rating may also be marked on the capacitor. Sometimes printing all the information on the capacitor is impractical. Similar to the color bands on resistors, capacitors also have color coding systems. The two most common color coding systems are JAN (Joint Army-Navy) and EIA (Electronics Industries Alliance). See Table 7.5 and Figure 7.8 for Color Codes for Molded Mica and Molded Paper Capacitors.

Table 7.5   Color Codes for Molded Mica and Molded Paper Capacitors

Type	Color	1st Digit	2nd Digit	Multiplier	Tolerance	Voltage Rating
	No color				± 20%	500
Molded paper	Silver			$\times 10^{-2}$	± 10%	2000
	Gold			$\times 10^{-1}$	± 5%	1000
JAN, mica	Black	0	0	$\times 10^{0}$	± 20%	
	Brown	1	1	$\times 10^{1}$	± 1%	100
	Red	2	2	$\times 10^{2}$	± 2%	200
	Orange	3	3	$\times 10^{3}$	± 3%	300
	Yellow	4	4	$\times 10^{4}$	± 4%	400
	Green	5	5	$\times 10^{5}$	± 5%	500
	Blue	6	6	$\times 10^{6}$	± 6%	600

Table continues below.

Table 7.5 Color Codes for Molded Mica and Molded Paper Capacitors *(continued)*

Type	Color	1st Digit	2nd Digit	Multiplier	Tolerance	Voltage Rating
	Violet	7	7	$\times 10^7$	± 7%	700
	Gray	8	8	$\times 10^8$	± 8%	800
EIA, mica	White	9	9	$\times 10^9$	± 9%	900

Note: The classification pertains to temperature coefficient or methods of testing
Note: JAN = Joint Army-Navy
Note: EIA = Electronics Industries Alliance (formerly RMA, RTMA, and RETMA)

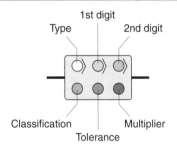

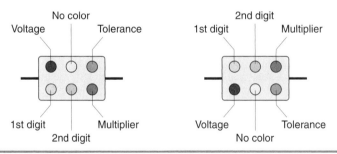

Figure 7.8 Mica capacitors.

ELECTRICAL FORMULAS AND TABLES

84

ELECTRICAL FORMULAS AND TABLES

Table 7.6 Preferred Values for Capacitors

pF	pF	pF	pF	μF	μF	μF	μF	μF	μF	μF
1.0	10	100	1000	0.01	0.1	1.0	10	100	1000	10,000
1.1	11	110	1100							
1.2	12	120	1200							
1.3	13	130	1300							
1.5	15	150	1500	0.015	0.15	1.5	15	150	1500	
1.6	16	160	1600							
1.8	18	180	1800							
2.0	20	200	2000							
2.2	22	220	2200	0.022	0.22	2.2	22	220	2200	
2.4	24	240	2400							
2.7	27	270	2700							
3.0	30	300	3000							
3.3	33	330	3300	0.033	0.33	3.3	33	330	3300	
3.6	36	360	3600							
3.9	39	390	3900							
4.3	43	430	4300							
4.7	47	470	4700	0.047	0.47	4.7	47	470	4700	
5.1	51	510	5100							
5.6	56	560	5600							
6.2	62	620	6200							
6.8	68	680	6800	0.068	0.68	6.8	68	680	6800	
7.5	75	750	7500							
8.2	82	820	8200							
9.1	91	910	9100							

Note: pF = picofarads = 1×10^{-12} farads
Note: μF = microfarads = 1×10^{-6} farads

POWER-FACTOR CORRECTION

Table 7.7 provides multipliers to determine the kVAR of capacitors necessary to raise the power factor. First, find the existing or original power factor percent. This is the column to the far left. Next, follow along the selected row until intersecting with the desired power factor column. This percentage is then used to find the capacitor kVAR required. Finally, multiply the load in kW by the selected percentage.

Example

The total plant load is 100 kW with an existing power factor of 70%. The desired power factor is 95%. The power-factor correction is 0.691. To determine the capacitor kVAR, multiply 0.691 by the total load.

Solution

0.691 × 100 = 69 kVAR of capacitor required

Table 7.7 Power-Factor Correction

Original Power Factor	Corrected Power Factor				
	80%	85%	90%	95%	100%
50	0.982	1.112	1.248	1.403	1.732
52	0.893	1.023	1.159	1.314	1.643
54	0.808	0.939	1.075	1.230	1.559
56	0.729	0.860	0.996	1.151	1.480
58	0.654	0.785	0.921	1.076	1.405
60	0.583	0.713	0.849	1.004	1.333
62	0.516	0.645	0.781	0.937	1.266
64	0.451	0.581	0.717	0.872	1.201
66	0.388	0.518	0.654	0.809	1.138
68	0.328	0.458	0.594	0.749	1.078
70	0.270	0.400	0.536	0.691	1.020
72	0.214	0.344	0.480	0.635	0.964
74	0.159	0.289	0.425	0.580	0.909
75	0.132	0.262	0.398	0.553	0.882
76	0.105	0.235	0.371	0.526	0.855
77	0.079	0.209	0.345	0.500	0.829
78	0.052	0.182	0.318	0.473	0.802
79	0.026	0.156	0.292	0.447	0.776
80	0.000	0.130	0.266	0.421	0.750
81		0.104	0.240	0.395	0.724
82		0.078	0.214	0.369	0.698
83		0.052	0.188	0.343	0.672
84		0.026	0.162	0.317	0.646
85		0.000	0.136	0.291	0.620
86			0.109	0.264	0.593
87			0.083	0.238	0.567
88			0.056	0.211	0.540
89			0.028	0.183	0.512
90			0.000	0.155	0.484
91				0.127	0.456
92				0.097	0.426
93				0.066	0.395
94				0.034	0.363
95				0.000	0.329
96					0.292
97					0.251
98					0.203
99					0.143

ELECTRICAL FORMULAS AND TABLES

ELECTRICAL FORMULAS AND TABLES

IMPEDANCE (Z)

Impedance is the total opposition to current flow in an electrical circuit. Impedance is measured in ohms. Impedance is represented by Z.

1. To find impedance when *voltage* and *current* are known:

$$\text{Impedance} = \frac{\text{Volts}}{\text{Amps}} \qquad \left[Z = \frac{E}{U} \right]$$

Example

What is the impedance in ohms of an ac circuit with 120 volts and 10 amps?

$$Z = \frac{120}{10} = 12 \text{ ohms}$$

2. To find impedance when *resistance* and *reactance* are known:

$$\text{Impedance} = \sqrt{\text{Resistance}^2 + \text{Reactance}^2} \qquad \left[Z = \sqrt{R^2 + X^2} \right]$$

Example

What is the impedance in ohms of an ac circuit with a resistance of 12 ohms and a reactance of 16 ohms?

$$Z = \sqrt{12^2 + 16^2} = \sqrt{144 + 256} = \sqrt{400} = 20 \text{ ohms}$$

3. To find impedance when *resistance, inductive reactance,* and *capacitive reactance* are known:

$$Z = \sqrt{R^2 + (X_L - X_C)^2}$$

Example

What is the impedance in ohms of an ac circuit with a resistance of 8 ohms, an inductive reactance of 50 ohms, and a capacitive reactance of 44 ohms?

$$Z = \sqrt{8^2 + (50 - 44)^2} = \sqrt{8^2 + 6^2} = \sqrt{64 + 36} = \sqrt{100} = 10 \text{ ohms}$$

REACTANCE (X)

Reactance is the opposition to alternating current due to inductance and capacitance. Reactance is represented by X.

1. To find inductive reactance:

Inductive Reactance = 2 × 3.1416 × Frequency × Inductance $[X_L = 2\pi FL]$

Example

What is the reactance of a circuit that has an inductance of 0.5 henrys on a 60-cycle ac circuit?

$$X_L = 2 \times 3.1416 \times 60 \times 0.5 = 188.5 \text{ ohms}$$

2. To find the inductance of the coil:

$$L \frac{X_L}{2 \times \pi \times F}$$

3. To find frequency (in hertz):

$$F = \frac{X_L}{2 \times \pi \times L}$$

4. To find capacitive reactance:

$$\text{Capacitive Reactance} = \frac{1}{2 \times 3.1416 \times \text{Frequency} \times \text{Capacitance}}$$

$$\left[X_c = \frac{1}{2\pi FC} \right]$$

Example

What is the reactance of a circuit that has a capacitance of 30 microfarads on a 60-cycle ac circuit?

$$X_c = \frac{1}{2 \times 3.1416 \times 60 \times 0.00003} = \frac{1}{0.01130976} = 88.4 \text{ ohms}$$

5. To find the capacitance of the capacitor:

$$C = \frac{1}{2 \times \pi \times F \times X_c}$$

6. To find frequency (in hertz):

$$F = \frac{1}{2 \times \pi \times C \times X_c}$$

ELECTRICAL FORMULAS AND TABLES

WATTS (W)

Power is the rate at which electrical energy is delivered and consumed. Power is measured in watts. The *National Electrical Code®* (*NEC®*) expresses power as VA or volt-amps.

1. To find watts when *amperes* and *ohms* are known,

   Watts = Amps^2 × Ohms [$W = I^2 \times R$]

   Example

   What is the wattage of a circuit with a current of 10 amps and a resistance of 12 ohms?

   W = 10 × 10 × 12 = 1200 watts

2. To find watts when *volts* and *ohms* are known:

   $$\text{Watts} = \frac{\text{Volts}^2}{\text{Ohms}} \quad \left[W = \frac{E^2}{R} \right]$$

   Example

   What is the wattage of a circuit with a source voltage of 120 volts and a resistance of 12 ohms?

   $$W = \frac{120 \times 120}{12} \times \frac{14,400}{12} = 1200 \text{ watts}$$

3. To find watts in a *single-phase* circuit when *volts* and *amperes* are known:

   Watts = Volts × Amps   [$W = E \times I$]

   Example

   What is the wattage of a circuit with a source voltage of 120 volts and a current of 10 amps?

   W = 120 × 10 = 1200 watts

4. To find watts in a *three-phase* circuit when *volts* and *amperes* are known:

   Watts = Volts × 1.732 × Amps   [$W = E \times 1.732 \times I$]

   Example

   What is the wattage of an electric heater with a current draw of 10 amps at 208 volts, three-phase?

   W = 208 × 1.732 × 10 = 3602 watts

   [Note: For two-phase (4-wire) voltage, use 2 instead of 1.732.]

5. To find watts in a *single-phase* circuit when volts, amperes, and *power factor* are known:

   Watts = Volts × Amps × Power Factor [$W = E \times I \times PF$]

   Example
   What is the wattage of a circuit supplied by a 240-volt, single-phase source, drawing 12.5 amperes, and operating at 80% power factor?

   $W = 240 \times 12.5 \times 0.80 = 2400$ watts

6. To find watts in a *three-phase* circuit when *volts, amperes,* and *power factor* are known:

   Watts = Volts × 1.732 × Amps × Power Factor   [$W = E \times 1.732 \times I \times PF$]

   Example
   What is the wattage of a circuit supplied by a 208-volt, three-phase source, drawing 25 amperes, and operating at 80% power factor?

   $W = 208 \times 1.732 \times 25 \times 0.80 = 7205$ watts

   [Note: For two-phase (4-wire) voltage, use 2 instead of 1.732.]

7. To find watts in a *single-phase* circuit when *volts, amperes, efficiency,* and *power factor* are known:

   Watts = Volts × Amps × Efficiency × Power Factor   [$W = E \times I \times Eff. \times PF$]

   Example
   What is the wattage of a circuit supplied by a 240-volt, single-phase source that is drawing 18.6 amperes? The efficiency is 70% and the power factor is 80%.

   $W = 240 \times 18.6 \times 0.70 \times 0.80 = 2500$ watts

8. To find watts in a *three-phase* circuit when *volts, amperes, efficiency,* and *power factor* are known:

   Watts = Volts × 1.732 × Amps × Efficiency × Power Factor

   [$W = E \times 1.732 \times I \times Eff. \times PF$]

ELECTRICAL FORMULAS AND TABLES

Example

What is the wattage of a circuit supplied by a 208-volt, three-phase source that is drawing 33.3 amperes? The efficiency is 75% and the power factor is 80%.

W = 208 × 1.732 × 33.3 × 0.75 × 0.80 = 7198 watts

[Note: For two-phase (4-wire) voltage, use 2 instead of 1.732.]

KILOWATTS

Since the prefix *kilo-* means one-thousand, the term *kilowatt* means one-thousand watts.

1. To find kilowatts in a *dc* or a *single-phase* ac circuit when *volts* and *amperes* are known:

$$kW = \frac{E \times I}{1000}$$

2. To find kilowatts in a *three-phase* circuit when *volts* and *amperes* are known:

$$kW = \frac{E \times 1.732 \times I}{1000}$$

[Note: For two-phase (4-wire) voltage, use 2 instead of 1.732.]

3. To find kilowatts in a *single-phase* circuit when *volts, amperes,* and *power factor* are known:

$$kW = \frac{E \times I \times PF}{1000}$$

4. To find kilowatts in a *three-phase* circuit when *volts, amperes,* and *power factor* are known:

$$kW = \frac{E \times 1.732 \times I \times PF}{1000}$$

[Note: For two-phase (4-wire) voltage, use 2 instead of 1.732.]

5. To find kilowatts in a *single-phase* circuit when *volts, amperes, efficiency,* and *power factor* are known:

$$kW = \frac{E \times I \times Eff. \times PF}{1000}$$

6. To find kilowatts in a *three-phase* circuit when *volts, amperes, efficiency,* and *power factor* are known:

$$kW = \frac{E \times 1.732 \times I \times \text{Eff.} \times PF}{1000}$$

[Note: For two-phase (4-wire) voltage, use 2 instead of 1.732.]

VOLT-AMPERES

Power supplied to the circuit by the source is the apparent power. Apparent power exists when voltage and current are out of phase with each other. Apparent power is also referred to as volt-amperes or VA. The *NEC®* expresses power as VA or volt-amps.

1. To find volt-amperes in a *single-phase* circuit when *volts* and *amperes* are known:

   $VA = E \times I$

2. To find volt-amperes in a *three-phase* circuit when *volts* and *amperes* are known:

   $VA = E \times 1.732 \times I$

[Note: For two-phase (4-wire) voltage, use 2 instead of 1.732.]

KILOVOLT-AMPERES

Since the prefix *kilo-* means one-thousand, the term *kilovolt-amperes* means one-thousand volt-amperes.

1. To find kilovolt-amperes in a *single-phase* circuit when *volts* and *amperes* are known:

   $$kVA = \frac{E \times I}{1000}$$

2. To find kilovolt-amperes in a *three-phase* circuit when *volts* and *amperes* are known:

   $$kVA = \frac{E \times 1.732 \times I}{1000}$$

ELECTRICAL FORMULAS AND TABLES

92

ELECTRICAL FORMULAS AND TABLES

EFFICIENCY

Efficiency is the ratio of input to output. Efficiency is expressed as a percentage or a decimal. Efficiency is always less than 100 percent.

1. To find efficiency in a *single-phase* circuit when *watts, volts, amperes,* and *power factor* are known:

$$\text{Efficiency} = \frac{\text{Watts}}{\text{Volts} \times \text{Amperes} \times \text{Power Factor}} \qquad \left[\textit{Eff.} = \frac{W}{E \times I \times PF} \right]$$

Example

What is the efficiency of a 2500-watt load supplied by a 240-volt, single-phase source that draws 18.6 amperes? The power factor is 80%.

$$\textit{Eff.} = \frac{2500}{240 \times 18.6 \times 0.80} = 0.70 = 70\%$$

2. To find efficiency in a *three-phase* circuit when *watts, volts, amperes,* and *power factor* are known:

$$\text{Efficiency} = \frac{\text{Watts}}{\text{Volts} \times 1.732 \times \text{Amperes} \times \text{Power Factor}}$$

$$\left[\textit{Eff.} = \frac{W}{E \times 1.732 \times I \times PF} \right]$$

Example

What is the efficiency of a 7200-watt load supplied by a 208-volt, three-phase source that draws 33.3 amperes? The power factor is 80%.

$$\textit{Eff.} = \frac{7200}{208 \times 1.732 \times 33.3 \times 0.80} = 0.75 = 75\%$$

[Note: For two-phase (4-wire) voltage, use 2 instead of 1.732.]

3. To find efficiency in a *dc* circuit when *horsepower, volts,* and *amperes* are known:

$$\text{Efficiency} = \frac{\text{Horsepower} \times 746}{\text{Volts} \times \text{Amperes}} \qquad \left[\textit{Eff.} = \frac{HP \times 746}{E \times I} \right]$$

Example

What is the efficiency of a 2-hp, 120-volt, dc motor that draws 17 amperes?

$$\textit{Eff.} = \frac{2 \times 746}{120 \times 17} = 0.73 = 73\%$$

4. To find efficiency in a *single-phase* circuit when *horsepower, volts, amperes,* and *power factor* are known:

$$\text{Efficiency} = \frac{\text{Horsepower} \times 746}{\text{Volts} \times \text{Amps} \times \text{Power Factor}} \qquad \left[\text{Eff.} = \frac{HP \times 746}{E \times I \times PF} \right]$$

Example

What is the efficiency of a 5-hp, 208-volt, single-phase motor that draws 32 amperes? The motor's power factor is 80%.

$$\text{Eff.} = \frac{5 \times 746}{208 \times 32 \times 0.80} = 0.70 = 70\%$$

5. To find efficiency in a *three-phase* circuit when *horsepower, volts, amperes,* and *power factor* are known:

$$\text{Efficiency} = \frac{\text{Horsepower} \times 746}{\text{Volts} \times 1.732 \times \text{Amps} \times \text{Power Factor}}$$

$$\left[\text{Eff.} = \frac{HP \times 746}{E \times 1.732 \times I \times PF} \right]$$

Example

What is the efficiency of a 5-hp, 208-volt, three-phase motor that draws 13.5 amperes? The motor's power factor is 90%.

$$\text{Eff.} = \frac{5 \times 746}{208 \times 1.732 \times 13.5 \times 0.90} = 0.85 = 85\%$$

[Note: For two-phase (4-wire) voltage, use 2 instead of 1.732.]

6. To find efficiency when *output* and *input* are known:

$$\text{Eff.} = \frac{\text{Output}}{\text{Input}}$$

POWER FACTOR

Power factor is the ratio between true power (measured in watts) and apparent power (measured in volt-amperes). Power factor can be 100 percent, but not more.

1. To find power factor in a *single-phase* circuit when *watts, volts,* and *amperes* are known:

$$\text{Power Factor} = \frac{\text{Watts}}{\text{Volts} \times \text{Amps}} \qquad \left[PF = \frac{W}{E \times I} \right]$$

ELECTRICAL FORMULAS AND TABLES

Example

What is the power factor of a 2400-watt load supplied by a 240-volt, single-phase source that draws 12.5 amperes?

$$PF = \frac{2400}{240 \times 12.5} = 0.8 = 80\%$$

2. To find power factor in a *three-phase* circuit when *watts, volts,* and *amperes* are known:

$$\text{Power Factor} = \frac{\text{Watts}}{\text{Volts} \times 1.732 \times \text{Amps}} \qquad \left[PF = \frac{W}{E \times 1.732 \times I} \right]$$

Example

What is the power factor of a 7200-watt load supplied by a 208-volt, three-phase source that draws 25 amperes?

$$PF = \frac{7200}{208 \times 1.732 \times 25} = 0.8 = 80\%$$

[Note: For two-phase (4-wire) voltage, use 2 instead of 1.732.]

3. To find power factor when true power (*watts*) and apparent power (*volt-amperes*) are known:

$$\text{Power Factor} = \frac{\text{True Power}}{\text{Apparent Power}} \qquad PF = \frac{\text{Watts}}{\text{Volt-Amps}}$$

$$\left[PF = \frac{W}{VA} \right]$$

Example

What is the power factor of a circuit that has an apparent power of 10,000 volt-amperes, but the true power of only 8500 watts?

$$PF = \frac{8500}{10,000} = 0.85 = 85\%$$

4. To find power factor in a *single-phase* circuit when *watts, volts, amperes,* and *efficiency* are known:

$$\text{Power Factor} = \frac{\text{Watts}}{\text{Volts} \times \text{Amperes} \times \text{Efficiency}} \qquad \left[PF = \frac{W}{E \times I \times Eff.} \right]$$

Example

What is the power factor of a 2500-watt load supplied by a 240-volt, single-phase source that draws 18.6 amperes? The efficiency is 70%.

$$PF = \frac{2500}{240 \times 18.6 \times 0.70} = 0.80 = 80\%$$

5. To find power factor in a *three-phase* circuit when *watts, volts, amperes,* and *efficiency* are known:

$$\text{Power Factor} = \frac{\text{Watts}}{\text{Volts} \times 1.732 \times \text{Amperes} \times \text{Efficiency}}$$

$$\left[PF = \frac{W}{E \times 1.732 \times I \times \textit{Eff.}} \right]$$

Example

What is the power factor of a 7200-watt load supplied by a 208-volt, three-phase source that draws 33.3 amperes? The efficiency is 75%.

$$PF = \frac{7200}{208 \times 1.732 \times 33.3 \times 0.75} = 0.80 = 80\%$$

[Note: For two-phase (4-wire) voltage, use 2 instead of 1.732.]

6. To find power factor in a *single-phase* circuit when *horsepower, volts, amperes,* and *efficiency* are known:

$$\text{Power Factor} = \frac{\text{Horsepower} \times 746}{\text{Volts} \times \text{Amps} \times \text{Efficiency}} \qquad \left[PF = \frac{HP \times 746}{E \times I \times \textit{Eff.}} \right]$$

Example

What is the power factor of a 5-hp, 208-volt, single-phase motor that draws 32 amperes? The motor's efficiency is 70%.

$$PF = \frac{5 \times 746}{208 \times 32 \times 0.70} = 0.80 = 80\%$$

7. To find power factor in a *three-phase* circuit when *horsepower, volts, amperes,* and *efficiency* are known:

$$\text{Power Factor} = \frac{\text{Horsepower} \times 746}{\text{Volts} \times 1.732 \times \text{Amps} \times \text{Efficiency}}$$

$$\left[PF = \frac{HP \times 746}{E \times 1.732 \times I \times \textit{Eff.}} \right]$$

Example

What is the power factor of a 5-hp, 208-volt, three-phase motor that draws 13.5 amperes? The motor's efficiency is 85%.

$$PF = \frac{5 \times 746}{208 \times 1.732 \times 13.5 \times 0.85} = 0.90 = 90\%$$

[Note: For two-phase (4-wire) voltage, use 2 instead of 1.732.]

8. To find power factor when *resistance* and *impedance* are known:

$$PF = \frac{R}{Z}$$

ELECTRICAL FORMULAS AND TABLES

96

ELECTRICAL FORMULAS AND TABLES

HORSEPOWER

Horsepower is a measurement of mechanical power. One horsepower is equal to 746 watts.

1. To find horsepower in a *dc* circuit when *volts, amperes,* and *efficiency* are known:

$$\text{Horsepower} = \frac{\text{Volts} \times \text{Amps} \times \text{Efficiency}}{746} \qquad \left[HP = \frac{E \times I \times \text{Eff.}}{746} \right]$$

Example

What is the horsepower of a 120-volt dc motor that draws 17 amperes? The motor's efficiency is 73%.

$$HP = \frac{120 \times 17 \times 0.73}{746} = 2 \text{ hp}$$

2. To find horsepower in a *single-phase* circuit when *volts, amperes, efficiency,* and *power factor* are known:

$$\text{Horsepower} = \frac{\text{Volts} \times \text{Amps} \times \text{Efficiency} \times \text{Power Factor}}{746}$$

$$\left[HP = \frac{E \times I \times \text{Eff.} \times PF}{746} \right]$$

Example

What is the horsepower of a 208-volt, single-phase (AC) motor that draws 32 amperes? The motor's efficiency is 70% and the power factor is 80%.

$$HP = \frac{208 \times 32 \times 0.70 \times 0.80}{746} = 5 \text{ hp}$$

3. To find horsepower in a *three-phase* circuit when *volts, amperes, efficiency,* and *power factor* are known:

$$\text{Horsepower} = \frac{\text{Volts} \times 1.732 \times \text{Amps} \times \text{Efficiency} \times \text{Power Factor}}{746}$$

$$\left[HP = \frac{E \times 1.732 \times I \times \text{Eff.} \times PF}{746} \right]$$

Example

What is the horsepower of a 208-volt, three-phase (AC) motor that draws 13.5 amperes? The motor's efficiency is 85% and the power factor is 90%.

$$HP = \frac{208 \times 1.732 \times 13.5 \times 0.85 \times 0.90}{746} = 5 \text{ hp}$$

[Note: For two-phase (4-wire) voltage, use 2 instead of 1.732.]

CHAPTER 8

ELECTRICAL PLAN SYMBOLS

Chapter 8 covers the electrical plan symbols associated with lighting outlets, switch outlets, and receptacle outlets (see Tables 8.1 through 8.3). Miscellaneous symbols are shown in Table 8.4.

Table 8.1 Lighting Outlets

Ceiling	Wall	Description
Ⓑ	─Ⓑ	Blanked Outlet
Ⓒ	─Ⓒ	Clock Outlet
Ⓓ	─Ⓓ	Drop Cord
Ⓕ	─Ⓕ	Fan Outlet
Ⓙ	─Ⓙ	Junction Box
Ⓛ	─Ⓛ	Lampholder
○$_{PS}$ Ⓛ$_{PS}$	─○$_{PS}$ ─Ⓛ$_{PS}$	Lampholder with Pull Switch
Ⓢ	─Ⓢ	Pull Switch
Ⓥ	─Ⓥ	Vapor Discharge Lamp Outlet
○ ⊠ ─○─	─○ ─⊠ ─○─	Surface or Pendant Luminaire (Fixture)
Ⓡ ▦ □○	─Ⓡ ─▦ ─□○	Recessed Luminaire (Fixture)
▭ ▭○	─▭ ─▭○	Surface Fluorescent
▭○$_R$	─▭○$_R$	Recessed Fluorescent
○▭▭		Surface or Pendant Continuous Row Fluorescent
○$_R$▭▭		Recessed Continuous Row Fluorescent

Continued

ELECTRICAL PLAN SYMBOLS

Table 8.1   Lighting Outlets (*continued*)

Ceiling	Wall	Description
├──────┤		Bare-Lamp Fluorescent Strips
├───┼───┤		Continuous Row Bare-Lamp Fluorescent Strips
◑ ◢		Luminaire (Fixture) Providing Emergency Illumination
Ⓧ	─Ⓧ	Surface or Pendant Exit Light
ⓇⓍ	─ⓇⓍ	Recessed Exit Light

Table 8.2   Switch Outlets

S	Single-Pole Switch
S_2	Double-Pole Switch
S_3	Three-Way Switch
S_4	Four-Way Switch
S_{DS}	Dimmer Switch
S_{DT}	Single-Pole Double-Throw Switch
S_D	Door Switch
S_K	Key-Operated Switch
S_{KP}	Key-Operated Switch with Pilot Lamp
S_P	Switch and Pilot Lamp
S_{CB}	Circuit-Breaker Switch
S_{LV}	Switch for Low-Voltage Switching System
S_{LM}	Master Switch for Low-Voltage Switching System
S_T	Time Switch
S_{WP}	Weather-Proof Single-Pole Switch
S_F	Fused Switch
S_{WPF}	Weather-Proof Fused Switch
S_{MC}	Momentary Contact Switch
S_{RC}	Remote-Control Switch

Table 8.3   Receptacle Outlets

─⊖ ═⊖$_1$	Single Receptacle
═⊖	Duplex Receptacle
═⊖$_3$	Triplex Receptacle

Table continues below.

Table 8.3 Receptacle Outlets (*continued*)

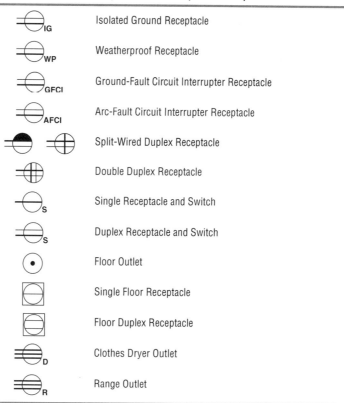

Symbol	Description
IG	Isolated Ground Receptacle
WP	Weatherproof Receptacle
GFCI	Ground-Fault Circuit Interrupter Receptacle
AFCI	Arc-Fault Circuit Interrupter Receptacle
	Split-Wired Duplex Receptacle
	Double Duplex Receptacle
S	Single Receptacle and Switch
S	Duplex Receptacle and Switch
	Floor Outlet
	Single Floor Receptacle
	Floor Duplex Receptacle
D	Clothes Dryer Outlet
R	Range Outlet

Table 8.4 Miscellaneous Symbols

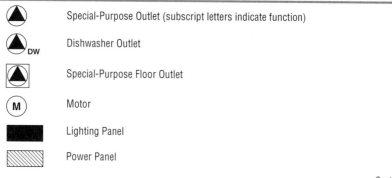

Symbol	Description
	Special-Purpose Outlet (subscript letters indicate function)
DW	Dishwasher Outlet
	Special-Purpose Floor Outlet
M	Motor
	Lighting Panel
	Power Panel

Continued

ELECTRICAL PLAN SYMBOLS

100

ELECTRICAL PLAN SYMBOLS

Table 8.4   Miscellaneous Symbols *(continued)*

	Disconnect Switch
	Thermostat
	Circuit Wiring Concealed in Wall or Ceiling
	Circuit Wiring Concealed in Floor
	Circuit Wiring Exposed
	Wiring Turned Up
	Wiring Turned Down
	Home Run to Panelboard (number of arrows indicate number of circuits)
	Conductors in Cable or Raceway (number of lines indicate number of conductors)
AFF	Above Finished Floor
AFG	Above Finished Grade
₵L	Centerline

CHAPTER 9
LOAD CALCULATIONS

Chapter 9 provides examples for calculating branch-circuit, feeder, and service loads for dwelling and non-dwelling units. Relevant requirements concerning loads have been directly extracted from the *NEC®*.

DWELLING LOAD

General Lighting and General-Use Receptacle Load

Using the outside dimensions, multiply the floor area for each floor by 3 volt-amperes per square foot (33 VA per square meter). Do not include open porches, garages, or unused or unfinished spaces not adaptable for future use.

220.12 Lighting Load for Specified Occupancies. A unit load of not less than that specified in Table 220.12 [shown in Exhibit 9.1] for occupancies specified therein shall constitute the minimum lighting load. The floor area for each floor shall be calculated from the outside dimensions of the building, dwelling unit, or other area involved.

220.14(J) Dwelling Occupancies. In one-family, two-family, and multifamily dwellings and in guest rooms or guest suites of hotels and motels, the outlets specified in (J)(1), (J)(2), and (J)(3) are included in the general lighting load calculations of 220.12. No additional load calculations shall be required for such outlets.

(1) All general-use receptacle outlets of 20-ampere rating or less, including receptacles connected to the circuits in 210.11(C)(3)
(2) The receptacle outlets specified in 210.52(E) and (G)
(3) The lighting outlets specified in 210.70(A) and (B)

Small Appliance Circuit Load

Include at least two small-appliance branch circuits per dwelling. The load must be calculated at 1500 volt-amperes for each 2-wire small-appliance branch circuit required.

220.52 Small Appliance and Laundry Loads—Dwelling Unit

(A) Small Appliance Circuit Load. In each dwelling unit, the load shall be calculated at 1500 volt-amperes for each 2-wire small-appliance branch circuit required by 210.11(C)(1). Where the load is subdivided through two or more feeders, the calculated load for each shall include not less than 1500 volt-amperes for each 2-wire

LOAD CALCULATIONS

EXHIBIT 9.1

NEC Table 220.12   General Lighting Loads by Occupancy

	Unit Load	
Type of Occupancy	Volt-Amperes per Square Meter	Volt-Amperes per Square Foot
Armories and auditoriums	11	1
Banks	39[b]	3½ [b]
Barber shops and beauty parlors	33	3
Churches	11	1
Clubs	22	2
Court rooms	22	2
Dwelling units[a]	33	3
Garages—commercial (storage)	6	½
Hospitals	22	2
Hotels and motels, including apartment houses without provision for cooking by tenants[a]	22	2
Industrial commercial (loft) buildings	22	2
Lodge rooms	17	1½
Office buildings	39[b]	3½ [b]
Restaurants	22	2
Schools	33	3
Stores	33	3
Warehouses (storage)	3	¼
In any of the preceding occupancies except one-family dwellings and individual dwelling units of two-family and multifamily dwellings:		
Assembly halls and auditoriums	11	1
Halls, corridors, closets, stairways	6	½
Storage spaces	3	¼

[a]See 220.14(J).

[b]See 220.14(K).

small-appliance branch circuit. These loads shall be permitted to be included with the general lighting load and subjected to the demand factors provided in Table 220.42 [shown here as Exhibit 9.2].

Exception: The individual branch circuit permitted by 210.52(B)(1), Exception No. 2, shall be permitted to be excluded from the calculation required by 220.52.

210.11(C) Dwelling Units.

(1) Small-Appliance Branch Circuits. In addition to the number of branch circuits required by other parts of this section, two or more 20-ampere small-appliance branch circuits shall be provided for all receptacle outlets specified by 210.52(B).

Laundry Circuit Load

Include at least one laundry branch circuit per dwelling unless meeting one of the exceptions. The load must be calculated at 1500 volt-amperes for each 2-wire laundry branch circuit required.

220.52(B) Laundry Circuit Load. A load of not less than 1500 volt-amperes shall be included for each 2-wire laundry branch circuit installed as required by 210.11(C)(2). This load shall be permitted to be included with the general lighting load and subjected to the demand factors provided in Table 220.42 [shown here as Exhibit 9.2].

210.11(C)(2) Laundry Branch Circuits. In addition to the number of branch circuits required by other parts of this section, at least one additional 20-ampere branch circuit shall be provided to supply the laundry receptacle outlet(s) required by 210.52(F). This circuit shall have no other outlets.

210.52(F) Laundry Areas. In dwelling units, at least one receptacle outlet shall be installed for the laundry.

EXHIBIT 9.2

NEC Table 220.42   Lighting Load Demand Factors

Type of Occupancy	Portion of Lighting Load to Which Demand Factor Applies (Volt-Amperes)	Demand Factor (Percent)
Dwelling units	First 3000 or less at	100
	From 3001 to 120,000 at	35
	Remainder over 120,000 at	25

[Note: Only the row pertaining to dwelling units is reproduced here.]

LOAD CALCULATIONS

104

LOAD CALCULATIONS

Exception No. 1: In a dwelling unit that is an apartment or living area in a multi-family building where laundry facilities are provided on the premises and are available to all building occupants, a laundry receptacle shall not be required.

Exception No. 2: In other than one-family dwellings where laundry facilities are not to be installed or permitted, a laundry receptacle shall not be required.

Lighting Load Demand Factors

Apply the demand factors from Exhibit 9.2 to that portion of the total branch-circuit load calculated for general illumination.

220.42 General Lighting. The demand factors specified in Table 220.42 [shown here as Exhibit 9.2 on p. 103] shall apply to that portion of the total branch-circuit load calculated for general illumination. They shall not be applied in determining the number of branch circuits for general illumination.

Appliance Load

Apply a demand factor of 75 percent to the nameplate rating load of four or more fastened-in-place appliances. Do not include electric ranges, clothes dryers, space-heating equipment, or air-conditioning equipment.

220.53 Appliance Load—Dwelling Unit(s). It shall be permissible to apply a demand factor of 75 percent to the nameplate rating load of four or more appliances fastened in place, other than electric ranges, clothes dryers, space-heating equipment, or air-conditioning equipment, that are served by the same feeder or service in a one-family, two-family, or multifamily dwelling.

Electric Clothes Dryers

Include 5000 watts (volt-amperes) or the nameplate rating, whichever is larger, for each household electric clothes dryer. Apply the demand factors from Exhibit 9.3 when there are five or more clothes dryers.

220.54 Electric Clothes Dryers—Dwelling Unit(s). The load for household electric clothes dryers in a dwelling unit(s) shall be either 5000 watts (volt-amperes) or the nameplate rating, whichever is larger, for each dryer served. The use of the demand factors in Table 220.54 [shown here as Exhibit 9.3 on p. 105] shall be permitted. Where two or more single-phase dryers are supplied by a 3-phase, 4-wire feeder or service, the total load shall be calculated on the basis of twice the maximum number connected between any two phases.

Electric Ranges and Other Cooking Appliances

Apply the demand factors from Exhibit 9.4 to household electric ranges, wall-mounted ovens, counter-mounted cooking units, and other household cooking appliances individually rated in excess of 1-3/4 kW (1750 watts).

EXHIBIT 9.3

NEC Table 220.54 Demand Factors for Household Electric Clothes Dryers

Number of Dryers	Demand Factor (Percent)
1–4	100%
5	85%
6	75%
7	65%
8	60%
9	55%
10	50%
11	47%
12–22	% = 47 – (number of dryers – 11)
23	35%
24–42	% = 35 – [0.5 × (number of dryers – 23)]
43 and over	25%

220.55 Electric Ranges and Other Cooking Appliances—Dwelling Unit(s). The load for household electric ranges, wall-mounted ovens, counter-mounted cooking units, and other household cooking appliances individually rated in excess of 1-3/4 kW shall be permitted to be calculated in accordance with Table 220.55 [shown here as Exhibit 9.4]. Kilovolt-amperes (kVA) shall be considered equivalent to kilowatts (kW) for loads calculated under this section.

Where two or more single-phase ranges are supplied by a 3-phase, 4-wire feeder or service, the total load shall be calculated on the basis of twice the maximum number connected between any two phases.

Noncoincident Loads

Compare the heating and air-conditioning load and include the larger. For heat pumps, include the compressor and the maximum amount of electric heat that is energized with the compressor. Include the air handler with either load.

220.60 Noncoincident Loads. Where it is unlikely that two or more noncoincident loads will be in use simultaneously, it shall be permissible to use only the largest load(s) that will be used at one time for calculating the total load of a feeder or service.

LOAD CALCULATIONS

106

LOAD CALCULATIONS

EXHIBIT 9.4

NEC Table 220.55   Demand Factors and Loads for Household
Electric Ranges, Wall-Mounted Ovens, Counter-Mounted Cooking Units,
and Other Household Cooking Appliances over 1-3/4 kW Rating (Column
C to be used in all cases except as otherwise permitted in Note 3.)

	Demand Factor (Percent) (See Notes)		
Number of Appliances	Column A (Less than 3½ kW Rating)	Column B (3½ kW to 8¾ kW Rating)	Column C Maximum Demand (kW) (See Notes) (Not over 12 kW Rating)
1	80	80	8
2	75	65	11
3	70	55	14
4	66	50	17
5	62	45	20
6	59	43	21
7	56	40	22
8	53	36	23
9	51	35	24
10	49	34	25
11	47	32	26
12	45	32	27
13	43	32	28
14	41	32	29
15	40	32	30
16	39	28	31
17	38	28	32
18	37	28	33
19	36	28	34
20	35	28	35
21	34	26	36
22	33	26	37
23	32	26	38
24	31	26	39
25	30	26	40
26–30	30	24	15 kW + 1 kW for each range
31–40	30	22	
41–50	30	20	25 kW + ¾ kW for each range
51–60	30	18	
61 and over	30	16	

1. Over 12 kW through 27 kW ranges all of same rating. For ranges individually rated more than 12 kW but not more than 27 kW, the maximum demand in Column C shall be increased 5 percent for each additional kilowatt of rating or major fraction thereof by which the rating of individual ranges exceeds 12 kW.

Table continues below.

EXHIBIT 9.4 *(continued)*

2. Over 8¾ kW through 27 kW ranges of unequal ratings. For ranges individually rated more than 8¾ kW and of different ratings, but none exceeding 27 kW, an average value of rating shall be calculated by adding together the ratings of all ranges to obtain the total connected load (using 12 kW for any range rated less than 12 kW) and dividing by the total number of ranges. Then the maximum demand in Column C shall be increased 5 percent for each kilowatt or major fraction thereof by which this average value exceeds 12 kW.

3. Over 1¾ kW through 8¾ kW. In lieu of the method provided in Column C, it shall be permissible to add the nameplate ratings of all household cooking appliances rated more than 1¾ kW but not more than 8¾ kW and multiply the sum by the demand factors specified in Column A or B for the given number of appliances. Where the rating of cooking appliances falls under both Column A and Column B, the demand factors for each column shall be applied to the appliances for that column, and the results added together.

4. Branch-Circuit Load. It shall be permissible to calculate the branch-circuit load for one range in accordance with Table 220.55. The branch-circuit load for one wall-mounted oven or one counter-mounted cooking unit shall be the nameplate rating of the appliance. The branch-circuit load for a counter-mounted cooking unit and not more than two wall-mounted ovens, all supplied from a single branch circuit and located in the same room, shall be calculated by adding the nameplate rating of the individual appliances and treating this total as equivalent to one range.

5. This table also applies to household cooking appliances rated over 1¾ kW and used in instructional programs.

Largest Motor

Multiply the largest motor by 25 percent and add to the load calculation. This applies to one motor only.

220.50 Motors. Motor loads shall be calculated in accordance with 430.24, 430.25, and 430.26 and with 440.6 for hermetic refrigerant motor compressors.

430.24 Several Motors or a Motor(s) and Other Load(s). Conductors supplying several motors, or a motor(s) and other load(s), shall have an ampacity not less than 125 percent of the full-load current rating of the highest rated motor plus the sum of the full-load current ratings of all the other motors in the group, as determined by 430.6(A), plus the ampacity required for the other loads.

Volt-Ampere Demand Load

Add the total volt-amperes from the above computations to find the minimum required demand load.

LOAD CALCULATIONS

OPTIONAL ONE-FAMILY DWELLING LOAD

Minimum Size Service

Regardless of the result of this calculation, the minimum size service is 100 amperes.

220.82(A) Feeder and Service Load. This section applies to a dwelling unit having the total connected load served by a single 120/240-volt or 208Y/120-volt set of 3-wire service or feeder conductors with an ampacity of 100 or greater. It shall be permissible to calculate the feeder and service loads in accordance with this section instead of the method specified in Part III of this article. The calculated load shall be the result of adding the loads from 220.82(B) and (C). Feeder and service-entrance conductors whose calculated load is determined by this optional calculation shall be permitted to have the neutral load determined by 220.61.

General Lighting and General-Use Receptacle Load

Using the outside dimensions, multiply the floor area for each floor by the 3 volt-ampere per square foot (33 VA per square meter). Do not include open porches, garages, or unused or unfinished spaces not adaptable for future use.

220.82(B)(1) 33 volt-amperes/m^2 or 3 volt-amperes/ft^2 for general lighting and general-use receptacles. The floor area for each floor shall be calculated from the outside dimensions of the dwelling unit. The calculated floor area shall not include open porches, garages, or unused or unfinished spaces not adaptable for future use.

Small Appliance and Laundry Circuit Load

Include at least two small-appliance branch circuits. The load must be calculated at 1500 volt-amperes for each 2-wire small-appliance branch circuit. Include at least one laundry branch circuit. The load must be calculated at 1500 volt-amperes for each 2-wire laundry branch circuit.

220.82(B)(2) 1500 volt-amperes for each 2-wire, 20-ampere small-appliance branch circuit and each laundry branch circuit specified in 220.52.

Appliance Load

Include the nameplate rating of all fastened-in-place appliances, including ranges, wall-mounted ovens, counter-mounted cooking units, clothes dryers, and water heaters.

220.82(B)(3) The nameplate rating of all appliances that are fastened in place, permanently connected, or located to be on a specific circuit, ranges, wall-mounted ovens, counter-mounted cooking units, clothes dryers, and water heaters.

Motors

Include the nameplate ampere or kVA rating of all motors and all low power factor loads. Do not include heating and air-conditioning loads here.

220.82(B)(4) The nameplate ampere or kVA rating of all motors and of all low-power-factor loads.

Demand for General Loads

The following demand factors apply to the general calculated loads in 220.82(B)(1) through (4). Include 100 percent of the first 10 kVA loads and 40 percent of the remaining loads.

220.82(B) General Loads. The general calculated load shall be not less than 100 percent of the first 10 kVA plus 40 percent of the remainder of the following loads: [220.82(B)(1)–220.82(B)(4).]

Heating and Air-Conditioning Load

Include the largest of the six selections in 220.82(C)(1) through (6).

220.82(C) Heating and Air-Conditioning Load. The largest of the following six selections (load in kVA) shall be included:

(1) 100 percent of the nameplate rating(s) of the air conditioning and cooling.
(2) 100 percent of the nameplate rating(s) of the heating when a heat pump is used without any supplemental electric heating.
(3) 100 percent of the nameplate ratings of electric thermal storage and other heating systems where the usual load is expected to be continuous at the full nameplate value. Systems qualifying under this selection shall not be calculated under any other selection in 220.30(C).
(4) 100 percent of the nameplate rating(s) of the heat pump compressor and 65 percent of the supplemental electric heating for central electric space heating systems. If the heat pump compressor is prevented from operating at the same time as the supplementary heat, it does not need to be added to the supplementary heat for the total central space heating load.
(5) 65 percent of the nameplate rating(s) of electric space heating if less than four separately controlled units.
(6) 40 percent of the nameplate rating(s) of electric space heating if four or more separately controlled units.

Volt-Ampere Demand Load

Add the total volt-amperes from the above computations to find the minimum required demand load.

NONDWELLING LOAD

Receptacle Outlets

Multiply each single or multiple receptacle on one yoke by 180 volt-amperes. Multiply each single piece of equipment comprised of four or more receptacles by 90 volt-amperes per receptacle.

LOAD CALCULATIONS

LOAD CALCULATIONS

220.14(I) Receptacle Outlets. Except as covered in 220.14(J) and (K), receptacle outlets shall be calculated at not less than 180 volt-amperes for each single or for each multiple receptacle on one yoke. A single piece of equipment consisting of a multiple receptacle comprised of four or more receptacles shall be calculated at not less than 90 volt-amperes per receptacle. This provision shall not be applicable to the receptacle outlets specified in 210.11(C)(1) and (C)(2).

Fixed Multioutlet Assemblies

Depending upon the use, calculate the loads in accordance with (1) or (2).

220.14(H) Fixed Multioutlet Assemblies. Fixed multioutlet assemblies used in other than dwelling units or the guest rooms or guest suites of hotels or motels shall be calculated in accordance with (H)(1) or (H)(2). For the purposes of this section, the calculation shall be permitted to be based on the portion that contains receptacle outlets.

(1) Where appliances are unlikely to be used simultaneously, each 1.5 m (5 ft) or fraction thereof of each separate and continuous length shall be considered as one outlet of not less than 180 volt-amperes.

(2) Where appliances are likely to be used simultaneously, each 300 mm (1 ft) or fraction thereof shall be considered as an outlet of not less than 180 volt-amperes.

Demand Factors for Nondwelling Receptacle Loads

Two options are available for computing nondwelling receptacle loads. The receptacle loads can be added to the lighting load and made subject to the demand factors given in Exhibit 9.6 on p. 112. Or the receptacle loads can be made subject to the demand factors given in Exhibit 9.5.

220.44 Receptacle Loads—Other Than Dwelling Units. Receptacle loads calculated in accordance with 220.14(H) and (I) shall be permitted to be made subject to the demand factors given in Table 220.42 [shown here as Exhibit 9.6 on p. 112] or Table 220.44 [shown here as Exhibit 9.5].

EXHIBIT 9.5

NEC **Table 220.44   Demand Factors for Non–dwelling Receptacle Loads**

Portion of Receptacle Load to Which Demand Factor Applies (Volt-Amperes)	Demand Factor (Percent)
First 10 kVA or less at	100
Remainder over 10 kVA at	50

Receptacle Loads for Banks and Office Buildings

If the actual number of general-purpose receptacle outlets is not known in banks or office buildings, include a unit load of 1 volt-ampere per square foot (11 volt-amperes per square meter) for general-purpose receptacles. When the number of receptacles is known, include the larger of (1) the calculated load from 220.14(K), or (2) 1 volt-ampere per square foot (11 volt-amperes per square meter).

> **220.14(K) Banks and Office Buildings.** In banks or office buildings, the receptacle loads shall be calculated to be the larger of (1) or (2):
>
> (1) The computed load from 220.14
> (2) 11 volt-amperes/m^2 or 1 volt-ampere/ft^2

General Lighting Load

Multiply the volt-ampere per square foot (or meter) for the type of occupancy by the total floor area. Calculate the floor area for each floor by using the outside dimensions of the building.

> **220.12 Lighting Load for Specified Occupancies.** A unit load of not less than that specified in Table 220.12 [shown here as Exhibit 9.1 on p. 102] for occupancies specified therein shall constitute the minimum lighting load. The floor area for each floor shall be calculated from the outside dimensions of the building, dwelling unit, or other area involved.

Lighting Load Demand Factors

Apply the demand factors from Exhibit 9.6 to certain portions of hospitals, hotels, motels, apartment houses without provision for cooking by tenants, and storage warehouses.

> **220.42 General Lighting.** The demand factors specified in Table 220.42 [shown here as Exhibit 9.6 on p. 112] shall apply to that portion of the total branch-circuit load calculated for general illumination. They shall not be applied in determining the number of branch circuits for general illumination.

Track Lighting

In addition to the general lighting load, include 150 volt-amperes for every 2 feet (600 mm) of lighting track or fraction thereof.

> **220.43(B) Track Lighting.** For track lighting in other than dwelling units or guest rooms or guest suites of hotels or motels, an additional load of 150 volt-amperes shall be included for every 600 mm (2 ft) of lighting track or fraction thereof. Where multicircuit track is installed, the load shall be considered to be divided equally between the track circuits.

LOAD CALCULATIONS

EXHIBIT 9.6

NEC Table 220.42   Lighting Load Demand Factors

Type of Occupancy	Portion of Lighting Load to Which Demand Factor Applies (Volt-Amperes)	Demand Factor (Percent)
Hospitals*	First 50,000 or less at	40
	Remainder over 50,000 at	20
Hotels and motels, including apartment houses without provision for cooking by tenants*	First 20,000 or less at	50
	From 20,001 to 100,000 at	40
	Remainder over 100,000 at	30
Warehouses (storage)	First 12,500 or less at	100
	Remainder over 12,500 at	50
All others	Total volt-amperes	100

*The demand factors of this table shall not apply to the calculated load of feeders or services supplying areas in hospitals, hotels, and motels where the entire lighting is likely to be used at one time, as in operationg rooms, ballrooms, or dining rooms.

[Note: Only rows pertaining to occupancies other than dwelling units are reproduced here.]

Sign and Outline Lighting

Where required, include at least 1200 volt-amperes for each required sign or outline lighting branch circuit.

220.14(F) Sign and Outline Lighting. Sign and outline lighting outlets shall be calculated at a minimum of 1200 volt-amperes for each required branch circuit specified in 600.5(A).

Show Windows

Include at least 200 volt-amperes per linear foot (660 volt-amperes per linear meter) of show window, measured horizontally along its base.

220.43(A) Show Windows. For show-window lighting, a load of not less than 660 volt-amperes/linear meter or 200 volt-amperes/linear foot shall be included for a show window, measured horizontally along its base.

Kitchen Equipment

Apply the demand factors from Exhibit 9.7 to commercial electric cooking equipment, dishwasher booster heaters, water heaters, and other kitchen equipment.

220.56 Kitchen Equipment—Other Than Dwelling Unit(s). It shall be permissible to calculate the load for commercial electric cooking equipment, dishwasher booster heaters, water heaters, and other kitchen equipment in accordance with Table 220.56 [shown here as Exhibit 9.7]. These demand factors shall be applied to all equipment that has either thermostatic control or intermittent use as kitchen equipment. These demand factors shall not apply to space-heating, ventilating, or air-conditioning equipment.

However, in no case shall the feeder or service calculated load be less than the sum of the largest two kitchen equipment loads.

Noncoincident Loads

Compare the heating and air-conditioning loads and include the larger. For heat pumps, include the compressor and the maximum amount of electric heat that is energized with the compressor. Include the air handler with either load.

220.60 Noncoincident Loads. Where it is unlikely that two or more noncoincident loads will be in use simultaneously, it shall be permissible to use only the largest load(s) that will be used at one time for calculating the total load of a feeder or service.

220.51 Fixed Electric Space Heating. Fixed electric space heating loads shall be calculated at 100 percent of the total connected load. However, in no case shall a feeder or service load current rating be less than the rating of the largest branch circuit supplied.

EXHIBIT 9.7

NEC Table 220.56 Demand Factors for Kitchen Equipment—Other Than Dwelling Unit(s)

Number of Units of Equipment	Demand Factor (Percent)
1	100
2	100
3	90
4	80
5	70
6 and over	65

LOAD CALCULATIONS

LOAD CALCULATIONS

Motor Loads on Services or Feeders

Add motor and refrigeration loads to the calculation. Multiply the largest motor by 25 percent and add to the load calculation.

220.50 Motors. Motor loads shall be calculated in accordance with 430.24, 430.25, and 430.26 and with 440.6 for hermetic refrigerant motor compressors.

430.24 Several Motors or a Motor(s) and Other Load(s). Conductors supplying several motors, or a motor(s) and other load(s), shall have an ampacity not less than 125 percent of the full-load current rating of the highest rated motor plus the sum of the full-load current ratings of all the other motors in the group, as determined by 430.6(A), plus the ampacity required for the other loads.

430.25 Multimotor and Combination-Load Equipment. The ampacity of the conductors supplying multimotor and combination-load equipment shall not be less than the minimum circuit ampacity marked on the equipment in accordance with 430.7(D). Where the equipment is not factory-wired and the individual nameplates are visible in accordance with 430.7(D)(2), the conductor ampacity shall be determined in accordance with 430.24.

430.26 Feeder Demand Factor. Where reduced heating of the conductors results from motors operating on duty-cycle, intermittently, or from all motors not operating at one time, the authority having jurisdiction may grant permission for feeder conductors to have an ampacity less than specified in 430.24, provided the conductors have sufficient ampacity for the maximum load determined in accordance with the sizes and number of motors supplied and the character of their loads and duties.

440.6(A) Hermetic Refrigerant Motor-Compressor. For a hermetic refrigerant motor-compressor, the rated-load current marked on the nameplate of the equipment in which the motor-compressor is employed shall be used in determining the rating or ampacity of the disconnecting means, the branch-circuit conductors, the controller, the branch-circuit short-circuit and ground-fault protection, and the separate motor overload protection. Where no rated-load current is shown on the equipment nameplate, the rated-load current shown on the compressor nameplate shall be used.

Exception No. 1: Where so marked, the branch-circuit selection current shall be used instead of the rated-load current to determine the rating or ampacity of the disconnecting means, the branch-circuit conductors, the controller, and the branch-circuit short-circuit and ground-fault protection.

Exception No. 2: For cord-and-plug-connected equipment, the nameplate marking shall be used in accordance with 440.22(B), Exception No. 2.

Volt-Ampere Demand Load

Add the total volt-amperes from the above computations to find the minimum required demand load.

CHAPTER 10
CLEARANCE AND COVER REQUIREMENTS

Chapter 10 provides specific clearance and cover requirements for electrical installations. The material in this chapter is extracted directly from the *NEC*® for quick and easy reference. This chapter makes it possible to quickly reference the clearance and cover requirements that are spread throughout various sections and chapters of the *NEC*.

WIRING AND WORKING SPACE CLEARANCES

Working Space Clearances around Electrical Equipment

110.26 Spaces About Electrical Equipment. Sufficient access and working space shall be provided and maintained about all electric equipment to permit ready and safe operation and maintenance of such equipment. Enclosures housing electrical apparatus that are controlled by a lock(s) shall be considered accessible to qualified persons.

(A) Working Space. Working space for equipment operating at 600 volts, nominal, or less to ground and likely to require examination, adjustment, servicing, or maintenance while energized shall comply with the dimensions of 110.26(A)(1), (A)(2), and (A)(3) or as required or permitted elsewhere in this *Code*.

(1) Depth of Working Space. The depth of the working space in the direction of live parts shall not be less than that specified in Table 110.26(A)(1) [shown here as Exhibit 10.1] unless the requirements of 110.26(A)(1)(a), (A)(1)(b), or (A)(1)(c) are met. Distances shall be measured from the exposed live parts or from the enclosure or opening if the live parts are enclosed.

(a) Dead-Front Assemblies. Working space shall not be required in the back or sides of assemblies, such as dead-front switchboards or motor control centers, where all connections and all renewable or adjustable parts, such as fuses or switches, are accessible from locations other than the back or sides. Where rear access is required to work on nonelectrical parts on the back of enclosed equipment, a minimum horizontal working space of 762 mm (30 in.) shall be provided.

(b) Low Voltage. By special permission, smaller working spaces shall be permitted where all uninsulated parts operate at not greater than 30 volts rms, 42 volts peak, or 60 volts dc.

(c) Existing Buildings. In existing buildings where electric equipment is being replaced, Condition 2 working clearance shall be permitted between dead-front switchboards, panelboards, or motor control centers located across the aisle from each other

CLEARANCE AND COVER REQUIREMENTS

EXHIBIT 10.1

NEC Table 110.26(A)(1)   Working Spaces

Nominal Voltage to Ground	Minimum Clear Distance		
	Condition 1	Condition 2	Condition 3
0–150	900 mm (3 ft)	900 mm (3 ft)	900 mm (3 ft)
151–600	900 mm (3 ft)	1.1 m (3½ ft)	1.2 m (4 ft)

Note: Where the conditions are as follows:

Condition 1—Exposed live parts on one side of the working space and no live or grounded parts on the other side of the working space, or exposed live parts on both sides of the working space that are effectively guarded by insulating materials.

Condition 2—Exposed live parts on one side of the working space and grounded parts on the other side of the working space. Concrete, brick, or tile walls shall be considered as grounded.

Condition 3—Exposed live parts on both sides of the working space.

where conditions of maintenance and supervision ensure that written procedures have been adopted to prohibit equipment on both sides of the aisle from being open at the same time. Qualified persons who are authorized will service the installation.

(2) Width of Working Space. The width of the working space in front of the electric equipment shall be the width of the equipment or 750 mm (30 in.), whichever is greater. In all cases, the work space shall permit at least a 90 degree opening of equipment doors or hinged panels.

(3) Height of Working Space. The work space shall be clear and extend from the grade, floor, or platform to the height required by 110.26(E). Within the height requirements of this section, other equipment that is associated with the electrical installation and is located above or below the electrical equipment shall be permitted to extend not more than 150 mm (6 in.) beyond the front of the electrical equipment.

(B) Clear Spaces. Working space required by this section shall not be used for storage. When normally enclosed live parts are exposed for inspection or servicing, the working space, if in a passageway or general open space, shall be suitably guarded.

(C) Entrance to Working Space.

(1) Minimum Required. At least one entrance of sufficient area shall be provided to give access to the working space about electric equipment.

(2) Large Equipment. For equipment rated 1200 amperes or more that contains overcurrent devices, switching devices, or control devices, there shall be one entrance to the required working space not less than 610 mm (24 in.) wide and 2.0 m (6-1/2 ft) high at each end of the working space. Where the entrance has a personnel door(s), the door(s) shall open in the direction of egress and be equipped with panic bars, pressure plates, or other devices that are normally latched but open under simple pressures.

A single entrance to the required working space shall be permitted where either of the conditions in 110.26(C)(2)(a) or (C)(2)(b) is met.

(a) Unobstructed Exit. Where the location permits a continuous and unobstructed way of exit travel, a single entrance to the working space shall be permitted.

(b) Extra Working Space. Where the depth of the working space is twice that required by 110.26(A)(1), a single entrance shall be permitted. It shall be located so that the distance from the equipment to the nearest edge of the entrance is not less than the minimum clear distance specified in Table 110.26(A)(1) for equipment operating at that voltage and in that condition.

(D) Illumination. Illumination shall be provided for all working spaces about service equipment, switchboards, panelboards, or motor control centers installed indoors. Additional lighting outlets shall not be required where the work space is illuminated by an adjacent light source or as permitted by 210.70(A)(1), Exception No. 1, for switched receptacles. In electrical equipment rooms, the illumination shall not be controlled by automatic means only.

(E) Headroom. The minimum headroom of working spaces about service equipment, switchboards, panelboards, or motor control centers shall be 2.0 m (6-1/2 ft). Where the electrical equipment exceeds 2.0 m (6-1/2 ft) in height, the minimum headroom shall not be less than the height of the equipment.

Exception: In existing dwelling units, service equipment or panelboards that do not exceed 200 amperes shall be permitted in spaces where the headroom is less than 2.0 m (6-1/2 ft).

(F) Dedicated Equipment Space. All switchboards, panelboards, distribution boards, and motor control centers shall be located in dedicated spaces and protected from damage.

Exception: Control equipment that by its very nature or because of other rules of the Code must be adjacent to or within sight of its operating machinery shall be permitted in those locations.

(1) Indoor. Indoor installations shall comply with 110.26(F)(1)(a) through (F)(1)(d).

(a) Dedicated Electrical Space. The space equal to the width and depth of the equipment and extending from the floor to a height of 1.8 m (6 ft) above the equipment or to the structural ceiling, whichever is lower, shall be dedicated to the electrical installation. No piping, ducts, leak protection apparatus, or equipment foreign to the electrical installation shall be located in this zone.

Exception: Suspended ceilings with removable panels shall be permitted within the 1.8-m (6-ft) zone.

(b) Foreign Systems. The area above the dedicated space required by 110.26(F)(1)(a) shall be permitted to contain foreign systems, provided protection is

118

CLEARANCE AND COVER REQUIREMENTS

installed to avoid damage to the electrical equipment from condensation, leaks, or breaks in such foreign systems.

(c) Sprinkler Protection. Sprinkler protection shall be permitted for the dedicated space where the piping complies with this section.

(d) Suspended Ceilings. A dropped, suspended, or similar ceiling that does not add strength to the building structure shall not be considered a structural ceiling.

(2) Outdoor. Outdoor electric equipment shall be installed in suitable enclosures and shall be protected from accidental contact by unauthorized personnel, or by vehicular traffic, or by accidental spillage or leakage from piping systems. The working clearance space shall include the zone described in 110.26(A). No architectural appurtenance or other equipment shall be located in this zone.

WORKING SPACE CLEARANCES (OVER 600 VOLTS, NOMINAL)

110.31 Enclosure for Electrical Installations. Electrical installations in a vault, room, or closet or in an area surrounded by a wall, screen, or fence, access to which is controlled by a lock(s) or other approved means, shall be considered to be accessible to qualified persons only. The type of enclosure used in a given case shall be designed and constructed according to the nature and degree of the hazard(s) associated with the installation.

For installations other than equipment as described in 110.31(D), a wall, screen, or fence shall be used to enclose an outdoor electrical installation to deter access by persons who are not qualified. A fence shall not be less than 2.1 m (7 ft) in height or a combination of 1.8 m (6 ft) or more of fence fabric and a 300-mm (1-ft) or more extension utilizing three or more strands of barbed wire or equivalent. The distance from the fence to live parts shall be not less than given in Table 110.31 [shown here as Exhibit 10.2].

110.34   Work Space and Guarding.

(A) Working Space. Except as elsewhere required or permitted in this *Code,* the minimum clear working space in the direction of access to live parts of electrical

---

EXHIBIT 10.2

NEC Table 110.31   Minimum Distance from Fence to Live Parts

Nominal Voltage	Minimum Distance to Live Parts	
	m	**ft**
601 – 13,799	3.05	10
13,800 – 230,000	4.57	15
Over 230,000	5.49	18

Note: For clearances of conductors for specific system voltages and typical BIL ratings, see ANSI C2-2002, *National Electrical Safety Code.*

EXHIBIT 10.3

NEC Table 110.34(A)   Minimum Depth of Clear Working Space at Electrical Equipment

Nominal Voltage to Ground	Minimum Clear Distance		
	Condition 1	Condition 2	Condition 3
601–2500 V	900 mm (3 ft)	1.2 m (4 ft)	1.5 m (5 ft)
2501–9000 V	1.2 m (4 ft)	1.5 m (5 ft)	1.8 m (6 ft)
9001–25,000 V	1.5 m (5 ft)	1.8 m (6 ft)	2.8 m (9 ft)
25,001V–75 kV	1.8 m (6 ft)	2.5 m (8 ft)	3.0 m (10 ft)
Above 75 kV	2.5 m (8 ft)	3.0 m (10 ft)	3.7 m (12 ft)

Note: Where the conditions are as follows:

Condition 1—Exposed live parts on one side of the working space and no live or grounded parts on the other side of the working space, or exposed live parts on both sides of the working space that are effectively guarded by insulating materials.

Condition 2—Exposed live parts on one side of the working space and grounded parts on the other side of the working space. Concrete, brick, or tile walls shall be considered as grounded.

Condition 3—Exposed live parts on both sides of the working space.

equipment shall not be less than specified in Table 110.34(A) [shown here as Exhibit 10.3]. Distances shall be measured from the live parts, if such are exposed, or from the enclosure front or opening if such are enclosed.

Exception: Working space shall not be required in back of equipment such as dead-front switchboards or control assemblies where there are no renewable or adjustable parts (such as fuses or switches) on the back and where all connections are accessible from locations other than the back. Where rear access is required to work on de-energized parts on the back of enclosed equipment, a minimum working space of 750 mm (30 in.) horizontally shall be provided.

(E) Elevation of Unguarded Live Parts. Unguarded live parts above working space shall be maintained at elevations not less than required by Table 110.34(E) [shown here as Exhibit 10.4].

EXHIBIT 10.4

NEC Table 110.34(E)   Elevation of Unguarded Live Parts Above Working Space

Nominal Voltage Between Phases	Elevation	
	m	ft
601–7500 V	2.8	9
7501–35,000 V	2.9	9½
Over 35 kV	2.9 m + 9.5 mm/kV above 35	9½ ft + 0.37 in./kV above 35

CLEARANCE AND COVER REQUIREMENTS

CLEARANCE AND COVER REQUIREMENTS

WIRING SPACE CLEARANCES
IN SWITCHBOARDS AND PANELBOARDS

See Exhibits 10.5 and 10.6.

Minimum Space Separation for Equipment, Over 600 Volts, Nominal

490.24 Minimum Space Separation. In field-fabricated installations, the minimum air separation between bare live conductors and between such conductors and adjacent grounded surfaces shall not be less than the values given in Table 490.24

EXHIBIT 10.5

NEC Table 408.5   Clearance for Conductors Entering Bus Enclosures

Conductor	Minimum Spacing Between Bottom of Enclosure and Busbars, Their Supports, or Other Obstructions	
	mm	in.
Insulated busbars, their supports, or other obstructions	200	8
Noninsulated busbars	250	10

EXHIBIT 10.6

NEC Table 408.56   Minimum Spacings Between Bare Metal Parts

Voltage	Opposite Polarity Where Mounted on the Same Surface		Opposite Polarity Where Held Free in Air		Live Parts to Ground*	
	mm	in.	mm	in.	mm	in.
Not over 125 volts, nominal	19.1	¾	12.7	½	12.7	½
Not over 250 volts, nominal	31.8	1¼	19.1	¾	12.7	½
Not over 600 volts, nominal	50.8	2	25.4	1	25.4	1

*For spacing between live parts and doors of cabinets, see 312.11(A)(1), (2), and (3).

EXHIBIT 10.7

NEC Table 490.24   Minimum Clearance of Live Parts*

Nominal Voltage Rating (kV)	Impulse Withstand, B.I.L (kV)		Minimum Clearance of Live Parts							
			Phase-to-Phase				Phase-to-Ground			
			Indoors		Outdoors		Indoors		Outdoors	
	Indoors	Outdoors	mm	in.	mm	in.	mm	in.	mm	in.
2.4–4.16	60	95	115	4.5	180	7	80	3.0	155	6
7.2	75	95	140	5.5	180	7	105	4.0	155	6
13.8	95	110	195	7.5	305	12	130	5.0	180	7
14.4	110	110	230	9.0	305	12	170	6.5	180	7
23	125	150	270	10.5	385	15	190	7.5	255	10
34.5	150	150	320	12.5	385	15	245	9.5	255	10
	200	200	460	18.0	460	18	335	13.0	335	13
46	—	200	—	—	460	18	—	—	335	13
	—	250	—	—	535	21	—	—	435	17
69	—	250	—	—	535	21	—	—	435	17
	—	350	—	—	790	31	—	—	635	25
115	—	550	—	—	1350	53	—	—	1070	42
138	—	550	—	—	1350	53	—	—	1070	42
	—	650	—	—	1605	63	—	—	1270	50
161	—	650	—	—	1605	63	—	—	1270	50
	—	750	—	—	1830	72	—	—	1475	58
230	—	750	—	—	1830	72	—	—	1475	58
	—	900	—	—	2265	89	—	—	1805	71
	—	1050	—	—	2670	105	—	—	2110	83

*The values given are the minimum clearance for rigid parts and bare conductors under favorable service conditions. They shall be increased for conductor movement or under unfavorable service conditions or wherever space limitations permit. The selection of the associated impulse withstand voltage for a particular system voltage is determined by the characteristics of the surge protective equipment.

[shown here as Exhibit 10.7]. These values shall not apply to interior portions or exterior terminals of equipment designed, manufactured, and tested in accordance with accepted national standards.

OVERHEAD CONDUCTOR CLEARANCES

Outside Branch Circuits and Feeders

225.18 Clearance from Ground. Overhead spans of open conductors and open multiconductor cables of not over 600 volts, nominal, shall have a clearance of not less than the following:

(1) 3.0 m (10 ft)—above finished grade, sidewalks, or from any platform or projection from which they might be reached where the voltage does not exceed 150 volts to ground and accessible to pedestrians only
(2) 3.7 m (12 ft)—over residential property and driveways, and those commercial areas not subject to truck traffic where the voltage does not exceed 300 volts to ground
(3) 4.5 m (15 ft)—for those areas listed in the 3.7-m (12-ft) classification where the voltage exceeds 300 volts to ground

CLEARANCE AND COVER REQUIREMENTS

CLEARANCE AND COVER REQUIREMENTS

(4) 5.5 m (18 ft)—over public streets, alleys, roads, parking areas subject to truck traffic, driveways on other than residential property, and other land traversed by vehicles, such as cultivated, grazing, forest, and orchard

225.19 Clearances from Buildings for Conductors of Not Over 600 Volts, Nominal.

(A) Above Roofs. Overhead spans of open conductors and open multiconductor cables shall have a vertical clearance of not less than 2.5 m (8 ft) above the roof surface. The vertical clearance above the roof level shall be maintained for a distance not less than 900 mm (3 ft) in all directions from the edge of the roof.

Exception No. 1: The area above a roof surface subject to pedestrian or vehicular traffic shall have a vertical clearance from the roof surface in accordance with the clearance requirements of 225.18.

Exception No. 2: Where the voltage between conductors does not exceed 300, and the roof has a slope of 100 mm in 300 mm (4 in. in 12 in.) or greater, a reduction in clearance to 900 mm (3 ft) shall be permitted.

Exception No 3: Where the voltage between conductors does not exceed 300, a reduction in clearance above only the overhanging portion of the roof to not less than 450 mm (18 in.) shall be permitted if (1) not more than 1.8 m (6 ft) of the conductors, 1.2 m (4 ft) horizontally, pass above the roof overhang and (2) they are terminated at a through-the-roof raceway or approved support.

Exception No 4: The requirement for maintaining the vertical clearance 900 mm (3 ft) from the edge of the roof shall not apply to the final conductor span where the conductors are attached to the side of a building.

(B) From Nonbuilding or Nonbridge Structures. From signs, chimneys, radio and television antennas, tanks, and other nonbuilding or nonbridge structures, clearances—vertical, diagonal, and horizontal—shall not be less than 900 mm (3 ft).

(C) Horizontal Clearances. Clearances shall not be less than 900 mm (3 ft).

(D) Final Spans. Final spans of feeders or branch circuits shall comply with 225.19(D)(1), (D)(2), and (D)(3).

(1) Clearance from Windows. Final spans to the building they supply, or from which they are fed, shall be permitted to be attached to the building, but they shall be kept not less than 900 mm (3 ft) from windows that are designed to be opened, and from doors, porches, balconies, ladders, stairs, fire escapes, or similar locations.

Exception: Conductors run above the top level of a window shall be permitted to be less than the 900-mm (3-ft) requirement.

(2) Vertical Clearance. The vertical clearance of final spans above, or within 900 mm (3 ft) measured horizontally of, platforms, projections, or surfaces from which they might be reached shall be maintained in accordance with 225.18.

(3) Building Openings. The overhead branch-circuit and feeder conductors shall not be installed beneath openings through which materials may be moved, such as openings in farm and commercial buildings, and shall not be installed where they obstruct entrance to these buildings' openings.

(E) Zone for Fire Ladders. Where buildings exceed three stories or 15 m (50 ft) in height, overhead lines shall be arranged, where practicable, so that a clear space (or zone) at least 1.8 m (6 ft) wide will be left either adjacent to the buildings or beginning not over 2.5 m (8 ft) from them to facilitate the raising of ladders when necessary for fire fighting.

Outside Branch Circuits and Feeders (Over 600 Volts)

225.60 Clearances over Roadways, Walkways, Rail, Water, and Open Land.

(A) 22 kV Nominal to Ground or Less. The clearances over roadways, walkways, rail, water, and open land for conductors and live parts up to 22 kV nominal to ground or less shall be not less than the values shown in Table 225.60 [shown here as Exhibit 10.8].

(B) Over 22 kV Nominal to Ground. Clearances for the categories shown in Table 225.60 shall be increased by 10 mm (0.4 in.) per kV above 22,000 volts.

(C) Special Cases. For special cases, such as where crossings will be made over lakes, rivers, or areas using large vehicles such as mining operations, specific designs shall be engineered considering the special circumstances and shall be approved by the authority having jurisdiction.

FPN: For additional information, see ANSI C2-2002, *National Electrical Safety Code*.

EXHIBIT 10.8

NEC Table 225.60   Clearances over Roadways, Walkways, Rail, Water, and Open Land

Location	Clearance	
	m	ft
Open land subject to vehicles, cultivation, or grazing	5.6	18.5
Roadways, driveways, parking lots, and alleys	5.6	18.5
Walkways	4.1	13.5
Rails	8.1	26.5
Spaces and ways for pedestrians and restricted traffic	4.4	14.5
Water areas not suitable for boating	5.2	17

CLEARANCE AND COVER REQUIREMENTS

CLEARANCE AND COVER REQUIREMENTS

EXHIBIT 10.9

NEC **Table 225.61   Clearances over Buildings and Other Structures**

Clearance from Conductors or Live Parts from:	Horizontal		Vertical	
	m	**ft**	**m**	**ft**
Building walls, projections, and windows	2.3	7.5	—	—
Balconies, catwalks, and similar areas accessible to people	2.3	7.5	4.1	13.5
Over or under roofs or projections not readily accessible to people	—	—	3.8	12.5
Over roofs accessible to vehicles but not trucks	—	—	4.1	13.5
Over roofs accessible to trucks	—	—	5.6	18.5
Other structures	2.3	7.5	—	—

225.61 Clearances over Buildings and Other Structures.

(A) 22 kV Nominal to Ground or Less. The clearances over buildings and other structures for conductors and live parts up to 22 kV, nominal, to ground or less shall be not less than the values shown in Table 225.61 [shown here as Exhibit 10.9].

(B) Over 22 kV Nominal to Ground. Clearances for the categories shown in Table 225.61 shall be increased by 10 mm (0.4 in.) per kV above 22,000 volts.

FPN: For additional information, see ANSI C2-2002, *National Electrical Safety Code.*

SERVICES

Service drop conductors shall not be readily accessible and shall comply with 230.24(A) through (D) for services not over 600 volts, nominal.

230.24 Clearances. Service-drop conductors shall not be readily accessible and shall comply with 230.24(A) through (D) for services not over 600 volts, nominal.

(A) Above Roofs. Conductors shall have a vertical clearance of not less than 2.5 m (8 ft) above the roof surface. The vertical clearance above the roof level shall be maintained for a distance of not less than 900 mm (3 ft) in all directions from the edge of the roof.

Exception No. 1: The area above a roof surface subject to pedestrian or vehicular traffic shall have a vertical clearance from the roof surface in accordance with the clearance requirements of 230.24(B).

Exception No. 2: Where the voltage between conductors does not exceed 300 and the roof has a slope of 100 mm in 300 mm (4 in. in 12 in.) or greater, a reduction in clearance to 900 mm (3 ft) shall be permitted.

Exception No. 3: Where the voltage between conductors does not exceed 300, a reduction in clearance above only the overhanging portion of the roof to not less than 450 mm (18 in.) shall be permitted if (1) not more than 1.8 m (6 ft) of service-drop conductors, 1.2 m (4 ft) horizontally, pass above the roof overhang, and (2) they are terminated at a through-the-roof raceway or approved support.

Exception: The requirement for maintaining the vertical clearance 900 mm (3 ft) from the edge of the roof shall not apply to the final conductor span where the service drop is attached to the side of a building.

MINIMUM COVER REQUIREMENTS

Underground Installations

See Exhibit 10.10.

CLEARANCE AND COVER REQUIREMENTS

EXHIBIT 10.10

NEC Table 300.5 Minimum Cover Requirements, 0 to 600 Volts, Nominal, Burial in Millimeters (Inches)

Type of Wiring Method or Circuit										
Location of Wiring Method or Circuit	Column 1 Direct Burial Cables or Conductors		Column 2 Rigid Metal Conduit or Intermediate Metal Conduit		Column 3 Nonmetallic Raceways Listed for Direct Burial Without Concrete Encasement or Other Approved Raceways		Column 4 Residential Branch Circuits Rated 120 Volts or Less with GFCI Protection and Maximum Overcurrent Protection of 20 Amperes		Column 5 Circuits for Control of Irrigation and Landscape Lighting Limited to Not More Than 30 Volts and Installed with Type UF or in Other Identified Cable or Raceway	
	mm	in.	mm	in.	mm	in.	mm	in.	mm	in.
All locations not specified below	600	24	150	6	450	18	300	12	150	6
In trench below 50-mm (2-in.) thick concrete or equivalent	450	18	150	6	300	12	150	6	150	6
Under a building	0 (in raceway)	0	0	0	0	0	0 (in raceway only)	0	0 (in raceway only)	0
Under minimum of 102-mm (4-in.) thick concrete exterior slab with no vehicular traffic and the slab extending not less than 152 mm (6 in.) beyond the underground installation	450	18	100	4	100	4	150 (direct burial) 100 (in raceway)	6 4	150	6
Under streets, highways, roads, alleys, driveways, and parking lots	600	24	600	24	600	24	600	24	600	24
One- and two-family dwelling driveways and outdoor parking areas, and used only for dwelling-related purposes	450	18	450	18	450	18	300	12	450	18
In or under airport runways, including adjacent areas where trespassing prohibited	450	18	450	18	450	18	450	18	450	18

Notes:
1. Cover is defined as the shortest distance in millimeters (inches) measured between a point on the top surface of any direct-buried conductor, cable, conduit, or other raceway and the top surface of finished grade, concrete, or similar cover.
2. Raceways approved for burial only where concrete encased shall require concrete envelope not less than 50 mm (2 in.) thick.
3. Lesser depths shall be permitted where cables and conductors rise for terminations or splices or where access is otherwise required.
4. Where one of the wiring method types listed in Columns 1–3 is used for one of the circuit types in Columns 4 and 5, the shallowest depth of burial shall be permitted.
5. Where solid rock prevents compliance with the cover depths specified in this table, the wiring shall be installed in metal or nonmetallic raceway permitted for direct burial. The raceways shall be covered by a minimum of 50 mm (2 in.) of concrete extending down to rock.

Underground Installations (Over 600 Volts, Nominal)

See Exhibit 10.11.

EXHIBIT 10.11

NEC Table 300.50 Minimum Cover[1] Requirements

	General Conditions (not otherwise specified)						Special Conditions (use if applicable)								
	(1) Direct-Buried Cables		(2) Rigid Nonmetallic Conduit[2]		(3) Rigid Metal Conduit and Intermediate Metal Conduit		(4) Raceways under buildings or exterior concrete slabs, 100 mm (4 in.) minimum thickness[3]		(5) Cables in airport runways or adjacent areas where trespass is prohibited		(6) Areas subject to vehicular traffic, such as thoroughfares and commercial parking areas	
Circuit Voltage	mm	in.	mm	in.	mm	in.	mm	in.	mm	in.	mm	in.
Over 600 V through 22 kV	750	30	450	18	150	6	100	4	450	18	600	24
Over 22 kV through 40 kV	900	36	600	24	150	6	100	4	450	18	600	24
Over 40 kV	1000	42	750	30	150	6	100	4	450	18	600	24

Notes:
1. Lesser depths shall be permitted where cables and conductors rise for terminations or splices or where access is otherwise required.
2. Where solid rock prevents compliance with the cover depths specified in this table, the wiring shall be installed in a metal or nonmetallic raceway permitted for direct burial. The raceways shall be covered by a minimum of 50 mm (2 in.) of concrete extending down to rock.

[1] Cover is defined as the shortest distance in millimeters (inches) measured between a point on the top surface of any direct-buried conductor, cable, conduit, or other raceway and the top surface of finished grade, concrete, or similar cover.

[2] Listed by a qualified testing agency as suitable for direct burial without encasement. All other nonmetallic systems shall require 50 mm (2 in.) of concrete or equivalent above conduit in addition to the table depth.

[3] The slab shall extend a minimum of 150 mm (6 in.) beyond the underground installation, and a warning ribbon or other effective means suitable for the conditions shall be placed above the underground installation.

CLEARANCE AND COVER REQUIREMENTS

CHAPTER 11
BOXES AND ENCLOSURES

Chapter 11 covers boxes and enclosures. Most of the material in this chapter is derived from Article 314 of the *NEC*®, which covers the installation and use of all boxes and conduit bodies used as outlet, device, junction, or pull boxes, depending on their use and handhole enclosures. Article 314 also includes installation requirements for fittings used to join raceways and to connect raceways and cables to boxes and conduit bodies.

BOX FILL CALCULATIONS

Boxes and conduit bodies containing conductors 18 AWG through 6 AWG must be sized in accordance with the provisions in 314.16. Boxes and conduit bodies must be of sufficient size to provide free space for all enclosed conductors. Standard size metal boxes are listed in Exhibit 11.1 [*NEC* Table 314.16(A)]. This exhibit shows the box trade size, minimum volume, and minimum number conductors 18 AWG through 6 AWG. Boxes and conduit bodies enclosing conductors 4 AWG or larger must also comply with the provisions of 314.28.

The volume of a wiring enclosure (box) must be the total volume of the assembled sections, and, where used, the space provided by plaster rings, domed covers, extension rings, and so forth, that are marked with their volume or are made from boxes the dimensions of which are listed in Exhibit 11.1. Boxes 100 cubic inches (1650 cubic centimeters) or less, except for those listed in Exhibit 11.1, and nonmetallic boxes must be durably and legibly marked by the manufacturer with their volume. Metal boxes listed in Exhibit 11.1 may or may not be marked with the cubic inch (cubic centimeter) volume.

The maximum numbers of conductors in Exhibit 11.1 are based on the same size conductors in a box that contains no internal cable clamps, support fittings, devices, or equipment. If the metal box contains fittings or devices, the number of conductors is reduced by the amount specified in Exhibit 11.2 [*NEC* Table 314.16(B)]. In no case can the volume of the box, as calculated in Exhibit 11.1, be less than the box fill as calculated in Exhibit 11.2.

130

BOXES AND ENCLOSURES

EXHIBIT 11.1

NEC Table 314.16(A)   Metal Boxes

Box Trade Size			Minimum Volume		Maximum Number of Conductors*						
mm	in.		cm³	in.³	18	16	14	12	10	8	6
100 × 32	(4 × 1¼)	round/octagonal	205	12.5	8	7	6	5	5	5	2
100 × 38	(4 × 1½)	round/octagonal	254	15.5	10	8	7	6	6	5	3
100 × 54	(4 × 2⅛)	round/octagonal	353	21.5	14	12	10	9	8	7	4
100 × 32	(4 × 1¼)	square	295	18.0	12	10	9	8	7	6	3
100 × 38	(4 × 1½)	square	344	21.0	14	12	10	9	8	7	4
100 × 54	(4 × 2⅛)	square	497	30.3	20	17	15	13	12	10	6
120 × 32	(4¹¹⁄₁₆ × 1¼)	square	418	25.5	17	14	12	11	10	8	5
120 × 38	(4¹¹⁄₁₆ × 1½)	square	484	29.5	19	16	14	13	11	9	5
120 × 54	(4¹¹⁄₁₆ × 2⅛)	square	689	42.0	28	24	21	18	16	14	8
75 × 50 × 38	(3 × 2 × 1½)	device	123	7.5	5	4	3	3	3	2	1
75 × 50 × 50	(3 × 2 × 2)	device	164	10.0	6	5	5	4	4	3	2
75 × 50 × 57	(3 × 2 × 2¼)	device	172	10.5	7	6	5	4	4	3	2
75 × 50 × 65	(3 × 2 × 2½)	device	205	12.5	8	7	6	5	5	4	2
75 × 50 × 70	(3 × 2 × 2¾)	device	230	14.0	9	8	7	6	5	4	2
75 × 50 × 90	(3 × 2 × 3½)	device	295	18.0	12	10	9	8	7	6	3
100 × 54 × 38	(4 × 2⅛ × 1½)	device	169	10.3	6	5	5	4	4	3	2
100 × 54 × 48	(4 × 2⅛ × 1⅞)	device	213	13.0	8	7	6	5	5	4	2
100 × 54 × 54	(4 × 2⅛ × 2⅛)	device	238	14.5	9	8	7	6	5	4	2
95 × 50 × 65	(3¾ × 2 × 2½)	masonry box/gang	230	14.0	9	8	7	6	5	4	2
95 × 50 × 90	(3¾ × 2 × 3½)	masonry box/gang	344	21.0	14	12	10	9	8	7	4
min. 44.5 depth	FS — single cover/gang (1¾)		221	13.5	9	7	6	6	5	4	2
min. 60.3 depth	FD — single cover/gang (2⅜)		295	18.0	12	10	9	8	7	6	3
min. 44.5 depth	FS — multiple cover/gang (1¾)		295	18.0	12	10	9	8	7	6	3
min. 60.3 depth	FD — multiple cover/gang (2⅜)		395	24.0	16	13	12	10	9	8	4

*Where no volume allowances are required by 314.16(B)(2) through (B)(5).

EXHIBIT 11.2

NEC Table 314.16(B)   Volume Allowance Required per Conductor

Size of Conductor (AWG)	Free Space Within Box for Each Conductor	
	cm³	in.³
18	24.6	1.50
16	28.7	1.75
14	32.8	2.00
12	36.9	2.25
10	41.0	2.50
8	49.2	3.00
6	81.9	5.00

Table 11.1 Box Fill Calculations

Items Within the Box	Determination of Conductor Volume Allowance	Number of Conductors Counted
Each conductor originating outside the box, and terminating or spliced within the box	Volume allowance for that specific conductor	1
Each conductor passing through the box without splice or termination (unbroken conductor)	Volume allowance for that specific conductor	1
A conductor, no part of which leaves the box (such as equipment bonding jumpers and pigtails)	Not counted	0
One or more internal cable clamps	Volume allowance for largest conductor in box	1
A cable connector with its clamping mechanism outside the box	Not counted	0
One or more luminaire (fixture) studs	Volume allowance for largest conductor in box	1
One or more hickeys	Volume allowance for largest conductor in box	1
Each yoke or strap containing one or more devices or equipment	Volume allowance for largest conductor in box	2
One or more equipment grounding conductors	Volume allowance for largest equipment grounding conductor in box	1
One or more isolated equipment grounding conductors	Volume allowance for largest isolated equipment grounding conductor in box	1
Small fittings such as locknuts and bushings	Not counted	0

Note: An equipment grounding conductor or conductors or not over four luminaire (fixture) wires smaller than 14 AWG, or both, shall be permitted to be omitted from the calculations where they enter a box from a domed luminaire (fixture) or similar canopy and terminate within that box. [314.16(B)(1) *Exception*]

Table 11.1 was constructed from the requirements in 314.16(B)(1) through (5).

Example 1

What is the minimum depth required for a 4-inch square metal box that will contain twelve 12 AWG conductors? The box will contain four grounded conductors, four ungrounded conductors, and four equipment grounding conductors. Four wire connectors will be used to terminate the conductors. The box will not contain internal cable clamps or devices.

BOXES AND ENCLOSURES

Solution

Step 1: Determine the number of conductors contributing to box fill.

Four 12 AWG grounded conductors = 4 conductors

Four 12 AWG ungrounded conductors = 4 conductors

Four 12 AWG equipment grounding conductors = 1 conductor

Step 2: Since the box will not contain internal cable clamps or devices, no additional volume allowance is required. No volume allowance is required for wire connectors. The total number of 12 AWG conductors contributing to this example's box fill is nine (4 + 4 + 1 = 9).

Step 3: Search Exhibit 11.1 to find a 4-inch square metal box that has enough depth to contain nine 12 AWG conductors. A 4-inch square box with a depth of 1-1/4 inches is too shallow for this example because the maximum number of 12 AWG conductors is only eight. The minimum depth 4-inch square metal box required for this example is 1-1/2 inches.

Example 2

What is the minimum depth required for a 3 × 2 metal device box that will contain two 14/2 (with ground) nonmetallic-sheathed cables, two internal cable clamps, and one receptacle?

Solution

Step 1: Determine the number of conductors contributing to box fill.

Two 14/2 cables (not counting grounds) = 4 conductors

Two 14 AWG grounds in the cables = 1 conductor

Step 2: Include the conductor volume allowance for other items contributing to box fill. Since this box will contain the same size conductors, calculate the volume allowance using 14 AWG conductors.

Two internal cable clamps = 1 conductor (14 AWG)

One receptacle = 2 conductors (14 AWG)

The total number of 14 AWG conductors contributing to this example's box fill is eight (4 + 1 + 1 + 2 = 8).

Step 3: Search Exhibit 11.1 to find a 3 × 2 metal device box that has enough depth to contain eight 14 AWG conductors. The minimum depth 3 × 2 metal device box for this example is 3-1/2 inches.

Example 3

Boxes not listed in Exhibit 11.1 or boxes containing different size conductors must be computed in accordance with Exhibit 11.2. What is the minimum cubic inch box required for a box that will contain the following items: one 14/2 (with ground) nonmetallic-sheathed cable, two 12/2 (with ground) nonmetallic-sheathed cables, one switch, and one receptacle? The 14 AWG conductors will be connected to the switch, and the 12 AWG conductors will be connected to the receptacle. All the equipment grounding conductors, along with two jumpers, will be tied together. A 14 AWG equipment grounding jumper will be connected to the switch. A 12 AWG equipment grounding jumper will be connected to the receptacle.

Solution

Step 1: Determine the number and size conductors contributing to box fill.

One 14/2 cable (not counting grounds) = 2 conductors (14 AWG)

Two 12/2 cables (not counting grounds) = 4 conductors (12 AWG)

Step 2: Although there are three equipment grounding conductors in the cables (one 14 and two 12s), only the largest is counted. No volume allowance is required for the equipment grounding jumpers because they do not leave the box.

Equipment grounds in the cables = 1 conductor (12 AWG)

Step 3: Include the conductor volume allowance for other items contributing to box fill.

Double the volume allowance for the conductors connected to the switch.

One switch = 2 conductors (14 AWG)

Double the volume allowance for the conductors connected to the receptacle.

One receptacle = 2 conductors (12 AWG)

Step 4: Calculate the cubic inch volume required for the conductors from Exhibit 11.2.

The amount of free space required within the box for each 14 AWG conductor is 2 cubic inches.

Four 14 AWG conductors = 8 in.^{3} (4 × 2)

The amount of free space required within the box for each 12 AWG conductor is 2.25 cubic inches.

Seven 12 AWG conductors = 15.75 in.^{3} (7 × 2.25)

The minimum volume allowance required for conductors and devices in this example is 23.75 cubic inches (8 + 15.75). Note, if the box contains internal cable clamps, another 2.25 cubic inches must be added to the calculation.

BOXES AND ENCLOSURES

CONDUIT BODIES

A conduit body is a separate portion of a conduit or tubing system that provides access through a removable cover(s) to the interior of the system at a junction of two or more sections of the system or at a terminal point of the system (see Table 11.2). Boxes such as FS and FD or larger cast or sheet metal boxes are not classified as conduit bodies.

(C) Conduit Bodies.

(1) General. Conduit bodies enclosing 6 AWG conductors or smaller, other than short-radius conduit bodies as described in 314.5, shall have a cross-sectional area not less than twice the cross-sectional area of the largest conduit or tubing to which it is attached. The maximum number of conductors permitted shall be the maximum number permitted by Table 1 of Chapter 9 for the conduit or tubing to which it is attached.

Table 11.2 Conduit Bodies

Conduit Bodies	Type	Conduit Bodies	Type
	C		L
	E		T
	LB		TA
	LL		TB
	LR		X

Source: Killark Division of Hubbell Incorporated (Delaware).

(2) **With Splices, Taps, or Devices.** Only those conduit bodies that are durably and legibly marked by the manufacturer with their volume shall be permitted to contain splices, taps, or devices. The maximum number of conductors shall be calculated in accordance with 314.16(B). Conduit bodies shall be supported in a rigid and secure manner.

CAST DEVICE BOXES

Cast device boxes are manufactured with various conduit entry configurations. The letters such as FS and FD indicate the number and location of conduit entries. Cast device boxes can be used as outlet, device, junction, and pull boxes. Table 11.3 provides images for the most common types of cast device boxes.

Table 11.3   Cast Device Boxes

Cast Device Box	Type	Depth	Cast Device Box	Type	Depth
	FS	Shallow		FSL	Shallow
	FD	Deep			
	FSC	Shallow		FSR	Shallow
	FDC	Deep			
	FSS	Shallow		FST	Shallow
	FDS	Deep		FDT	Deep
	FSCC	Shallow		FSX	Shallow
	FDCC	Deep		FDX	Deep

Continued

BOXES AND ENCLOSURES

BOXES AND ENCLOSURES

Table 11.3   Cast Device Boxes *(continued)*

Cast Device Box	Type	Depth	Cast Device Box	Type	Depth
	FSA	Shallow		FSCT	Shallow
	FDA	Deep		FDCT	Deep
	FSLB	Shallow		FSOC	Deep
	FSCA	Shallow			

Source: Killark Division of Hubbell Incorporated (Delaware).

PULL AND JUNCTION BOX CALCULATIONS

Boxes and conduit bodies used as pull or junction boxes must comply with 314.28(A) through (D). Boxes and conduit bodies containing conductors 18 AWG through 6 AWG are calculated from the sizes and numbers of conductors. Boxes and conduit bodies containing conductors larger than 6 AWG (under 600 volts) are calculated from the sizes and numbers of raceways or conduits.

314.28 Pull and Junction Boxes and Conduit Bodies. Boxes and conduit bodies used as pull or junction boxes shall comply with 314.28(A) through (D).

(A) Minimum Size. For raceways containing conductors of 4 AWG or larger, and for cables containing conductors of 4 AWG or larger, the minimum dimensions of pull or junction boxes installed in a raceway or cable run shall comply with (A)(1) through (A)(3). Where an enclosure dimension is to be calculated based on the diameter of entering raceways, the diameter shall be the metric designator (trade size) expressed in the units of measurement employed.

Straight Pulls

Pull and junction boxes containing straight pulls must be computed in accordance with 314.28(A)(1). The calculation is based on the largest raceway entering the box. The length must not be less than eight times the trade size (metric designator) of the largest raceway. Unless a raceway enters the back of the box, the depth is not specified. The depth of the box must be large enough to provide proper installation of the raceway (or cable), including locknuts and bushings.

(1) Straight Pulls. In straight pulls, the length of the box shall not be less than eight times the metric designator (trade size) of the largest raceway.

Example

A junction box is needed for 10 raceway entries. The right side of the box will have a 2-inch and a 3-inch conduit. The left side will have two 3-inch conduits. The top of the box will have two 4-inch conduits. The bottom of the box will have four 2-inch conduits. If all pulls are straight pulls, what is the minimum size junction box required?

Solution

Since the largest conduit entering the left/right side is 3 inches, multiply eight by three ($8 \times 3 = 24$). The minimum left/right dimension is 24 inches. Since the largest conduit entering the top or bottom is 4 inches, multiply eight by four ($8 \times 4 = 32$). The minimum top/bottom dimension is 32 inches.

Angle or U Pulls

Angle or U pulls must be calculated in accordance with 314.28(A)(2). To find the minimum dimension for a box that contains angle pulls, start with the wall where the raceways enter the box, and find the distance to the opposite wall. First, multiply the largest raceway by six. Add to that number the sum of the diameters of all other raceway entries in the same row on the same wall of the box. For boxes containing more than one row, calculate each row individually. The single row providing the maximum distance is the minimum dimension required for that side of the box. The distance between raceway entries enclosing the same conductor cannot be less than six times the trade size (metric designator) of the largest raceway.

(2) Angle or U Pulls. Where splices or where angle or U pulls are made, the distance between each raceway entry inside the box and the opposite wall of the box shall not be less than six times the metric designator (trade size) of the largest raceway in a row. This distance shall be increased for additional entries by the amount of the sum of the diameters of all other raceway entries in the same row on the same wall of the box. Each row shall be calculated individually, and the single row that provides the maximum distance shall be used.

Exception: Where a raceway or cable entry is in the wall of a box or conduit body opposite a removable cover, the distance from that wall to the cover shall be permitted to comply with the distance required for one wire per terminal in Table 312.6(A).

The distance between raceway entries enclosing the same conductor shall not be less than six times the metric designator (trade size) of the larger raceway.

When transposing cable size into raceway size in 314.28(A)(1) and (A)(2), the minimum metric designator (trade size) raceway required for the number and size of conductors in the cable shall be used.

Example

A junction box is needed for six raceway entries. The right side of the box will have a row with a 2-inch, a 3-inch, and a 4-inch conduit. The top of the box will have a

BOXES AND ENCLOSURES

row with a 2-inch and two 3-inch conduits. The bottom and left side will not have any conduit entries. What is the minimum size junction box required?

Solution

The largest conduit entering the right side is 4 inches; therefore multiply six by four (6 × 4 = 24). Add to that number the other conduits on the same side of the box (24 + 3 + 2 = 29). The minimum left/right dimension is 29 inches. The largest conduit entering the top is 3 inches. Therefore, multiply six by three (6 × 3 = 18). Add to that number the other conduits on the same side (18 + 3 + 2 = 23). The minimum top/bottom dimension is 23 inches.

NEMA ENCLOSURES

Enclosures are designed to meet different classes of exposure. Table 11.4 is a quick reference table providing the National Electrical Manufacturers Association (NEMA) types of enclosures. This table is useful for selecting the appropriate enclosure for the conditions.

Table 11.4   NEMA Enclosure Types

Type	Use	Enclosure Types Constructed to Provide a Degree of Protection Against:	May Also Be Marked
1	Indoor		Indoor Use Only
2	Indoor	dripping and light splashing of liquids	Driptight
3	Indoor or Outdoor	rain, sleet, snow, and windblown dust; and that will be undamaged by the external formation of ice on the enclosure	Raintight; Dusttight
3R	Indoor or Outdoor	rain, sleet, and snow; and that will be undamaged by the external formation of ice on the enclosure	Rainproof
3S	Indoor or Outdoor	rain, sleet, snow, and windblown dust; and in which the external mechanism(s) remain operable when ice laden	Raintight; Dusttight
4	Indoor or Outdoor	rain, sleet, snow, windblown dust, splashing water, and hose-directed water; and that will be undamaged by the external formation of ice on the enclosure	Raintight; Watertight

Table continues below.

Table 11.4 NEMA Enclosure Types *(continued)*

Type	Use	Enclosure Types Constructed to Provide a Degree of Protection Against:	May Also Be Marked
4X	Indoor or Outdoor	rain, sleet, snow, windblown dust, splashing water, hose-directed water, and corrosion; and that will be undamaged by the external formation of ice on the enclosure	Raintight; Watertight; Corrosion Resistant
5	Indoor	settling airborne dust, lint, fibers, and flyings; and to provide a degree of protection against dripping and light splashing of liquids	Driptight; Dusttight
6	Indoor or Outdoor	hose-directed water and the entry of water during occasional temporary submersion at a limited depth; and that will be undamaged by the external formation of ice on the enclosure	Raintight; Watertight
6P	Indoor or Outdoor	hose-directed water and the entry of water during prolonged submersion at a limited depth; and that will be undamaged by the external formation of ice on the enclosure	Raintight; Watertight; Corrosion Resistant
12	Indoor	circulating dust, lint, fibers, and flyings; and against dripping and light splashing of liquids (without knockouts)	Driptight
12K	Indoor	circulating dust, lint, fibers, and flyings; and against dripping and light splashing of liquids (with knockouts)	Driptight; Dusttight
13	Indoor	circulating dust, lint, fibers, and flyings; and against the spraying, splashing, and seepage of water, oil, and noncorrosive coolants	Driptight; Dusttight

Note: All enclosures are constructed to provide a degree of protection to personnel against incidental contact with the enclosed equipment when doors of covers are closed and in place. All type enclosures provide a degree of protection against ordinary corrosion and against a limited amount of falling dirt.

Source: NEMA, National Electrical Manufacturers Association.

CHAPTER 12
RACEWAYS AND CABLE TRAYS

Chapter 12 covers raceway and cable tray components, such as metric designators and trade sizes, bends, supports, numbers of conductors, and fill tables.

METRIC DESIGNATORS AND TRADE SIZES

Metric designators and trade sizes for conduit, tubing, and associated fittings and accessories are shown in Exhibit 12.1 [*NEC* Table 300.1(C)].

EXHIBIT 12.1

NEC Table 300.1(C)   Metric Designator and Trade Sizes

Metric Designator	Trade Size
12	⅜
16	½
21	¾
27	1
35	1¼
41	1½
53	2
63	2½
78	3
91	3½
103	4
129	5
155	6

Note: The metric designators and trade sizes are for identification purposes only and are not actual dimensions.

BENDS

Bends must be made so the conduit or tubing is not damaged and the internal diameter is not effectively reduced. The bend is measured from the radius of the curve to the centerline.

In *NEC* Table 2, Chapter 9 (Exhibit 12.2), the column titled "One Shot and Full Shoe Benders" applies to intermediate metal conduit (IMC), rigid metal conduit (RMC), and electrical metallic tubing (EMT).

RACEWAYS AND CABLE TRAYS

The column titled "Other Bends" applies to flexible metal conduit (FMC), liquidtight flexible metal conduit (LFMC), rigid nonmetallic conduit (RNC), liquidtight flexible nonmetallic conduit (LFNC), and electrical nonmetallic tubing (ENT).

EXHIBIT 12.2

NEC Table 2   Radius of Conduit and Tubing Bends

Conduit or Tubing Size		One Shot and Full Shoe Benders		Other Bends	
Metric Designator	Trade Size	mm	in.	mm	in.
16	½	101.6	4	101.6	4
21	¾	114.3	4½	127	5
27	1	146.05	5¾	152.4	6
35	1¼	184.15	7¼	203.2	8
41	1½	209.55	8¼	254	10
53	2	241.3	9½	304.8	12
63	2½	266.7	10½	381	15
78	3	330.2	13	457.2	18
91	3½	381	15	533.4	21
103	4	406.4	16	609.6	24
129	5	609.6	24	762	30
155	6	762	30	914.4	36

SUPPORTS FOR INTERMEDIATE METAL CONDUIT AND RIGID METAL CONDUIT

The distance between supports for straight runs of conduit is permitted in accordance with *NEC* Table 344.30(B)(2) (Exhibit 12.3), provided the conduit is made up with threaded couplings and such supports prevent transmission of stress to termination where conduit is deflected between supports.

EXHIBIT 12.3

NEC Table 344.30(B)(2)   Supports for Rigid Metal Conduit

Conduit Size		Maximum Distance Between Rigid Metal Conduit Supports	
Metric Designator	Trade Size	m	ft
16–21	½–¾	3.0	10
27	1	3.7	12
35–41	1¼–1½	4.3	14
53–63	2–2½	4.9	16
78 and larger	3 and larger	6.1	20

CONDUCTORS IN FLEXIBLE METAL CONDUIT

The number of conductors in flexible metal conduit (FMC) must not exceed that permitted by the percentage fill specified in Table 1, Chapter 9, or as permitted in *NEC* Table 348.22 (Exhibit 12.4).

EXHIBIT 12.4

NEC Table 348.22   Maximum Number of Insulated Conductors in Metric Designator 12 (Trade Size 3/8) Flexible Metal Conduit*

	Types RFH-2, SF-2		Types TF, XHHW, TW		Types TFN, THHN, THWN		Types FEP, FEBP, PF, PGF	
Size (AWG)	Fittings Inside Conduit	Fittings Outside Conduit	Fittings Inside Conduit	Fittings Outside Conduit	Fittings Inside Conduit	Fittings Outside Conduit	Fittings Inside Conduit	Fittings Outside Conduit
18	2	3	3	5	5	8	5	8
16	1	2	3	4	4	6	4	6
14	1	2	2	3	3	4	3	4
12	—	—	2	2	2	3	2	3
10	—	—	1	1	1	1	1	2

*In addition, one covered or bare equipment grounding conductor of the same size shall be permitted.

RACEWAYS AND CABLE TRAYS

SUPPORT AND EXPANSION CHARACTERISTICS OF RIGID NONMETALLIC CONDUIT

Rigid nonmetallic conduit must be supported as required in *NEC* Table 352.30(B) (Exhibit 12.5).

EXHIBIT 12.5

NEC Table 352.30(B) Support of Rigid Nonmetallic Conduit (RNC)

Conduit Size		Maximum Spacing Between Supports	
Metric Designator	Trade Size	mm or m	ft
16–27	½–1	900 mm	3
35–53	1¼–2	1.5 m	5
63–78	2½–3	1.8 m	6
91–129	3½–5	2.1 m	7
155	6	2.5 m	8

Expansion fittings for rigid nonmetallic conduit must be provided to compensate for thermal expansion and contraction where the length change, in accordance with *NEC* Table 352.44(A) or (B) (Exhibits 12.6 and 12.7), is expected to be 6 mm (1/4 in.) or greater in a straight run between securely mounted items such as boxes, cabinets, elbows, or other conduit terminations.

EXHIBIT 12.6

***NEC* Table 352.44(A)   Expansion Characteristics of PVC Rigid Nonmetallic Conduit Coefficient of Thermal Expansion = 6.084×10^{-5} mm/mm/°C (3.38×10^{-5} in./in./°F)**

Temperature Change (°C)	Length Change of PVC Conduit (mm/m)	Temperature Change (°F)	Length Change of PVC Conduit (in./100 ft)	Temperature Change (°F)	Length Change of PVC Conduit (in./100 ft)
5	0.30	5	0.20	105	4.26
10	0.61	10	0.41	110	4.46
15	0.91	15	0.61	115	4.66
20	1.22	20	0.81	120	4.87
25	1.52	25	1.01	125	5.07
30	1.83	30	1.22	130	5.27
35	2.13	35	1.42	135	5.48
40	2.43	40	1.62	140	5.68
45	2.74	45	1.83	145	5.88
50	3.04	50	2.03	150	6.08
55	3.35	55	2.23	155	6.29
60	3.65	60	2.43	160	6.49
65	3.95	65	2.64	165	6.69
70	4.26	70	2.84	170	6.90
75	4.56	75	3.04	175	7.10
80	4.87	80	3.24	180	7.30
85	5.17	85	3.45	185	7.50
90	5.48	90	3.65	190	7.71
95	5.78	95	3.85	195	7.91
100	6.08	100	4.06	200	8.11

EXHIBIT 12.7

***NEC* Table 352.44(B)   Expansion Characteristics of Reinforced Thermosetting Resin Conduit (RTRC) Coefficient of Thermal Expansion = 2.7×10^{-5} mm/mm/°C (1.5×10^{-5} in./in./°F)**

Temperature Change (°C)	Length Change of RTRC Conduit (mm/m)	Temperature Change (°F)	Length Change of RTRC Conduit (in./100 ft)	Temperature Change (°F)	Length Change of RTRC Conduit (in./100 ft)
5	0.14	5	0.09	105	1.89
10	0.27	10	0.18	110	1.98
15	0.41	15	0.27	115	2.07
20	0.54	20	0.36	120	2.16
25	0.68	25	0.45	125	2.25
30	0.81	30	0.54	130	2.34
35	0.95	35	0.63	135	2.43
40	1.08	40	0.72	140	2.52
45	1.22	45	0.81	145	2.61
50	1.35	50	0.90	150	2.70
55	1.49	55	0.99	155	2.79
60	1.62	60	1.08	160	2.88
65	1.76	65	1.17	165	2.97
70	1.89	70	1.26	170	3.06
75	2.03	75	1.35	175	3.15
80	2.16	80	1.44	180	3.24
85	2.30	85	1.53	185	3.33
90	2.43	90	1.62	190	3.42
95	2.57	95	1.71	195	3.51
100	2.70	100	1.80	200	3.60

RACEWAYS AND CABLE TRAYS

SURFACE NONMETALLIC RACEWAYS

Figure 12.1 provides conductor fill information for specific surface metal raceways.

Figure 12.1 Conductor fill table for various surface metal raceways.

Type of Raceway	Wire Size Gauge No.	Number of Wires — Types RHH, RHW	Type THW	Type TW	Types THHN, THWN
No. 200 (½ in. × 11/32 in.)	12	—	2	3	3
	14	—	2	3	5
No. 500 (¾ in. × 17/32 in.)	8	—	—	2	2
	10	2	2	3	4
	12	2	3	4	7
	14	2	4	6	9
No. 700 (¾ in. × 21/32 in.)	6	—	—	—	2
	8	—	2	2	3
	10	2	3	4	5
	12	2	4	6	8
	14	3	5	7	11
No. 1500 (1 9/16 in. × 11/32 in.)	6	—	—	—	2
	8	—	—	2	3
	10	2	3	4	5
	12	2	3	5	7
	14	2	4	6	10
No. 2000[a] (1 3/32 in. × ¾ in.)	12	—	—	7	7
	14	—	—	7	7
No. 2100[a] (7/8 in. × 1¼ in.)	6	2	4 6	4 8	6
	8	4	10	14	10
	10	7	13	19	17
	12	8	15	24	28
	14	10			37
No. 2200[a] (2 3/8 in. × ¾ in.)	6	5 8	7	3[b] 7	11
	8	13	11	7[b] 14	19
	10	15	19	10[b] 26	32
	12	18	23	10[b] 34	51
	14		29	10[b] 44	69
No. 2600 (2 7/32 in. × 23/32 in.)	6	2	3	3 7	5 9
	8	4	5	12	15
	10	6	9	16	24
	12	7	11	21	33
	14	9	14		
G-3000 (2 ¾ in. × 1 17/32 in.)	6	4[b] 11	6[b] 19	6[b] 17	6[b] 27
	8	6[b] 18	8[b] 26	8[b] 34	8[b] 44
	10	10[b] 30	10[b] 45	10[b] 62	10[b] 76
	12	14[b] 36	18[b] 55	18[b] 81	18[b] 119
	14	16[b] 42	26[b] 67	26[b] 103	26[b] 160
G-4000, with divider (4 ¾ in. × 1 ¾ in.)	2	— 7 8	— 10	— 10	— 12
	3	— 9	— 11	— 11	— 15
	4	— 12	— 13	— 13	— 17
	6	4[b] 19	4[b] 18	4[b] 18	7[b] 28
	8	7[b] 32	7[b] 28	7[b] 36	8[b] 47
	10	11[b] 39	11[b] 48	11[b] 66	15[b] 81
	12	15[b] 45	15[b] 59	15[b] 86	24[b] 128
	14	17[b]	17[b] 72	17[b] 110	32[b] 171

Figure continues below.

Figure 12.1 Conductor fill table for various surface metal raceways.

Type of Raceway	Wire Size Gauge No.	Number of Wires			
		Types RHH, RHW	Type THW	Type TW	Types THHN, THWN
G-4000, without divider (1¾ in. × 4¾ in.)	2	— 14	— 20	— 20	— 25
	3	— 16	— 23	— 23	— 30
	4	— 18	— 27	— 27	— 35
	6	8[b] 24	8[b] 36	8[b] 36	10[b] 57
	8	10[b] 39	10[b] 57	10[b] 78	15[b] 94
	10	15[b] 65	15[b] 96	12[b] 133	18[b] 163
	12	21[b] 78	21[b] 119	16[b] 174	34[b] 256
	14	21[b] 91	21[b] 145	17[b] 222	34[b] 344
G-6000 (4¾ in. × 3⁹⁄₁₆ in.)	2/0	10[b] 17	12[b] 22	12[b] 22	15[b] 27
	1/0	11[b] 20	14[b] 26	14[b] 26	18[b] 33
	1 2	12[b] 23	17[b] 31	17[b] 31	21[b] 39
	3 4	16[b] 30	23[b] 43	23[b] 43	29[b] 53
	6 8	19[b] 34	27[b] 50	27[b] 50	34[b] 63
	10	21[b] 39	32[b] 58	32[b] 58	40[b] 74
	12	27[b] 51	42[b] 77	42[b] 77	66[b] 122
	14	40[b] 74	57[b] 106	73[b] 134	92[b] 169
		75[b] 137	111[b] 203	154[b] 282	187[b] 343
		90[b] 164	137[b] 252	200[b] 386	295[b] 540
		105[b] 193	167[b] 307	255[b] 469	396[b] 726

[a]Figures for Nos. 2000, 2100, 2200, G-3000, G-4000, and G-6000 are without receptacles, except where noted.
[b]With receptacles.
Source: Redrawn from The Wiremold Co.

RACEWAYS AND CABLE TRAYS

RACEWAYS AND CABLE TRAYS

CABLE TRAYS

Exhibit 12.8 shows the minimum cross-sectional area of metal required for cable trays that are used as equipment grounding conductors. Exhibits 12.9 to 12.12 contain allowable fill for multiconductor and single-conductor cables installed in cable trays.

EXHIBIT 12.8

NEC Table 392.7(B)   Metal Area Requirements for
Cable Trays Used as Equipment Grounding Conductor

Maximum Fuse Ampere Rating, Circuit Breaker Ampere Trip Setting, or Circuit Breaker Protective Relay Ampere Trip Setting for Ground-Fault Protection of Any Cable Circuit in the Cable Tray System	Minimum Cross-Sectional Area of Metal[a]			
	Steel Cable Trays		Aluminum Cable Trays	
	mm^2	in.^{2}	mm^2	in.^{2}
60	129	0.20	129	0.20
100	258	0.40	129	0.20
200	451.5	0.70	129	0.20
400	645	1.00	258	0.40
600	967.5	1.50[b]	258	0.40
1000	—	—	387	0.60
1200	—	—	645	1.00
1600	—	—	967.5	1.50
2000	—	—	1290	2.00[b]

[a]Total cross-sectional area of both side rails for ladder or trough cable trays; or the minimum cross-sectional area of metal in channel cable trays or cable trays of one-piece construction.

[b]Steel cable trays shall not be used as equipment grounding conductors for circuits with ground-fault protection above 600 amperes. Aluminum cable trays shall not be used as equipment grounding conductors for circuits with ground-fault protection above 2000 amperes.

EXHIBIT 12.9

NEC Table 392.9 Allowable Cable Fill Area for Multiconductor Cables in Ladder, Ventilated Trough, or Solid Bottom Cable Trays for Cables Rated 2000 Volts or Less

		Maximum Allowable Fill Area for Multiconductor Cables						
		Ladder or Ventilated Trough Cable Trays, 392.9(A)				Solid Bottom Cable Trays, 392.9(C)		
Inside Width of Cable Tray		Column 1 Applicable for 392.9(A)(2) Only		Column 2[a] Applicable for 392.9(A)(3) Only		Column 3 Applicable for 392.9(C)(2) Only		Column 4[a] Applicable for 392.9(C)(3) Only
mm	in.	mm^2	in.^{2}	mm^2	in.^{2}	mm^2	in.^{2}	mm^2	in.^{2}
150	6.0	4,500	7.0	4,500 − (30 Sd)[b]	7 − (1.2 Sd)[b]	3,500	5.5	3,500 − (25 Sd[b])	5.5 − Sd[b]
225	9.0	6,800	10.5	6,800 − (30 Sd)	10.5 − (1.2 Sd)	5,100	8.0	5,100 − (25 Sd)	8.0 − Sd
300	12.0	9,000	14.0	9,000 − (30 Sd)	14 − (1.2 Sd)	7,100	11.0	7,100 − (25 Sd)	11.0 − Sd
450	18.0	13,500	21.0	13,500 − (30 Sd)	21 − (1.2 Sd)	10,600	16.5	10,600 − (25 Sd)	16.5 − Sd
600	24.0	18,000	28.0	18,000 − (30 Sd)	28 − (1.2 Sd)	14,200	22.0	14,200 − (25 Sd)	22.0 − Sd
750	30.0	22,500	35.0	22,500 − (30 Sd)	35 − (1.2 Sd)	17,700	27.5	17,700 − (25 Sd)	27.5 − Sd
900	36.0	27,000	42.0	27,000 − (30 Sd)	42 − (1.2 Sd)	21,300	33.0	21,300 − (25 Sd)	33.0 − Sd

[a] The maximum allowable fill areas in Columns 2 and 4 shall be calculated. For example, the maximum allowable fill in mm^2 for a 150-mm wide cable tray in Column 2 shall be 4500 minus (30 multiplied by Sd) [the maximum allowable fill, in square inches, for a 6-in. wide cable tray in Column 2 shall be 7 minus (1.2 multiplied by Sd)].
[b] The term *Sd* in Columns 2 and 4 is equal to the sum of the diameters, in mm, of all cables 107.2 mm (in inches, of all 4/0 AWG) and larger multiconductor cables in the same cable tray with smaller cables.

EXHIBIT 12.10

NEC Table 392.9(E) Allowable Cable Fill Area for Multiconductor Cables in Ventilated Channel Cable Trays for Cables Rated 2000 Volts or Less

Inside Width of Cable Tray		Maximum Allowable Fill Area for Multiconductor Cables			
		Column 1 One Cable		Column 2 More Than One Cable	
mm	in.	mm^2	in.^{2}	mm^2	in.^{2}
75	3	1500	2.3	850	1.3
100	4	2900	4.5	1600	2.5
150	6	4500	7.0	2450	3.8

RACEWAYS AND CABLE TRAYS

150

RACEWAYS AND CABLE TRAYS

EXHIBIT 12.11

NEC Table 392.9(F)   Allowable Cable Fill Area for Multiconductor Cables
in Solid Bottom Channel Cable Trays for Cables Rated 2000 Volts or Less

Inside Width of Cable Tray		Column 1 One Cable		Column 2 More Than One Cable	
mm	in.	mm^2	in.^{2}	mm^2	in.^{2}
50	2	850	1.3	500	0.8
75	3	1300	2.0	700	1.1
100	4	2400	3.7	1400	2.1
150	6	3600	5.5	2100	3.2

EXHIBIT 12.12

NEC Table 392.10(A)   Allowable Cable Fill Area for Single-Conductor Cables in
Ladder or Ventilated Trough Cable Trays for Cables Rated 2000 Volts or Less

Inside Width of Cable Tray		Maximum Allowable Fill Area for Single-Conductor Cables in Ladder or Ventilated Trough Cable Trays			
		Column 1 Applicable for 392.10(A)(2) Only		Column 2[a] Applicable for 392.10(A)(3) Only	
mm	in.	mm^2	in.^{2}	mm^2	in.^{2}
150	6	4,200	6.5	4,200 – (28 Sd)[b]	6.5 – (1.1 Sd)[b]
225	9	6,100	9.5	6,100 – (28 Sd)	9.5 – (1.1 Sd)
300	12	8,400	13.0	8,400 – (28 Sd)	13.0 – (1.1 Sd)
450	18	12,600	19.5	12,600 – (28 Sd)	19.5 – (1.1 Sd)
600	24	16,800	26.0	16,800 – (28 Sd)	26.0 – (1.1 Sd)
750	30	21,000	32.5	21,000 – (28 Sd)	32.5 – (1.1 Sd)
900	36	25,200	39.0	25,200 – (28 Sd)	39.0 – (1.1 Sd)

[a] The maximum allowable fill areas in Column 2 shall be calculated. For example, the maximum allowable fill, in mm^2, for a 150-mm wide cable tray in Column 2 shall be 4200 minus (28 multiplied by Sd) [the maximum allowable fill, in square inches, for a 6-in. wide cable tray in Column 2 shall be 6.5 minus (1.1 multiplied by Sd)].

[b] The term Sd in Column 2 is equal to the sum of the diameters, in mm, of all cables 507 mm^2 (in inches, of all 1000 kcmil) and larger single-conductor cables in the same ladder or ventilated trough cable tray with small cables.

CONDUIT AND TUBING FILL TABLES

Table 1 from *NEC* Chapter 9 (Exhibit 12.13) sets forth the absolute maximum fill permitted for conduit or tubing.

EXHIBIT 12.13

NEC Table 1   Percent of Cross Section of Conduit and Tubing for Conductors

Number of Conductors	All Conductor Types
1	53
2	31
Over 2	40

FPN No. 1: Table 1 is based on common conditions of proper cabling and alignment of conductors where the length of the pull and the number of bends are within reasonable limits. It should be recognized that, for certain conditions, a larger size conduit or a lesser conduit fill should be considered.

FPN No. 2: When pulling three conductors or cables into a raceway, if the ratio of the raceway (inside diameter) to the conductor or cable (outside diameter) is between 2.8 and 3.2, jamming can occur. While jamming can occur when pulling four or more conductors or cables into a raceway, the probability is very low.

Chapter 9 contains a number of tables pertaining to conductors, conduit, and tubing. (See Chapter 9's **Notes to Tables** for specifications pertaining to the tables.)

Table 4 (Exhibit 12.14) contains internal diameters and areas for different conduit and tubing types. This electrical reference includes all but two of the twelve tables. (See Chapter 9 in the *NEC* for Type A and Type EB, PVC conduit.) Tables 4 and 5 are used to find the maximum number of conductors permitted in conduit and tubing.

Table 5 (Exhibit 12.15) contains approximate diameters and areas for different conductor types. This table has been condensed from Table 5 in the *NEC* to include the most common conductor types. For a complete list of conductor types, see Table 5 in the *NEC*. Tables 4 and 5 are used to find the maximum number of conductors permitted in conduit and tubing.

EXHIBIT 12.14

NEC Table 4 [Abridged] Dimensions and Percent Area of Conduit and Tubing
(Areas of Conduit or Tubing for the Combinations of Wires Permitted in Table 1, Chapter 9)

Article 358—Electrical Metallic Tubing (EMT)

Metric Designator	Trade Size	Nominal Internal Diameter		Total Area 100%		60%		1 Wire 53%		2 Wires 31%		Over 2 Wires 40%	
		mm	in.	mm²	in.²	mm²	in.²	mm²	in.²	mm²	in.²	mm²	in.²
16	½	15.8	0.622	196	0.304	118	0.182	104	0.161	61	0.094	78	0.122
21	¾	20.9	0.824	343	0.533	206	0.320	182	0.283	106	0.165	137	0.213
27	1	26.6	1.049	556	0.864	333	0.519	295	0.458	172	0.268	222	0.346
35	1¼	35.1	1.380	968	1.496	581	0.897	513	0.793	300	0.464	387	0.598
41	1½	40.9	1.610	1314	2.036	788	1.221	696	1.079	407	0.631	526	0.814
53	2	52.5	2.067	2165	3.356	1299	2.013	1147	1.778	671	1.040	866	1.342
63	2½	69.4	2.731	3783	5.858	2270	3.515	2005	3.105	1173	1.816	1513	2.343
78	3	85.2	3.356	5701	8.846	3421	5.307	3022	4.688	1767	2.742	2280	3.538
91	3½	97.4	3.834	7451	11.545	4471	6.927	3949	6.119	2310	3.579	2980	4.618
103	4	110.1	4.334	9521	14.753	5712	8.852	5046	7.819	2951	4.573	3808	5.901

Continued

EXHIBIT 12.14 (continued)

NEC Table 4 [Abridged] Dimensions and Percent Area of Conduit and Tubing
(Areas of Conduit or Tubing for the Combinations of Wires Permitted in Table 1, Chapter 9)

Article 362—Electrical Nonmetallic Tubing (ENT)

Metric Designator	Trade Size	Nominal Internal Diameter		Total Area 100%		60%		1 Wire 53%		2 Wires 31%		Over 2 Wires 40%	
		mm	in.	mm²	in.²	mm²	in.²	mm²	in.²	mm²	in.²	mm²	in.²
16	½	14.2	0.560	158	0.246	95	0.148	84	0.131	49	0.076	63	0.099
21	¾	19.3	0.760	293	0.454	176	0.272	155	0.240	91	0.141	117	0.181
27	1	25.4	1.000	507	0.785	304	0.471	269	0.416	157	0.243	203	0.314
35	1¼	34.0	1.340	908	1.410	545	0.846	481	0.747	281	0.437	363	0.564
41	1½	39.9	1.570	1250	1.936	750	1.162	663	1.026	388	0.600	500	0.774
53	2	51.3	2.020	2067	3.205	1240	1.923	1095	1.699	641	0.993	827	1.282
63	2½	—	—	—	—	—	—	—	—	—	—	—	—
78	3	—	—	—	—	—	—	—	—	—	—	—	—
91	3½	—	—	—	—	—	—	—	—	—	—	—	—

Continued

EXHIBIT 12.14 (continued)

NEC Table 4 [Abridged] Dimensions and Percent Area of Conduit and Tubing
(Areas of Conduit or Tubing for the Combinations of Wires Permitted in Table 1, Chapter 9)

Article 348—Flexible Metal Conduit (FMC)

Metric Designator	Trade Size	Nominal Internal Diameter		Total Area 100%		60%		1 Wire 53%		2 Wires 31%		Over 2 Wires 40%	
		mm	in.	mm²	in.²	mm²	in.²	mm²	in.²	mm²	in.²	mm²	in.²
12	3/8	9.7	0.384	74	0.116	44	0.069	39	0.061	23	0.036	30	0.046
16	1/2	16.1	0.635	204	0.317	122	0.190	108	0.168	63	0.098	81	0.127
21	3/4	20.9	0.824	343	0.533	206	0.320	182	0.283	106	0.165	137	0.213
27	1	25.9	1.020	527	0.817	316	0.490	279	0.433	163	0.253	211	0.327
35	1¼	32.4	1.275	824	1.277	495	0.766	437	0.677	256	0.396	330	0.511
41	1½	39.1	1.538	1201	1.858	720	1.115	636	0.985	372	0.576	480	0.743
53	2	51.8	2.040	2107	3.269	1264	1.961	1117	1.732	653	1.013	843	1.307
63	2½	63.5	2.500	3167	4.909	1900	2.945	1678	2.602	982	1.522	1267	1.963
78	3	76.2	3.000	4560	7.069	2736	4.241	2417	3.746	1414	2.191	1824	2.827
91	3½	88.9	3.500	6207	9.621	3724	5.773	3290	5.099	1924	2.983	2483	3.848
103	4	101.6	4.000	8107	12.566	4864	7.540	4297	6.660	2513	3.896	3243	5.027

Continued

EXHIBIT 12.14 (continued)

NEC Table 4 [Abridged] Dimensions and Percent Area of Conduit and Tubing
(Areas of Conduit or Tubing for the Combinations of Wires Permitted in Table 1, Chapter 9)

Article 342—Intermediate Metal Conduit (IMC)

Metric Designator	Trade Size	Nominal Internal Diameter		Total Area 100%		60%		1 Wire 53%		2 Wires 31%		Over 2 Wires 40%	
		mm	in.	mm²	in.²	mm²	in.²	mm²	in.²	mm²	in.²	mm²	in.²
12	⅜	—	—	—	—	—	—	—	—	—	—	—	—
16	½	16.8	0.660	222	0.342	133	0.205	117	0.181	69	0.106	89	0.137
21	¾	21.9	0.864	377	0.586	226	0.352	200	0.311	117	0.182	151	0.235
27	1	28.1	1.105	620	0.959	372	0.575	329	0.508	192	0.297	248	0.384
35	1¼	36.8	1.448	1064	1.647	638	0.988	564	0.873	330	0.510	425	0.659
41	1½	42.7	1.683	1432	2.225	859	1.335	759	1.179	444	0.690	573	0.890
53	2	54.6	2.150	2341	3.630	1405	2.178	1241	1.924	726	1.125	937	1.452
63	2½	64.9	2.557	3308	5.135	1985	3.081	1753	2.722	1026	1.592	1323	2.054
78	3	80.7	3.176	5115	7.922	3069	4.753	2711	4.199	1586	2.456	2046	3.169
91	3½	93.2	3.671	6822	10.584	4093	6.351	3616	5.610	2115	3.281	2729	4.234
103	4	105.4	4.166	8725	13.631	5235	8.179	4624	7.224	2705	4.226	3490	5.452

Continued

EXHIBIT 12.14 (continued)

NEC Table 4 [Abridged] Dimensions and Percent Area of Conduit and Tubing
(Areas of Conduit or Tubing for the Combinations of Wires Permitted in Table 1, Chapter 9)

Article 356—Liquidtight Flexible Nonmetallic Conduit (LFNC-B*)

Metric Designator	Trade Size	Nominal Internal Diameter		Total Area 100%		60%		1 Wire 53%		2 Wires 31%		Over 2 Wires 40%	
		mm	in.	mm²	in.²	mm²	in.²	mm²	in.²	mm²	in.²	mm²	in.²
12	3/8	12.5	0.494	123	0.192	74	0.115	65	0.102	38	0.059	49	0.077
16	1/2	16.1	0.632	204	0.314	122	0.188	108	0.166	63	0.097	81	0.125
21	3/4	21.1	0.830	350	0.541	210	0.325	185	0.287	108	0.168	140	0.216
27	1	26.8	1.054	564	0.873	338	0.524	299	0.462	175	0.270	226	0.349
35	1 1/4	35.4	1.395	984	1.528	591	0.917	522	0.810	305	0.474	394	0.611
41	1 1/2	40.3	1.588	1276	1.981	765	1.188	676	1.050	395	0.614	510	0.792
53	2	51.6	2.033	2091	3.246	1255	1.948	1108	1.720	648	1.006	836	1.298

*Corresponds to 356.2(2)

EXHIBIT 12.14 *(continued)*

NEC Table 4 [Abridged] Dimensions and Percent Area of Conduit and Tubing
(Areas of Conduit or Tubing for the Combinations of Wires Permitted in Table 1, Chapter 9)

Article 356—Liquidtight Flexible Nonmetallic Conduit (LFNC-A*)

Metric Designator	Trade Size	Nominal Internal Diameter		Total Area 100%		1 Wire 53%		2 Wires 31%		Over 2 Wires 40%	
		mm	in.	mm²	in.²	mm²	in.²	mm²	in.²	mm²	in.²
12	3/8	12.6	0.495	125	0.192	66	0.102	39	0.060	50	0.077
16	1/2	16.0	0.630	201	0.312	107	0.165	62	0.097	80	0.125
21	3/4	21.0	0.825	346	0.535	184	0.283	107	0.166	139	0.214
27	1	26.5	1.043	552	0.854	292	0.453	171	0.265	221	0.342
35	1 1/4	35.1	1.383	968	1.502	513	0.796	300	0.466	387	0.601
41	1 1/2	40.7	1.603	1301	2.018	690	1.070	403	0.626	520	0.807
53	2	52.4	2.063	2157	3.343	1143	1.772	669	1.036	863	1.337

*Corresponds to 356.2(1)

Continued

EXHIBIT 12.14 (continued)

NEC Table 4 [Abridged] Dimensions and Percent Area of Conduit and Tubing
(Areas of Conduit or Tubing for the Combinations of Wires Permitted in Table 1, Chapter 9)

Article 350—Liquidtight Flexible Metal Conduit (LFMC)

Metric Designator	Trade Size	Nominal Internal Diameter		Total Area 100%		60%		1 Wire 53%		2 Wires 31%		Over 2 Wires 40%	
		mm	in.	mm²	in.²	mm²	in.²	mm²	in.²	mm²	in.²	mm²	in.²
12	3/8	12.5	0.494	123	0.192	74	0.115	65	0.102	38	0.059	49	0.077
16	1/2	16.1	0.632	204	0.314	122	0.188	108	0.166	63	0.097	81	0.125
21	3/4	21.1	0.830	350	0.541	210	0.325	185	0.287	108	0.168	140	0.216
27	1	26.8	1.054	564	0.873	338	0.524	299	0.462	175	0.270	226	0.349
35	1 1/4	35.4	1.395	984	1.528	591	0.917	522	0.810	305	0.474	394	0.611
41	1 1/2	40.3	1.588	1276	1.981	765	1.188	676	1.050	395	0.614	510	0.792
53	2	51.6	2.033	2091	3.246	1255	1.948	1108	1.720	648	1.006	836	1.298
63	2 1/2	63.3	2.493	3147	4.881	1888	2.929	1668	2.587	976	1.513	1259	1.953
78	3	78.4	3.085	4827	7.475	2896	4.485	2559	3.962	1497	2.317	1931	2.990
91	3 1/2	89.4	3.520	6277	9.731	3766	5.839	3327	5.158	1946	3.017	2511	3.893
103	4	102.1	4.020	8187	12.692	4912	7.615	4339	6.727	2538	3.935	3275	5.077
129	5	—	—	—	—	—	—	—	—	—	—	—	—
155	6	—	—	—	—	—	—	—	—	—	—	—	—

Continued

EXHIBIT 12.14 (continued)

NEC Table 4 [Abridged]   Dimensions and Percent Area of Conduit and Tubing
(Areas of Conduit or Tubing for the Combinations of Wires Permitted in Table 1, Chapter 9)

Article 344—Rigid Metal Conduit (RMC)

Metric Designator	Trade Size	Nominal Internal Diameter		Total Area 100%		1 Wire 60%		1 Wire 53%		2 Wires 31%		Over 2 Wires 40%	
		mm	in.	mm²	in.²	mm²	in.²	mm²	in.²	mm²	in.²	mm²	in.²
12	⅜	—	—	—	—	—	—	—	—	—	—	—	—
16	½	16.1	0.632	204	0.314	122	0.188	108	0.166	63	0.097	81	0.125
21	¾	21.2	0.836	353	0.549	212	0.329	187	0.291	109	0.170	141	0.220
27	1	27.0	1.063	573	0.887	344	0.532	303	0.470	177	0.275	229	0.355
35	1¼	35.4	1.394	984	1.526	591	0.916	522	0.809	305	0.473	394	0.610
41	1½	41.2	1.624	1333	2.071	800	1.243	707	1.098	413	0.642	533	0.829
53	2	52.9	2.083	2198	3.408	1319	2.045	1165	1.806	681	1.056	879	1.363
63	2½	63.2	2.489	3137	4.866	1882	2.919	1663	2.579	972	1.508	1255	1.946
78	3	78.5	3.090	4840	7.499	2904	4.499	2565	3.974	1500	2.325	1936	3.000
91	3½	90.7	3.570	6461	10.010	3877	6.006	3424	5.305	2003	3.103	2584	4.004
103	4	102.9	4.050	8316	12.882	4990	7.729	4408	6.828	2578	3.994	3326	5.153
129	5	128.9	5.073	13050	20.212	7830	12.127	6916	10.713	4045	6.266	5220	8.085
155	6	154.8	6.093	18821	29.158	11292	17.495	9975	15.454	5834	9.039	7528	11.663

Continued

EXHIBIT 12.14 (continued)

NEC Table 4 [Abridged] Dimensions and Percent Area of Conduit and Tubing
(Areas of Conduit or Tubing for the Combinations of Wires Permitted in Table 1, Chapter 9)

Article 352—Rigid PVC Conduit (RNC), Schedule 80

Metric Designator	Trade Size	Nominal Internal Diameter		Total Area 100%		60%		1 Wire 53%		2 Wires 31%		Over 2 Wires 40%	
		mm	in.	mm²	in.²	mm²	in.²	mm²	in.²	mm²	in.²	mm²	in.²
12	3/8	—	—	—	—	—	—	—	—	—	—	—	—
16	1/2	13.4	0.526	141	0.217	85	0.130	75	0.115	44	0.067	56	0.087
21	3/4	18.3	0.722	263	0.409	158	0.246	139	0.217	82	0.127	105	0.164
27	1	23.8	0.936	445	0.688	267	0.413	236	0.365	138	0.213	178	0.275
35	1¼	31.9	1.255	799	1.237	480	0.742	424	0.656	248	0.383	320	0.495
41	1½	37.5	1.476	1104	1.711	663	1.027	585	0.907	342	0.530	442	0.684
53	2	48.6	1.913	1855	2.874	1113	1.725	983	1.523	575	0.891	742	1.150
63	2½	58.2	2.290	2660	4.119	1596	2.471	1410	2.183	825	1.277	1064	1.647
78	3	72.7	2.864	4151	6.442	2491	3.865	2200	3.414	1287	1.997	1660	2.577
91	3½	84.5	3.326	5608	8.688	3365	5.213	2972	4.605	1738	2.693	2243	3.475
103	4	96.2	3.786	7268	11.258	4361	6.755	3852	5.967	2253	3.490	2907	4.503
129	5	121.1	4.768	11518	17.855	6911	10.713	6105	9.463	3571	5.535	4607	7.142
155	6	145.0	5.709	16513	25.598	9908	15.359	8752	13.567	5119	7.935	6605	10.239

Continued

EXHIBIT 12.14 (continued)

NEC Table 4 [Abridged]   Dimensions and Percent Area of Conduit and Tubing
(Areas of Conduit or Tubing for the Combinations of Wires Permitted in Table 1, Chapter 9)

Articles 352 and 353—Rigid PVC Conduit (RNC), Schedule 40, and HDPE Conduit

Metric Designator	Trade Size	Nominal Internal Diameter		Total Area 100%		60%		1 Wire 53%		2 Wires 31%		Over 2 Wires 40%	
		mm	in.	mm²	in.²	mm²	in.²	mm²	in.²	mm²	in.²	mm²	in.²
12	⅜	—	—	—	—	—	—	—	—	—	—	—	—
16	½	15.3	0.602	184	0.285	110	0.171	97	0.151	57	0.088	74	0.114
21	¾	20.4	0.804	327	0.508	196	0.305	173	0.269	101	0.157	131	0.203
27	1	26.1	1.029	535	0.832	321	0.499	284	0.441	166	0.258	214	0.333
35	1¼	34.5	1.360	935	1.453	561	0.872	495	0.770	290	0.450	374	0.581
41	1½	40.4	1.590	1282	1.986	769	1.191	679	1.052	397	0.616	513	0.794
53	2	52.0	2.047	2124	3.291	1274	1.975	1126	1.744	658	1.020	849	1.316
63	2½	62.1	2.445	3029	4.695	1817	2.817	1605	2.488	939	1.455	1212	1.878
78	3	77.3	3.042	4693	7.268	2816	4.361	2487	3.852	1455	2.253	1877	2.907
91	3½	89.4	3.521	6277	9.737	3766	5.842	3327	5.161	1946	3.018	2511	3.895
103	4	101.5	3.998	8091	12.554	4855	7.532	4288	6.654	2508	3.892	3237	5.022
129	5	127.4	5.016	12748	19.761	7649	11.856	6756	10.473	3952	6.126	5099	7.904
155	6	153.2	6.031	18433	28.567	11060	17.140	9770	15.141	5714	8.856	7373	11.427

Continued

EXHIBIT 12.15

NEC Table 5 [Abridged] Dimensions of Insulated Conductors and Fixture Wires

Type	Size (AWG or kcmil)	Approximate Diameter		Approximate Area	
		mm	in.	mm^2	in.^{2}
TFN, TFFN	18	2.134	0.084	3.548	0.0055
	16	2.438	0.096	4.645	0.0072
THHN, THWN, THWN-2	14	2.819	0.111	6.258	0.0097
	12	3.302	0.130	8.581	0.0133
	10	4.166	0.164	13.61	0.0211
	8	5.486	0.216	23.61	0.0366
	6	6.452	0.254	32.71	0.0507
	4	8.230	0.324	53.16	0.0824
	3	8.941	0.352	62.77	0.0973
	2	9.754	0.384	74.71	0.1158
	1	11.33	0.446	100.8	0.1562
	1/0	12.34	0.486	119.7	0.1855
	2/0	13.51	0.532	143.4	0.2223
	3/0	14.83	0.584	172.8	0.2679
	4/0	16.31	0.642	208.8	0.3237
	250	18.06	0.711	256.1	0.3970
	300	19.46	0.766	297.3	0.4608

Continued

EXHIBIT 12.15 (continued)

NEC Table 5 [Abridged] Dimensions of Insulated Conductors and Fixture Wires

Type	Size (AWG or kcmil)	Approximate Diameter		Approximate Area	
		mm	in.	mm²	in.²
THHN, THWN, THWN-2	350	20.75	0.817	338.2	0.5242
	400	21.95	0.864	378.3	0.5863
	500	24.10	0.949	456.3	0.7073
	600	26.70	1.051	559.7	0.8676
	700	28.50	1.122	637.9	0.9887
	750	29.36	1.156	677.2	1.0496
	800	30.18	1.188	715.2	1.1085
	900	31.80	1.252	794.3	1.2311
	1000	33.27	1.310	869.5	1.3478
XHHW, ZW, XHHW-2, XHH	14	3.378	0.133	8.968	0.0139
	12	3.861	0.152	11.68	0.0181
	10	4.470	0.176	15.68	0.0243
	8	5.994	0.236	28.19	0.0437
	6	6.960	0.274	38.06	0.0590
	4	8.179	0.322	52.52	0.0814
	3	8.890	0.350	62.06	0.0962
	2	9.703	0.382	73.94	0.1146
XHHW, XHHW-2, XHH	1	11.23	0.442	98.97	0.1534
	1/0	12.24	0.482	117.7	0.1825
	2/0	13.41	0.528	141.3	0.2190

Continued

RACEWAYS AND CABLE TRAYS

EXHIBIT 12.15 (continued)

NEC Table 5 [Abridged] Dimensions of Insulated Conductors and Fixture Wires

Type	Size (AWG or kcmil)	Approximate Diameter		Approximate Area	
		mm	in.	mm²	in.²
XHHW, XHHW-2, XHH	3/0	14.73	0.58	170.5	0.2642
	4/0	16.21	0.638	206.3	0.3197
	250	17.91	0.705	251.9	0.3904
	300	19.30	0.76	292.6	0.4536
	350	20.60	0.811	333.3	0.5166
	400	21.79	0.858	373.0	0.5782
	500	23.95	0.943	450.6	0.6984
	600	26.75	1.053	561.9	0.8709
	700	28.55	1.124	640.2	0.9923
	750	29.41	1.158	679.5	1.0532
	800	30.23	1.190	717.5	1.1122
	900	31.85	1.254	796.8	1.2351
	1000	33.32	1.312	872.2	1.3519
	1250	37.57	1.479	1108	1.7180
	1500	40.69	1.602	1300	2.0157
	1750	43.59	1.716	1492	2.3127
	2000	46.28	1.822	1682	2.6073

ANNEX C TABLES—CONDUIT AND TUBING FILL FOR CONDUCTORS AND FIXTURE WIRES OF THE SAME SIZE

NEC Tables C1 through C10 are shown here as Exhibits 12.16 through 12.25 to provide the reader with a quick reference to find the maximum number of conductors (or fixture wires) of the same size permitted in conduit and tubing.

Annex C Tables (Abridged)

C1—Electrical Metallic Tubing (EMT) p. 166
C2—Electrical Nonmetallic Tubing (ENT) p. 167
C3—Flexible Metal Conduit (FMC) p. 169
C4—Intermediate Metal Conduit (IMC) p. 170
C5—Liquidtight Flexible Nonmetallic Conduit (Type LFNC-B) p. 171
C6—Liquidtight Flexible Nonmetallic Conduit (Type LFNC-A) p. 173
C7—Liquidtight Flexible Metal Conduit (LFML) p. 174
C8—Rigid Metal Conduit (RMC) p. 175
C9—Rigid PVC Conduit, Schedule 80 p. 177
C10—Rigid PVC Conduit, Schedule 40 and HDPE Conduit p. 178

Note: Refer to the "A" tables in Annex C of the *NEC* for compact stranded conductors.

RACEWAYS AND CABLE TRAYS

EXHIBIT 12.16

NEC Table C1 [Abridged]   Maximum Number of Conductors or Fixture
Wires in Electrical Metallic Tubing (EMT) *(Based on Table 1, Chapter 9)*

		CONDUCTORS									
	Conductor Size	Metric Designator (Trade Size)									
Type	(AWG kcmil)	16 (½)	21 (¾)	27 (1)	35 (1¼)	41 (1½)	53 (2)	63 (2½)	78 (3)	91 (3½)	103 (4)
TFN,	18	22	38	63	108	148	244	–	–	–	–
TFFN	16	17	29	48	83	113	186	–	–	–	–
THHN,	14	12	22	35	61	84	138	241	364	476	608
THWN,	12	9	16	26	45	61	101	176	266	347	443
THWN-2	10	5	10	16	28	38	63	111	167	219	279
	8	3	6	9	16	22	36	64	96	126	161
	6	2	4	7	12	16	26	46	69	91	116
	4	1	2	4	7	10	16	28	43	56	71
	3	1	1	3	6	8	13	24	36	47	60
	2	1	1	3	5	7	11	20	30	40	51
	1	1	1	1	4	5	8	15	22	29	37
	1/0	1	1	1	3	4	7	12	19	25	32
	2/0	0	1	1	2	3	6	10	16	20	26
	3/0	0	1	1	1	3	5	8	13	17	22
	4/0	0	1	1	1	2	4	7	11	14	18
	250	0	0	1	1	1	3	6	9	11	15
	300	0	0	1	1	1	3	5	7	10	13
	350	0	0	1	1	1	2	4	6	9	11
	400	0	0	0	1	1	1	4	6	8	10
	500	0	0	0	1	1	1	3	5	6	8
	600	0	0	0	1	1	1	2	4	5	7
	700	0	0	0	1	1	1	2	3	4	6
	750	0	0	0	0	1	1	1	3	4	5
	800	0	0	0	0	1	1	1	3	4	5
	900	0	0	0	0	1	1	1	3	3	4
	1000	0	0	0	0	1	1	1	2	3	4
XHH,	14	8	15	25	43	58	96	168	254	332	424
XHHW,	12	6	11	19	33	45	74	129	195	255	326
XHHW-2,	10	5	8	14	24	33	55	96	145	190	243
ZW	8	2	5	8	13	18	30	53	81	105	135
	6	1	3	6	10	14	22	39	60	78	100
	4	1	2	4	7	10	16	28	43	56	72
	3	1	1	3	6	8	14	24	36	48	61
	2	1	1	3	5	7	11	20	31	40	51
XHH,	1	1	1	1	4	5	8	15	23	30	38
XHHW,	1/0	1	1	1	3	4	7	13	19	25	32
XHHW-2	2/0	0	1	1	2	3	6	10	16	21	27
	3/0	0	1	1	1	3	5	9	13	17	22
	4/0	0	1	1	1	2	4	7	11	14	18
	250	0	0	1	1	1	3	6	9	12	15
	300	0	0	1	1	1	3	5	8	10	13
	350	0	0	1	1	1	2	4	7	9	11
	400	0	0	0	1	1	1	4	6	8	10
	500	0	0	0	1	1	1	3	5	6	8

Table continues below.

EXHIBIT 12.16 (continued)

NEC Table C1 [Abridged]   Maximum Number of Conductors or Fixture Wires in Electrical Metallic Tubing (EMT) *(Based on Table 1, Chapter 9)*

	CONDUCTORS									
	Conductor Size	Metric Designator (Trade Size)								
Type	(AWG kcmil)	16 (½)	21 (¾)	27 (1)	35 (1¼)	41 (1½)	53 (2)	63 (2½)	78 (3)	91 (3½)	103 (4)
XHH, XHHW, XHHW-2	600	0	0	0	1	1	1	2	4	5	6
	700	0	0	0	0	1	1	2	3	4	6
	750	0	0	0	0	1	1	1	3	4	5
	800	0	0	0	0	1	1	1	3	4	5
	900	0	0	0	0	1	1	1	3	3	4
	1000	0	0	0	0	0	1	1	2	3	4
	1250	0	0	0	0	0	1	1	1	2	3
	1500	0	0	0	0	0	1	1	1	1	3
	1750	0	0	0	0	0	0	1	1	1	2
	2000	0	0	0	0	0	0	1	1	1	1

EXHIBIT 12.17

NEC Table C2 [Abridged]   Maximum Number of Conductors or Fixture Wires in Electrical Nonmetallic Tubing (ENT) *(Based on Table 1, Chapter 9)*

	CONDUCTORS						
	Conductor Size	Metric Designator (Trade Size)					
Type	(AWG/ kcmil)	16 (½)	21 (¾)	27 (1)	35 (1¼)	41 (1½)	53 (2)
TFN, TFFN	18	18	33	57	102	141	233
	16	13	25	43	78	107	178
THHN, THWN, THWN-2	14	10	18	32	58	80	132
	12	7	13	23	42	58	96
	10	4	8	15	26	36	60
	8	2	5	8	15	21	35
	6	1	3	6	11	15	25
	4	1	1	4	7	9	15
	3	1	1	3	5	8	13
	2	1	1	2	5	6	11
	1	1	1	1	3	5	8
	1/0	0	1	1	3	4	7
	2/0	0	1	1	2	3	5
	3/0	0	1	1	1	3	4
	4/0	0	0	1	1	2	4
	250	0	0	1	1	1	3
	300	0	0	1	1	1	2
	350	0	0	0	1	1	2
	400	0	0	0	1	1	1
	500	0	0	0	1	1	1
	600	0	0	0	0	1	1
	700	0	0	0	0	1	1
	750	0	0	0	0	1	1

Continued

168

RACEWAYS AND CABLE TRAYS

EXHIBIT 12.17 *(continued)*

NEC Table C2 [Abridged]   Maximum Number of Conductors or Fixture
Wires in Electrical Nonmetallic Tubing (ENT) *(Based on Table 1, Chapter 9)*

		CONDUCTORS					
	Conductor Size	Metric Designator (Trade Size)					
Type	(AWG/ kcmil)	16 (½)	21 (¾)	27 (1)	35 (1¼)	41 (1½)	53 (2)
THHN,	800	0	0	0	0	1	1
THWN,	900	0	0	0	0	1	1
THWN-2	1000	0	0	0	0	0	1
XHH,	14	7	13	22	40	55	92
XHHW,	12	5	10	17	31	42	71
XHHW-2,	10	4	7	13	23	32	52
ZW	8	1	4	7	13	17	29
	6	1	3	5	9	13	21
	4	1	1	4	7	9	15
	3	1	1	3	6	8	13
	2	1	1	2	5	6	11
XHH, XHHW,	1	1	1	1	3	5	8
XHHW-2	1/0	0	1	1	3	4	7
	2/0	0	1	1	2	3	6
	3/0	0	1	1	1	3	5
	4/0	0	0	1	1	2	4
	250	0	0	1	1	1	3
	300	0	0	1	1	1	3
	350	0	0	1	1	1	2
	400	0	0	0	1	1	1
	500	0	0	0	1	1	1
	600	0	0	0	1	1	1
	700	0	0	0	0	1	1
	750	0	0	0	0	1	1
	800	0	0	0	0	1	1
	900	0	0	0	0	1	1
XHH,	1000	0	0	0	0	0	1
XHHW,	1250	0	0	0	0	0	1
XHHW-2	1500	0	0	0	0	0	1
	1750	0	0	0	0	0	0
	2000	0	0	0	0	0	0

EXHIBIT 12.18

NEC Table C3 [Abridged]   Maximum Number of Conductors or Fixture Wires in Flexible Metal Conduit (FMC) *(Based on Table 1, Chapter 9)*

		CONDUCTORS								
	Conductor Size	Metric Designator (Trade Size)								
Type	(AWG/ kcmil)	16 (½)	21 (¾)	27 (1)	35 (1¼)	41 (1½)	53 (2)	63 (2½)	78 (3)	91 (3½)	103 (4)
TFN,	18	23	38	59	93	135	237	–	–	–	–
TFFN	16	17	29	45	71	103	181	–	–	–	–
THHN,	14	13	22	33	52	76	134	202	291	396	518
THWN,	12	9	16	24	38	56	98	147	212	289	378
THWN-2	10	6	10	15	24	35	62	93	134	182	238
	8	3	6	9	14	20	35	53	77	105	137
	6	2	4	6	10	14	25	38	55	76	99
	4	1	2	4	6	9	16	24	34	46	61
	3	1	1	3	5	7	13	20	29	39	51
	2	1	1	3	4	6	11	17	24	33	43
	1	1	1	1	3	4	8	12	18	24	32
	1/0	1	1	1	2	4	7	10	15	20	27
	2/0	0	1	1	1	3	6	9	12	17	22
	3/0	0	1	1	1	2	5	7	10	14	18
	4/0	0	1	1	1	1	4	6	8	12	15
	250	0	0	1	1	1	3	5	7	9	12
	300	0	0	1	1	1	3	4	6	8	11
	350	0	0	1	1	1	2	3	5	7	9
	400	0	0	0	1	1	1	3	5	6	8
	500	0	0	0	1	1	1	2	4	5	7
	600	0	0	0	0	1	1	1	3	4	5
	700	0	0	0	0	1	1	1	3	4	5
	750	0	0	0	0	1	1	1	2	3	4
	800	0	0	0	0	1	1	1	2	3	4
	900	0	0	0	0	0	1	1	1	3	4
	1000	0	0	0	0	0	1	1	1	3	3
XHH,	14	9	15	23	36	53	94	141	203	277	361
XHHW,	12	7	11	18	28	41	72	108	156	212	277
XHHW-2,	10	5	8	13	21	30	54	81	116	158	207
ZW	8	3	5	7	11	17	30	45	64	88	115
	6	1	3	5	8	12	22	33	48	65	85
	4	1	2	4	6	9	16	24	34	47	61
	3	1	1	3	5	7	13	20	29	40	52
	2	1	1	3	4	6	11	17	24	33	44
XHH,	1	1	1	1	3	5	8	13	18	25	32
XHHW,	1/0	1	1	1	2	4	7	10	15	21	27
XHHW-2	2/0	0	1	1	2	3	6	9	13	17	23
	3/0	0	1	1	1	3	5	7	10	14	19
	4/0	0	1	1	1	2	4	6	9	12	15
	250	0	0	1	1	1	3	5	7	10	13
	300	0	0	1	1	1	3	4	6	8	11
	350	0	0	1	1	1	2	4	5	7	9
	400	0	0	0	1	1	1	3	5	6	8
	500	0	0	0	1	1	1	3	4	5	7
	600	0	0	0	0	1	1	1	3	4	5
	700	0	0	0	0	1	1	1	3	4	5
	750	0	0	0	0	1	1	1	2	3	4
	800	0	0	0	0	1	1	1	2	3	4
	900	0	0	0	0	0	1	1	1	3	4

Continued

170

RACEWAYS AND CABLE TRAYS

EXHIBIT 12.18 *(continued)*

NEC Table C [Abridged]   Maximum Number of Conductors or Fixture Wires in Flexible Metal Conduit (FMC) *(Based on Table 1, Chapter 9)*

		CONDUCTORS									
	Conductor Size	Metric Designator (Trade Size)									
Type	(AWG/ kcmil)	16 (½)	21 (¾)	27 (1)	35 (1¼)	41 (1½)	53 (2)	63 (2½)	78 (3)	91 (3½)	103 (4)
XHH, XHHW, XHHW-2	1000	0	0	0	0	0	1	1	1	3	3
	1250	0	0	0	0	0	1	1	1	1	3
	1500	0	0	0	0	0	1	1	1	1	2
	1750	0	0	0	0	0	0	1	1	1	1
	2000	0	0	0	0	0	0	1	1	1	1

EXHIBIT 12.19

NEC Table C4 [Abridged]   Maximum Number of Conductors or Fixture Wires in Intermediate Metal Conduit (IMC) *(Based on Table 1, Chapter 9)*

		CONDUCTORS									
	Conductor Size	Metric Designator (Trade Size)									
Type	(AWG/ kcmil)	16 (½)	21 (¾)	27 (1)	35 (1¼)	41 (1½)	53 (2)	63 (2½)	78 (3)	91 (3½)	103 (4)
TFN, TFFN	18	25	42	69	191	161	264	–	–	–	–
	16	19	32	53	91	123	201	–	–	–	–
THHN, THWN, THWN-2	14	14	24	39	68	91	149	211	326	436	562
	12	10	17	29	49	67	109	154	238	318	410
	10	6	11	18	31	42	68	97	150	200	258
	8	3	6	10	18	24	39	56	86	115	149
	6	2	4	7	13	17	28	40	62	83	107
	4	1	3	4	8	10	17	25	38	51	66
	3	1	2	4	6	9	15	21	32	43	56
	2	1	1	3	5	7	12	17	27	36	47
	1	1	1	2	4	5	9	13	20	27	35
	1/0	1	1	1	3	4	8	11	17	23	29
	2/0	1	1	1	3	4	6	9	14	19	24
	3/0	0	1	1	2	3	5	7	12	16	20
	4/0	0	1	1	1	2	4	6	9	13	17
	250	0	0	1	1	1	3	5	8	10	13
	300	0	0	1	1	1	3	4	7	9	12
	350	0	0	1	1	1	2	4	6	8	10
	400	0	0	1	1	1	2	3	5	7	9
	500	0	0	0	1	1	1	3	4	6	7
	600	0	0	0	1	1	1	2	3	5	6
	700	0	0	0	1	1	1	1	3	4	5
	750	0	0	0	1	1	1	1	3	4	5
	800	0	0	0	0	1	1	1	3	4	5
	900	0	0	0	0	1	1	1	2	3	4
	1000	0	0	0	0	1	1	1	2	3	4

Table continues below.

EXHIBIT 12.19 *(continued)*

NEC Table C4 [Abridged]   Maximum Number of Conductors or Fixture Wires in Intermediate Metal Conduit (IMC) *(Based on Table 1, Chapter 9)*

		CONDUCTORS								
	Conductor Size	Metric Designator (Trade Size)								
Type	(AWG/ kcmil)	16 (½)	21 (¾)	27 (1)	35 (1¼)	41 (1½)	53 (2)	63 (2½)	78 (3)	91 (3½)	103 (4)
XHH,	14	10	17	27	47	64	104	147	228	304	392
XHHW,	12	7	13	21	36	49	80	113	175	234	301
XHHW-2,	10	5	9	15	27	36	59	84	130	174	224
ZW	8	3	5	8	15	20	33	47	72	97	124
	6	1	4	6	11	15	24	35	53	71	92
	4	1	3	4	8	11	18	25	39	52	67
	3	1	2	4	7	9	15	21	33	44	56
	2	1	1	3	5	7	12	18	27	37	47
XHH,	1	1	1	2	4	5	9	13	20	27	35
XHHW,	1/0	1	1	1	3	5	8	11	17	23	30
XHHW-2	2/0	1	1	1	3	4	6	9	14	19	25
	3/0	0	1	1	2	3	5	7	12	16	20
	4/0	0	1	1	1	2	4	6	10	13	17
	250	0	0	1	1	1	3	5	8	11	14
	300	0	0	1	1	1	3	4	7	9	12
	350	0	0	1	1	1	3	4	6	8	10
	400	0	0	1	1	1	2	3	5	7	9
	500	0	0	0	1	1	1	3	4	6	8
	600	0	0	0	1	1	1	2	3	5	6
	700	0	0	0	1	1	1	1	3	4	5
	750	0	0	0	1	1	1	1	3	4	5
	800	0	0	0	0	1	1	1	3	4	5
	900	0	0	0	0	1	1	1	2	3	4
	1000	0	0	0	0	1	1	1	2	3	4
	1250	0	0	0	0	0	1	1	1	2	3
	1500	0	0	0	0	0	1	1	1	1	2
	1750	0	0	0	0	0	1	1	1	1	2
	2000	0	0	0	0	0	0	1	1	1	1

EXHIBIT 12.20

NEC Table C5 [Abridged]   Maximum Number of Conductors or Fixture Wires in Liquidtight Flexible Nonmetallic Conduit (Type LFNC-B*) *(Based on Table 1, Chapter 9)*

		CONDUCTORS						
	Conductor Size	Metric Designator (Trade Size)						
Type	(AWG/ kcmil)	12 (⅜)	16 (½)	21 (¾)	27 (1)	35 (1¼)	41 (1½)	53 (2)
TFN,	18	14	23	39	63	111	144	236
TFFN	16	10	17	30	48	85	110	180
THHN,	14	8	13	22	36	63	81	133
THWN,	12	5	9	16	26	46	59	97
THWN-2	10	3	6	10	16	29	37	61
	8	1	3	6	9	16	21	35
	6	1	2	4	7	12	15	25

Continued

172

RACEWAYS AND CABLE TRAYS

EXHIBIT 12.20 (continued)

NEC Table C5 [Abridged]   Maximum Number of Conductors or Fixture Wires in
Liquidtight Flexible Nonmetallic Conduit (Type LFNC-B*) *(Based on Table 1, Chapter 9)*

		CONDUCTORS						
	Conductor Size	Metric Designator (Trade Size)						
Type	(AWG/ kcmil)	12 (⅜)	16 (½)	21 (¾)	27 (1)	35 (1¼)	41 (1½)	53 (2)
THHN,	4	1	1	2	4	7	9	15
THWN,	3	1	1	1	3	6	8	13
THWN-2	2	1	1	1	3	5	7	11
	1	0	1	1	1	4	5	8
	1/0	0	1	1	1	3	4	7
	2/0	0	0	1	1	2	3	6
	3/0	0	0	1	1	1	3	5
	4/0	0	0	1	1	1	2	4
	250	0	0	0	1	1	1	3
	300	0	0	0	1	1	1	3
	350	0	0	0	1	1	1	2
	400	0	0	0	0	1	1	1
	500	0	0	0	0	1	1	1
	600	0	0	0	0	1	1	1
	700	0	0	0	0	1	1	1
	750	0	0	0	0	0	1	1
	800	0	0	0	0	0	1	1
	900	0	0	0	0	0	1	1
	1000	0	0	0	0	0	0	1
XHH,	14	5	9	15	25	44	57	93
XHHW,	12	4	7	12	19	33	43	71
XHHW-2,	10	3	5	9	14	25	32	53
ZW	8	1	3	5	8	14	18	29
	6	1	1	3	6	10	13	22
	4	1	1	2	4	7	9	16
	3	1	1	1	3	6	8	13
	2	1	1	1	3	5	7	11
XHH,	1	0	1	1	1	4	5	8
XHHW,	1/0	0	1	1	1	3	4	7
XHHW-2	2/0	0	0	1	1	2	3	6
	3/0	0	0	1	1	1	3	5
	4/0	0	0	1	1	1	2	4
	250	0	0	0	1	1	1	3
	300	0	0	0	1	1	1	3
	350	0	0	0	1	1	1	2
	400	0	0	0	0	1	1	1
	500	0	0	0	0	1	1	1
	600	0	0	0	0	1	1	1
	700	0	0	0	0	1	1	1
	750	0	0	0	0	0	1	1
	800	0	0	0	0	0	1	1
	900	0	0	0	0	0	1	1
	1000	0	0	0	0	0	0	1
	1250	0	0	0	0	0	0	1
	1500	0	0	0	0	0	0	1
	1750	0	0	0	0	0	0	0
	2000	0	0	0	0	0	0	0

EXHIBIT 12.21

NEC Table C6 [Abridged]   Maximum Number of Conductors or Fixture Wires in Liquidtight Flexible Nonmetallic Conduit (Type LFNC-A*) *(Based on Table 1, Chapter 9)*

		CONDUCTORS						
	Conductor Size	Metric Designator (Trade Size)						
Type	(AWG/ kcmil)	12 (⅜)	16 (½)	21 (¾)	27 (1)	35 (1¼)	41 (1½)	53 (2)
TFN, TFFN	18	14	22	39	62	109	146	243
	16	10	17	29	47	83	112	185
THHN, THWN, THWN-2	14	8	13	22	35	62	83	137
	12	5	9	16	25	45	60	100
	10	3	6	10	16	28	38	63
	8	1	3	6	9	16	22	36
	6	1	2	4	6	12	16	26
	4	1	1	2	4	7	9	16
	3	1	1	1	3	6	8	13
	2	1	1	1	3	5	7	11
	1	0	1	1	1	4	5	8
	1/0	0	1	1	1	3	4	7
	2/0	0	0	1	1	2	3	6
	3/0	0	0	1	1	1	3	5
	4/0	0	0	1	1	1	2	4
	250	0	0	0	1	1	1	3
	300	0	0	0	1	1	1	3
	350	0	0	0	1	1	1	2
	400	0	0	0	0	1	1	1
	500	0	0	0	0	1	1	1
	600	0	0	0	0	1	1	1
	700	0	0	0	0	1	1	1
	750	0	0	0	0	0	1	1
	800	0	0	0	0	0	1	1
	900	0	0	0	0	0	1	1
	1000	0	0	0	0	0	0	1
XHH, XHHW, XHHW-2, ZW	14	5	9	15	24	43	58	96
	12	4	7	12	19	33	44	74
	10	3	5	9	14	24	33	55
	8	1	3	5	8	13	18	30
	6	1	1	3	5	10	13	22
	4	1	1	2	4	7	10	16
	3	1	1	1	3	6	8	14
	2	1	1	1	3	5	7	11
XHH, XHHW, XHHW-2	1	0	1	1	1	4	5	8
	1/0	0	1	1	1	3	4	7
	2/0	0	0	1	1	2	3	6
	3/0	0	0	1	1	1	3	5
	4/0	0	0	1	1	1	2	4
	250	0	0	0	1	1	1	3
	300	0	0	0	1	1	1	3
	350	0	0	0	1	1	1	2
	400	0	0	0	0	1	1	1
	500	0	0	0	0	1	1	1
	600	0	0	0	0	1	1	1
	700	0	0	0	0	1	1	1
	750	0	0	0	0	0	1	1
	800	0	0	0	0	0	1	1
	900	0	0	0	0	0	1	1

Continued

174

RACEWAYS AND CABLE TRAYS

EXHIBIT 12.21 *(continued)*

NEC Table C6 [Abridged]   Maximum Number of Conductors or Fixture Wires in Liquidtight Flexible Nonmetallic Conduit (Type LFNC-A*) *(Based on Table 1, Chapter 9)*

		CONDUCTORS						
	Conductor Size	Metric Designator (Trade Size)						
Type	(AWG/ kcmil)	12 (⅜)	16 (½)	21 (¾)	27 (1)	35 (1¼)	41 (1½)	53 (2)
XHH,	1000	0	0	0	0	0	0	1
XHHW,	1250	0	0	0	0	0	0	1
XHHW-2	1500	0	0	0	0	0	0	1
	1750	0	0	0	0	0	0	0
	2000	0	0	0	0	0	0	0

EXHIBIT 12.22

NEC Table C7 [Abridged]   Maximum Number of Conductors or Fixture Wires in Liquidtight Flexible Metal Conduit (LFMC) *(Based on Table 1, Chapter 9)*

		CONDUCTORS								
	Conductor Size	Metric Designator (Trade Size)								
Type	(AWG/ kcmil)	16 (½)	21 (¾)	27 (1)	35 (1¼)	41 (1½)	53 (2)	63 (2½)	78 (3)	91 (3½)	103 (4)
TFN,	18	23	39	63	111	144	236	–	–	–	–
TFFN	16	17	30	48	85	110	180	–	–	–	–
THHN,	14	13	22	36	63	81	133	201	308	401	523
THWN,	12	9	16	26	46	59	97	146	225	292	381
THWN-2	10	6	10	16	29	37	61	92	141	184	240
	8	3	6	9	16	21	35	53	81	106	138
	6	2	4	7	12	15	25	38	59	76	100
	4	1	2	4	7	9	15	23	36	47	61
	3	1	1	3	6	8	13	20	30	40	52
	2	1	1	3	5	7	11	17	26	33	44
	1	1	1	1	4	5	8	12	19	25	32
	1/0	1	1	1	3	4	7	10	16	21	27
	2/0	0	1	1	2	3	6	8	13	17	23
	3/0	0	1	1	1	3	5	7	11	14	19
	4/0	0	1	1	1	2	4	6	9	12	15
	250	0	0	1	1	1	3	5	7	10	12
	300	0	0	1	1	1	3	4	6	8	11
	350	0	0	1	1	1	2	3	5	7	9
	400	0	0	0	1	1	1	3	5	6	8
	500	0	0	0	1	1	1	2	4	5	7
	600	0	0	0	1	1	1	1	3	4	6
	700	0	0	0	1	1	1	1	3	4	5
	750	0	0	0	0	1	1	1	3	3	5
	800	0	0	0	0	1	1	1	2	3	4
	900	0	0	0	0	1	1	1	2	3	4
	1000	0	0	0	0	0	1	1	1	3	3

Table continues below.

EXHIBIT 12.22 *(continued)*

NEC Table C7 [Abridged]   Maximum Number of Conductors or Fixture Wires in Liquidtight Flexible Metal Conduit (LFMC) *(Based on Table 1, Chapter 9)*

		CONDUCTORS								
	Conductor Size	Metric Designator (Trade Size)								
Type	(AWG/ kcmil)	16 (½)	21 (¾)	27 (1)	35 (1¼)	41 (1½)	53 (2)	63 (2½)	78 (3)	91 (3½)	103 (4)
XHH, XHHW, XHHW-2, ZW	14	9	15	25	44	57	93	140	215	280	365
	12	7	12	19	33	43	71	108	165	215	280
	10	5	9	14	25	32	53	80	123	160	209
	8	3	5	8	14	18	29	44	68	89	116
	6	1	3	6	10	13	22	33	50	66	86
	4	1	2	4	7	9	16	24	36	48	62
	3	1	1	3	6	8	13	20	31	40	52
	2	1	1	3	5	7	11	17	26	34	44
	1	1	1	1	4	5	8	12	19	25	33
XHH, XHHW, XHHW-2	1/0	1	1	1	3	4	7	10	16	21	28
	2/0	0	1	1	2	3	6	9	13	17	23
	3/0	0	1	1	1	3	5	7	11	14	19
	4/0	0	1	1	1	2	4	6	9	12	16
	250	0	0	1	1	1	3	5	7	10	13
	300	0	0	1	1	1	3	4	6	8	11
	350	0	0	1	1	1	2	3	5	7	10
	400	0	0	0	1	1	1	3	5	6	8
	500	0	0	0	1	1	1	2	4	5	7
	600	0	0	0	1	1	1	1	3	4	6
	700	0	0	0	1	1	1	1	3	4	5
	750	0	0	0	0	1	1	1	3	3	5
	800	0	0	0	0	1	1	1	2	3	4
	900	0	0	0	0	1	1	1	2	3	4
	1000	0	0	0	0	0	1	1	1	3	3
	1250	0	0	0	0	0	1	1	1	1	3
	1500	0	0	0	0	0	0	1	1	1	2
	1750	0	0	0	0	0	0	1	1	1	2
	2000	0	0	0	0	0	0	1	1	1	2

EXHIBIT 12.23

NEC Table C8 [Abridged]   Maximum Number of Conductors or Fixture Wires in Rigid Metal Conduit (RMC) *(Based on Table 1, Chapter 9)*

		CONDUCTORS										
	Conductor Size	Metric Designator (Trade Size)										
Type	(AWG/ kcmil)	16 (½)	21 (¾)	27 (1)	35 (1¼)	41 (1½)	53 (2)	63 (2½)	78 (3)	91 (3½)	103 (4)	129 (5)	155 (6)
TFN,	18	23	40	64	111	150	248	–	–	–	–	–	–
TFFN	16	17	30	49	84	115	189	–	–	–	–	–	–
THHN, THWN, THWN-2	14	13	22	36	63	85	140	200	309	412	531	833	1202
	12	9	16	26	46	62	102	146	225	301	387	608	877
	10	6	10	17	29	39	64	92	142	189	244	383	552
	8	3	6	9	16	22	37	53	82	109	140	221	318
	6	2	4	7	12	16	27	38	59	79	101	159	230

Continued

176

RACEWAYS AND CABLE TRAYS

EXHIBIT 12.23 *(continued)*

NEC Table C8 [Abridged]   Maximum Number of Conductors or Fixture Wires in
Rigid Metal Conduit (RMC) *(Based on Table 1, Chapter 9)*

		CONDUCTORS											
	Conductor Size	Metric Designator (Trade Size)											
Type	(AWG/ kcmil)	16 (½)	21 (¾)	27 (1)	35 (1¼)	41 (1½)	53 (2)	63 (2½)	78 (3)	91 (3½)	103 (4)	129 (5)	155 (6)
THHN,	4	1	2	4	7	10	16	23	36	48	62	98	141
THWN,	3	1	1	3	6	8	14	20	31	41	53	83	120
THWN-2	2	1	1	3	5	7	11	17	26	34	44	70	100
	1	1	1	1	4	5	8	12	19	25	33	51	74
	1/0	1	1	1	3	4	7	10	16	21	27	43	63
	2/0	0	1	1	2	3	6	8	13	18	23	36	52
	3/0	0	1	1	1	3	5	7	11	15	19	30	43
	4/0	0	1	1	1	2	4	6	9	12	16	25	36
	250	0	0	1	1	1	3	5	7	10	13	20	29
	300	0	0	1	1	1	3	4	6	8	11	17	25
	350	0	0	1	1	1	2	3	5	7	10	15	22
	400	0	0	1	1	1	2	3	5	7	8	13	20
	500	0	0	0	1	1	1	2	4	5	7	11	16
	600	0	0	0	1	1	1	1	3	4	6	9	13
	700	0	0	0	1	1	1	1	3	4	5	8	11
	750	0	0	0	0	1	1	1	3	4	5	7	11
	800	0	0	0	0	1	1	1	2	3	4	7	10
	900	0	0	0	0	1	1	1	2	3	4	6	9
	1000	0	0	0	0	1	1	1	1	3	4	6	8
XHH,	14	9	15	25	44	59	98	140	216	288	370	581	839
XHHW,	12	7	12	19	33	45	75	107	165	221	284	446	644
XHHW-2,	10	5	9	14	25	34	56	80	123	164	212	332	480
ZW	8	3	5	8	14	19	31	44	68	91	118	185	267
	6	1	3	6	10	14	23	33	51	68	87	137	197
	4	1	2	4	7	10	16	24	37	49	63	99	143
	3	1	1	3	6	8	14	20	31	41	53	84	121
	2	1	1	3	5	7	12	17	26	35	45	70	101
	1	1	1	1	4	5	9	12	19	26	33	52	76
XHH,	1/0	1	1	1	3	4	7	10	16	22	28	44	64
XHHW,	2/0	0	1	1	2	3	6	9	13	18	23	37	53
XHHW-2	3/0	0	1	1	1	3	5	7	11	15	19	30	44
	4/0	0	1	1	1	2	4	6	9	12	16	25	36
	250	0	0	1	1	1	3	5	7	10	13	20	30
	300	0	0	1	1	1	3	4	6	9	11	18	25
	350	0	0	1	1	1	2	3	6	7	10	15	22
	400	0	0	1	1	1	2	3	5	7	9	14	20
	500	0	0	0	1	1	1	2	4	5	7	11	16
	600	0	0	0	1	1	1	1	3	4	6	9	13
	700	0	0	0	1	1	1	1	3	4	5	8	11
	750	0	0	0	0	1	1	1	3	4	5	7	11
	800	0	0	0	0	1	1	1	2	3	4	7	10
	900	0	0	0	0	1	1	1	2	3	4	6	9
	1000	0	0	0	0	1	1	1	1	3	4	6	8
	1250	0	0	0	0	0	1	1	1	2	3	4	6
	1500	0	0	0	0	0	1	1	1	1	2	4	5
	1750	0	0	0	0	0	0	1	1	1	1	3	5
	2000	0	0	0	0	0	0	1	1	1	1	3	4

EXHIBIT 12.24

NEC Table C9 [Abridged]   Maximum Number of Conductors or Fixture Wires in Rigid PVC Conduit, Schedule 80 *(Based on Table 1, Chapter 9)*

		CONDUCTORS										
	Conductor Size	Metric Designator (Trade Size)										
Type	(AWG/ kcmil)	16 (½)	21 (¾)	27 (1)	35 (1¼)	41 (1½)	53 (2)	63 (2½)	78 (3)	91 (3½)	103 (4)	129 (5)	155 (6)
TFN,	18	16	29	50	90	124	209	–	–	–	–	–	–
TFFN	16	12	22	38	68	95	159	–	–	–	–	–	–
THHN,	14	9	17	28	51	70	118	170	265	358	464	736	1055
THWN,	12	6	12	20	37	51	86	124	193	261	338	537	770
THWN-2	10	4	7	13	23	32	54	78	122	164	213	338	485
	8	2	4	7	13	18	31	45	70	95	123	195	279
	6	1	3	5	9	13	22	32	51	68	89	141	202
	4	1	1	3	6	8	14	20	31	42	54	86	124
	3	1	1	3	5	7	12	17	26	35	46	73	105
	2	1	1	2	4	6	10	14	22	30	39	61	88
	1	0	1	1	3	4	7	10	16	22	29	45	65
	1/0	0	1	1	2	3	6	9	14	18	24	38	55
	2/0	0	1	1	1	3	5	7	11	15	20	32	46
	3/0	0	1	1	1	2	4	6	9	13	17	26	38
	4/0	0	0	1	1	1	3	5	8	10	14	22	31
	250	0	0	1	1	1	3	4	6	8	11	18	25
	300	0	0	0	1	1	2	3	5	7	9	15	22
	350	0	0	0	1	1	1	3	5	6	8	13	19
	400	0	0	0	1	1	1	3	4	6	7	12	17
	500	0	0	0	1	1	1	2	3	5	6	10	14
	600	0	0	0	0	1	1	1	3	4	5	8	12
	700	0	0	0	0	1	1	1	2	3	4	7	10
	750	0	0	0	0	1	1	1	2	3	4	7	9
	800	0	0	0	0	0	1	1	2	3	4	6	9
	900	0	0	0	0	0	1	1	1	3	3	6	8
	1000	0	0	0	0	0	1	1	1	2	3	5	7
XHH,	14	6	11	20	35	49	82	118	185	250	324	514	736
XHHW,	12	5	9	15	27	38	63	91	142	192	248	394	565
XHHW-2,	10	3	6	11	20	28	47	67	106	143	185	294	421
ZW	8	1	3	6	11	15	26	37	59	79	103	163	234
	6	1	2	4	8	11	19	28	43	59	76	121	173
	4	1	1	3	6	8	14	20	31	42	55	87	125
	3	1	1	3	5	7	12	17	26	36	47	74	106
	2	1	1	2	4	6	10	14	22	30	39	62	89
XHH,	1	0	1	1	3	4	7	10	16	22	29	46	66
XHHW,	1/0	0	1	1	2	3	6	9	14	19	24	39	56
XHHW-2	2/0	0	1	1	1	3	5	7	11	16	20	32	46
	3/0	0	1	1	1	2	4	6	9	13	17	27	38
	4/0	0	0	1	1	1	3	5	8	11	14	22	32
	250	0	0	1	1	1	3	4	6	9	11	18	26
	300	0	0	1	1	1	2	3	5	7	10	15	22
	350	0	0	0	1	1	1	3	5	6	8	14	20
	400	0	0	0	1	1	1	3	4	6	7	12	17
	500	0	0	0	1	1	1	2	3	5	6	10	14

Continued

178

RACEWAYS AND CABLE TRAYS

EXHIBIT 12.24 *(continued)*

NEC Table C9 [Abridged]   Maximum Number of Conductors or Fixture
Wires in Rigid PVC Conduit, Schedule 80 *(Based on Table 1, Chapter 9)*

		CONDUCTORS										
	Conductor Size	Metric Designator (Trade Size)										
Type	(AWG/ kcmil)	16 (½)	21 (¾)	27 (1)	35 (1¼)	41 (1½)	53 (2)	63 (2½)	78 (3)	91 (3½)	103 (4)	129 (5)	155 (6)
XHH,	600	0	0	0	0	1	1	1	3	4	5	8	11
XHHW,	700	0	0	0	0	1	1	1	2	3	4	7	10
XHHW-2	750	0	0	0	0	1	1	1	2	3	4	6	9
	800	0	0	0	0	1	1	1	1	3	4	6	9
	900	0	0	0	0	0	1	1	—	3	3	5	8
	1000	0	0	0	0	0	1	1	1	2	3	5	7
	1250	0	0	0	0	0	1	1	1	1	2	4	6
	1500	0	0	0	0	0	0	1	1	1	1	3	5
	1750	0	0	0	0	0	0	1	1	1	1	3	4
	2000	0	0	0	0	0	0	1	1	1	1	2	4

EXHIBIT 12.25

NEC Table C10 [Abridged]   Maximum Number of Conductors or Fixture Wires in
Rigid PVC Conduit, Schedule 40 and HDPE Conduit *(Based on Table 1, Chapter 9)*

		CONDUCTORS										
	Conductor Size	Metric Designator (Trade Size)										
Type	(AWG/ kcmil)	16 (½)	21 (¾)	27 (1)	35 (1¼)	41 (1½)	53 (2)	63 (2½)	78 (3)	91 (3½)	103 (4)	129 (5)	155 (6)
TFN,	18	20	37	60	105	144	239	–	–	–	–	–	–
TFFN	16	16	28	46	80	110	183	–	–	–	–	–	–
THHN,	14	11	21	34	60	82	135	193	299	401	517	815	1178
THWN,	12	8	15	25	43	59	99	141	218	293	377	594	859
THWN-2	10	5	9	15	27	37	62	89	137	184	238	374	541
	8	3	5	9	16	21	36	51	79	106	137	216	312
	6	1	4	6	11	15	26	37	57	77	99	156	225
	4	1	2	4	7	9	16	22	35	47	61	96	138
	3	1	1	3	6	8	13	19	30	40	51	81	117
	2	1	1	3	5	7	11	16	25	33	43	68	98
	1	1	1	1	3	5	8	12	18	25	32	50	73
	1/0	1	1	1	3	4	7	10	15	21	27	42	61
	2/0	0	1	1	2	3	6	8	13	17	22	35	51
	3/0	0	1	1	1	3	5	7	11	14	18	29	42
	4/0	0	1	1	1	2	4	6	9	12	15	24	35
	250	0	0	1	1	1	3	4	7	10	12	20	28
	300	0	0	1	1	1	3	4	6	8	11	17	24

Table continues below.

EXHIBIT 12.25 *(continued)*

NEC Table C10 [Abridged] Maximum Number of Conductors or Fixture Wires in Rigid PVC Conduit, Schedule 40 and HDPE Conduit *(Based on Table 1, Chapter 9)*

		CONDUCTORS										
	Conductor Size				Metric Designator (Trade Size)							
Type	(AWG/kcmil)	16 (½)	21 (¾)	27 (1)	35 (1¼)	41 (1½)	53 (2)	63 (2½)	78 (3)	91 (3½)	103 (4)	129 (5)	155 (6)
THHN, THWN, THWN-2	350	0	0	1	1	1	2	3	5	7	9	15	21
	400	0	0	0	1	1	1	3	5	6	8	13	19
	500	0	0	0	1	1	1	2	4	5	7	11	16
	600	0	0	0	1	1	1	1	3	4	5	9	13
	700	0	0	0	0	1	1	1	3	4	5	8	11
	750	0	0	0	0	1	1	1	2	3	4	7	11
	800	0	0	0	0	1	1	1	2	3	4	7	10
	900	0	0	0	0	1	1	1	2	3	4	6	9
	1000	0	0	0	0	0	1	1	1	3	3	6	8
XHH, XHHW, XHHW-2, ZW	14	8	14	24	42	57	94	135	209	280	361	568	822
	12	6	11	18	32	44	72	103	160	215	277	436	631
	10	4	8	13	24	32	54	77	119	160	206	325	470
	8	2	4	7	13	18	30	43	66	89	115	181	261
	6	1	3	5	10	13	22	32	49	66	85	134	193
	4	1	2	4	7	9	16	23	35	48	61	97	140
	3	1	1	3	6	8	13	19	30	40	52	82	118
	2	1	1	3	5	7	11	16	25	34	44	69	99
XHH, XHHW, XHHW-2	1	1	1	1	3	5	8	12	19	25	32	51	74
	1/0	1	1	1	3	4	7	10	16	21	27	43	62
	2/0	0	1	1	2	3	6	8	13	17	23	36	52
	3/0	0	1	1	1	3	5	7	11	14	19	30	43
	4/0	0	1	1	1	2	4	6	9	12	15	24	35
	250	0	0	1	1	1	3	5	7	10	13	20	29
	300	0	0	1	1	1	3	4	6	8	11	17	25
	350	0	0	1	1	1	2	3	5	7	9	15	22
	400	0	0	0	1	1	1	3	5	6	8	13	19
	500	0	0	0	1	1	1	2	4	5	7	11	16
	600	0	0	0	1	1	1	1	3	4	5	9	13
	700	0	0	0	0	1	1	1	3	4	5	8	11
	750	0	0	0	0	1	1	1	2	3	4	7	11
	800	0	0	0	0	1	1	1	2	3	4	7	10
	900	0	0	0	0	1	1	1	2	3	4	6	9
	1000	0	0	0	0	0	1	1	1	3	3	6	8
	1250	0	0	0	0	0	1	1	1	1	3	4	6
	1500	0	0	0	0	0	1	1	1	1	2	4	5
	1750	0	0	0	0	0	0	1	1	1	1	3	5
	2000	0	0	0	0	0	0	1	1	1	1	3	4

CHAPTER 13
CONDUIT BENDING

Chapter 13 covers different methods for bending conduit, including bending 90 degree stubs, back-to-back bends, offset bends, 3-point saddles, segment bends, and concentric bends. The chapter concludes with a table of natural trigonometric functions.

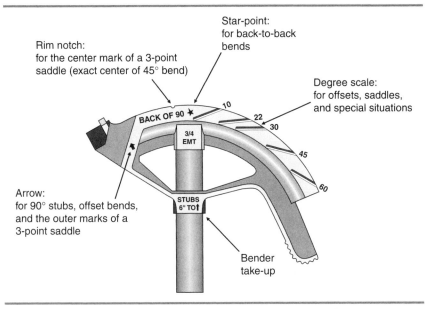

Figure 13.1   Typical markings on a conduit bender.
Bender courtesy of Ideal Industries.

BENDING 90 DEGREE STUBS

A 90 degree stub is a conduit that is bent at a 90 degree angle. It is also referred to as a ninety, stub, stub-up, and 90 degree bend. This is the most common type of bend. Since the bender draws the conduit up, the length between the mark and the floor must be deducted from the desired length. The length between the mark and

182

CONDUIT BENDING

the floor is called "take-up," as shown in Figure 13.2. As the conduit size increases, the take-up increases. The take-up length is stamped into most hand benders.

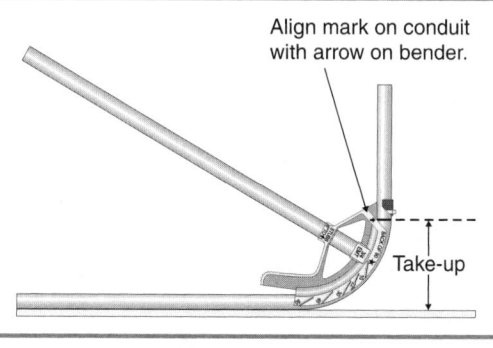

Align mark on conduit with arrow on bender.

Take-up

Figure 13.2   Take-up in 90 degree stub.

See Table 13.1 for take-up lengths for common 90 degree conduit bends. The steps for creating a 90 degree stub are as follows:

Step 1: Determine the length required for the 90 degree bend.

Step 2: Deduct the take-up length and mark the conduit.

Step 3: Position the bender so the arrow lines up with the mark and bend the conduit to a 90 degree angle.

Example

A 90 degree stub is needed in a 3/4-in. electrical metallic tubing (EMT). The required length is 10 in. Since the take-up for 3/4-in. EMT is 6 in., deduct 6 from 10 and mark the conduit 4 in. from the end (see Figure 13.3). Place the bender over the conduit, as shown in Figure 13.4, and align the arrow with the mark on the conduit. Bend the conduit to a 90 degree angle (see Figure 13.5).

There is an alternative method for bending 90 degree stubs. Instead of deducting the take-up length, place the tape measure (or folding rule) beyond the end of the conduit by the amount of the take-up length (see Figure 13.6). Either the conduit can be marked, or the arrow on the bender can be lined up with the required

Table 13.1   90 Degree Stubs

Bender	Take-Up
1/2-in. EMT	5 in.
3/4-in. EMT (1/2 in. rigid)	6 in.
1-in. EMT (3/4 in. rigid)	8 in.
1-1/4-in. EMT (1 in. rigid)	11 in.

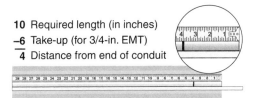

10 Required length (in inches)
−6 Take-up (for 3/4-in. EMT)
―――
 4 Distance from end of conduit

Mark the conduit 4 in. from the end.

Figure 13.3   Determining the required length.

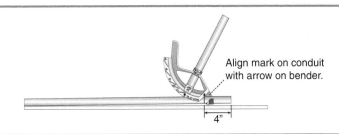

Align mark on conduit with arrow on bender.

Figure 13.4   Aligning the bender.

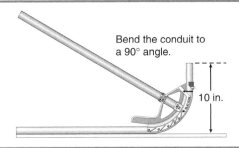

Bend the conduit to a 90° angle.

10 in.

Figure 13.5   Bending the conduit to 90 degrees.

Extend the measuring device beyond the conduit by the amount of the take-up.

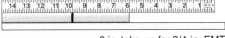

6 in. take-up for 3/4-in. EMT

Figure 13.6   Allowing for take-up length.

CONDUIT BENDING

184

CONDUIT BENDING

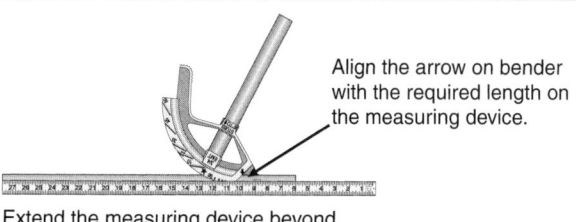

Align the arrow on bender with the required length on the measuring device.

Extend the measuring device beyond the conduit by the amount of the take-up.

Figure 13.7   Aligning the bender.

length on the tape measure (see Figure 13.7). For example, the length needed for a 90 degree stub in 3/4-in. EMT is 9 in. Since the take-up for 3/4-in. EMT is 6 in., place the tape measure 6 in. beyond the end of the conduit. Position the bender so the arrow lines up with the mark and bend the conduit (see Figure 13.5).

BACK-TO-BACK BENDS

Although a back-to-back bend results in a U shape in a single length of conduit, this bending method can be employed in many different applications. This type of bend is useful when installing a run of conduit and a 90 degree angle is needed. The marking on the bender for this bend is the back-to-back symbol. The back-to-back

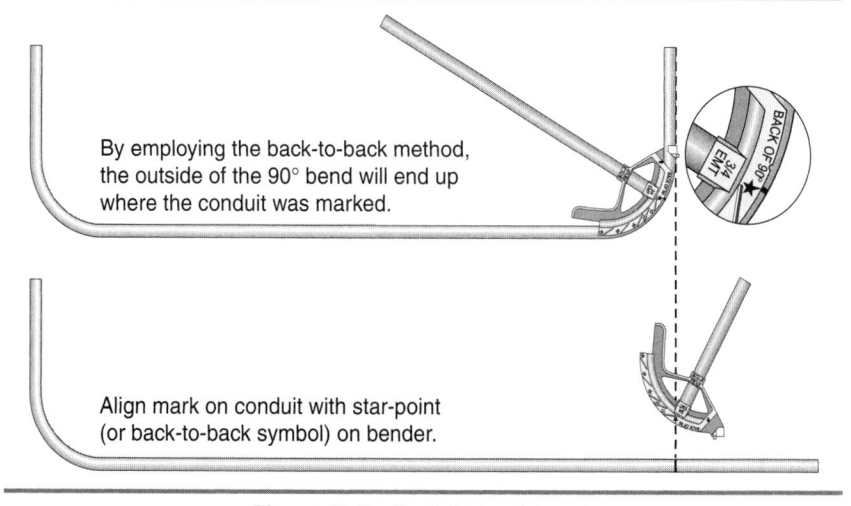

By employing the back-to-back method, the outside of the 90° bend will end up where the conduit was marked.

Align mark on conduit with star-point (or back-to-back symbol) on bender.

Figure 13.8   Back-to-back bend.

symbol can be a star-point, line, letter, or some other symbol (see detail in Figure 13.8). By employing the back-to-back method, the outside of the 90 degree bend will end up where the conduit was marked.

The steps for creating a back-to-back bend are as follows:

Step 1: Determine the length required from a point on the conduit to the back of the 90 degree bend and mark the conduit.

Step 2: Position the bender so the star-point (or back-to-back symbol) lines up with the mark on the conduit and bend it to a 90 degree angle.

Example

As a run of conduit is being installed, a 90 degree change in direction is required. The distance to the back of the 90 degree angle is 50 in. Use the back-to-back method for this installation. First, mark the conduit at 50 in., as shown in Figure 13.9. Next, position the bender so the star-point lines up with the mark on the conduit. Finally, bend a 90 degree angle and install the conduit (see Figure 13.10).

OFFSET BENDS

An offset bend is two bends in opposite directions. An offset bend is employed to change the plane (level) of the conduit (see Figure 13.11). While offset bends can be any angle, the most common include 10, 22-1/2, 30, 45, and 60 degrees. From these angles, the most common is 30 degrees. The angle of the first bend should be the same as the angle of the second bend. For example, if the angle of the first bend is 45 degrees, the angle of the second bend should also be 45 degrees.

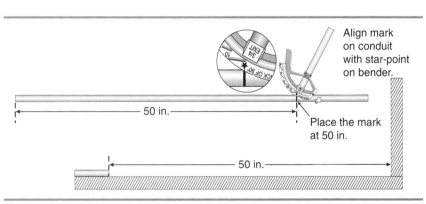

Figure 13.9   Marking the conduit at 50 in. and aligning with star-point symbol.

CONDUIT BENDING

185

186

CONDUIT BENDING

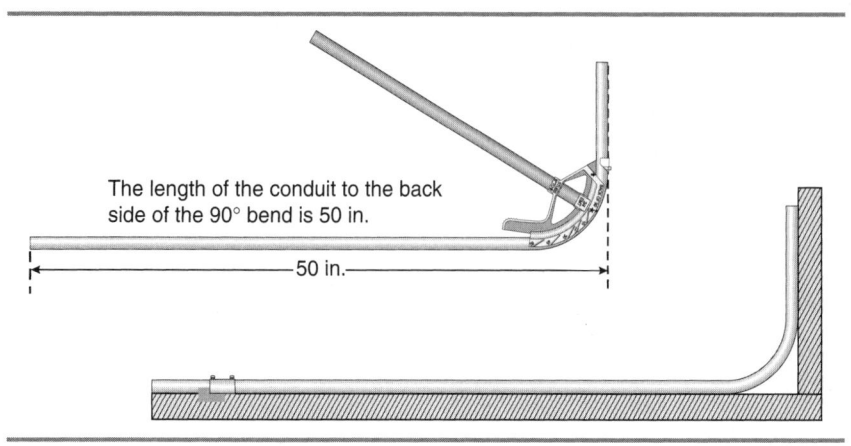

The length of the conduit to the back side of the 90° bend is 50 in.

50 in.

Figure 13.10   Bending the conduit.

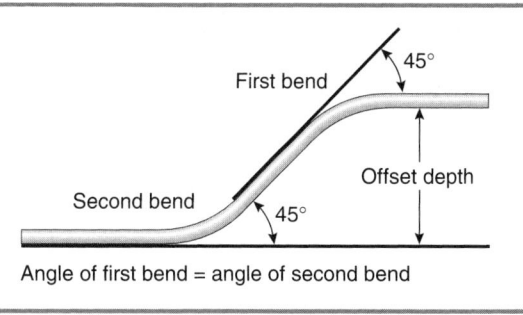

45°

First bend

Offset depth

Second bend

45°

Angle of first bend = angle of second bend

Figure 13.11   Offset bend.

Before bending an offset, choose the offset angle (see Figure 13.12). Shallow bends make wire pulling easier, but require more space. Steeper bends make wire pulling more difficult, but require less space. The shrink per inch of offset depth is also greater with steeper bends.

The distance between bends depends upon the offset depth and the offset angle. Each offset angle has a different multiplier. The offset depth × the multiplier (for the offset angle) will provide the distance between bends (see Table 13.2).

The length of the conduit (along the plane) will shorten because of the bends. This is called *conduit shrink*. Conduit shrink per inch increases as the degree of offset angle increases. For example, while 10 degree offset angles shrink 1/16 in. per inch of offset depth, 60 degree angles shrink 1/2 in. for each inch of offset depth. The shrink per inch multiplied by the offset depth will provide the total length of conduit shrink. Table 13.3 shows conduit shrink for some common offset angles.

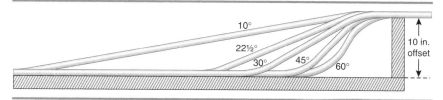

Figure 13.12 Offset angles.

Conduit shrink should be considered when installing conduit toward an obstruction. Measure to the obstruction, add to that measurement the amount of conduit shrink, and mark the conduit. Deduct the distance between bends and mark the conduit again. Bend and install the conduit. Conduit shrink can be ignored when installing conduit away from an obstruction.

The steps for creating a back-to-back bend are as follows:

Step 1: Determine the offset depth and choose an offset angle.
Step 2: Measure to the obstruction.
Step 3: Add the conduit shrink to the measurement and mark the conduit.
Step 4: Determine the distance between bends and mark the conduit again.
Step 5: Align arrow on bender with first mark and make first bend.

Table 13.2 Distance Between Bends

Offset Angle	Multiplier
10° × 10°	5.8
22-1/2° × 22-1/2°	2.6
30° × 30°	2
45° × 45°	1.4
60° × 60°	1.2

Offset depth × multiplier = distance between bends

Table 13.3 Conduit Shrink

Offset Angle	Per Inch of Offset Depth
10° × 10°	1/16 in.
22-1/2° × 22-1/2°	3/16 in.
30° × 30°	1/4 in.
45° × 45°	3/8 in.
60° × 60°	1/2 in.

CONDUIT BENDING

188

CONDUIT BENDING

Step 6: Flip conduit 90 degrees, slide the bender down, and align arrow with second mark. Bend the offsetting angle.

Example

An offset is needed to clear an obstacle. The distance to the obstruction is 50 in., and the offset depth is 8 in. The offset angle selected for this installation is 30 degrees. The conduit shrink will be 2 in. (1/4 × 8). Since the distance to the obstruction is 50 inches, mark the conduit at 52 in. (50 + 2) (see Figure 13.13). The distance between bends is 16 in. (2 × 8). Make the second mark on the conduit 16 in. before the first mark (see Figure 13.14). Bend and install the conduit (see Figure 13.15 on p. 238).

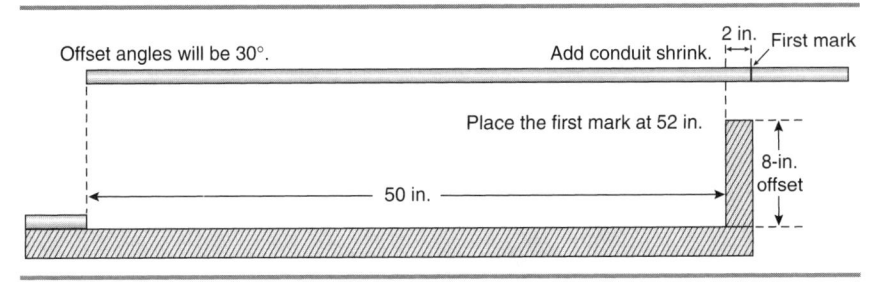

Figure 13.13   The first mark at 52 in.

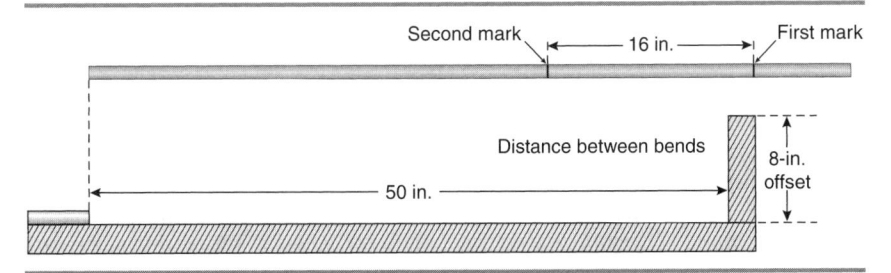

Figure 13.14   The second mark at 16 in.

Tables for Offset Bends

Tables 13.4 through 13.8 contain useful measurements for making offset bends. Determine the offset depth and choose an offset angle. The middle column shows the distance (in inches) between bends, the right column shows the amount of conduit shrink.

Table 13.4 10 Degree Angle of Offset Bends

Offset Depth in Inches	Distance Between Bends (Depth x 5.8)	Conduit Shrink (1/16 in. per Inch of Offset Depth)
3/8	2-1/8	—
1/2	2-7/8	1/16
3/4	4-3/8	1/16
1	5-3/4	1/16
1-1/2	8-3/4	1/8
2	11-5/8	1/8
2-1/2	14-1/2	3/16
3	17-3/8	3/16
3-1/2	20-1/4	1/4
4	23-1/4	1/4
4-1/2	26-1/8	5/16
5	29	5/16
5-1/2	31-7/8	3/8
6	34-3/4	3/8
6-1/2	37-3/4	7/16
7	40-5/8	7/16
7-1/2	43-1/2	1/2
8	46-3/8	1/2
8-1/2	49-1/4	9/16
9	52-1/4	9/16
9-1/2	55-1/8	5/8
10	58	5/8
10-1/2	60-7/8	11/16
11	63-3/4	11/16
11-1/2	66-3/4	3/4
12	69-5/8	3/4

Note: All dimensions are shown in inches.
Note: Distance between bends is rounded to the nearest eighth.
Note: Conduit shrink is rounded to the nearest sixteenth.
Note: The conduit will shorten 1/16 in. per inch of offset depth.

Table 13.5 22-1/2 Degree Angle of Offset Bends

Offset Depth in Inches	Distance Between Bends (Depth x 2.6)	Conduit Shrink (3/16 in. per Inch of Offset Depth)
1	2-5/8	3/16
1-1/2	3-7/8	5/16
2	5-1/4	3/8
2-1/2	6-1/2	1/2
3	7-3/4	9/16

Continued

CONDUIT BENDING

CONDUIT BENDING

Table 13.5   22-1/2 Degree Angle of Offset Bends *(continued)*

Offset Depth in Inches	Distance Between Bends (Depth x 2.6)	Conduit Shrink (3/16 in. per Inch of Offset Depth)
3-1/2	9-1/8	11/16
4	10-3/8	3/4
4-1/2	11-3/4	7/8
5	13	15/16
5-1/2	14-1/4	1-1/16
6	15-5/8	1-1/8
6-1/2	16-7/8	1-1/4
7	18-1/4	1-5/16
7-1/2	19-1/2	1-7/16
8	20-3/4	1-1/2
8-1/2	22-1/8	1-5/8
9	23-3/8	1-11/16
9-1/2	24-3/4	1-13/16
10	26	1-7/8
10-1/2	27-1/4	2
11	28-5/8	2-1/16
11-1/2	29-7/8	2-3/16
12	31-1/4	2-1/4
12-1/2	32-1/2	2-3/8
13	33-3/4	2-7/16
13-1/2	35-1/8	2-9/16
14	36-3/8	2-5/8
14-1/2	37-3/4	2-3/4
15	39	2-13/16
15-1/2	40-1/4	2-15/16
16	41-5/8	3
16-1/2	42-7/8	3-1/8
17	44-1/4	3-3/16
17-1/2	45-1/2	3-5/16
18	46-3/4	3-3/8
18-1/2	48-1/8	3-1/2
19	49-3/8	3-9/16
19-1/2	50-3/4	3-11/16
20	52	3-3/4
20-1/2	53-1/4	3-7/8
21	54-5/8	3-15/16
21-1/2	55-7/8	4-1/16
22	57-1/4	4-1/8
22-1/2	58-1/2	4-1/4
23	59-3/4	4-5/16

Table continues below.

Table 13.5 22-1/2 Degree Angle of Offset Bends *(continued)*

Offset Depth in Inches	Distance Between Bends (Depth x 2.6)	Conduit Shrink (3/16 in. per Inch of Offset Depth)
23-1/2	61-1/8	4-7/16
24	62-3/8	4-1/2

Note: All dimensions are shown in inches.
Note: Distance between bends is rounded to the nearest eighth.
Note: Conduit shrink is rounded to the nearest sixteenth.
Note: The conduit will shorten 3/16 in. per inch of offset depth.

Table 13.6 30 Degree Angle of Offset Bends

Offset Depth in Inches	Distance Between Bends (Depth x 2)	Conduit Shrink (1/4 in. per Inch of Offset Depth)
3	6	3/4
3-1/2	7	7/8
4	8	1
4-1/2	9	1-1/8
5	10	1-1/4
5-1/2	11	1-3/8
6	12	1-1/2
6-1/2	13	1-5/8
7	14	1-3/4
7-1/2	15	1-7/8
8	16	2
8-1/2	17	2-1/8
9	18	2-1/4
9-1/2	19	2-3/8
10	20	2-1/2
10-1/2	21	2-5/8
11	22	2-3/4
11-1/2	23	2-7/8
12	24	3
12-1/2	25	3-1/8
13	26	3-1/4
13-1/2	27	3-3/8
14	28	3-1/2
14-1/2	29	3-5/8
15	30	3-3/4
15-1/2	31	3-7/8
16	32	4
16-1/2	33	4-1/8

Continued

CONDUIT BENDING

Table 13.6   30 Degree Angle of Offset Bends   *(continued)*

Offset Depth in Inches	Distance Between Bends (Depth x 2)	Conduit Shrink (1/4 in. per Inch of Offset Depth)
17	34	4-1/4
17-1/2	35	4-3/8
18	36	4-1/2
18-1/2	37	4-5/8
19	38	4-3/4
19-1/2	39	4-7/8
20	40	5
20-1/2	41	5-1/8
21	42	5-1/4
21-1/2	43	5-3/8
22	44	5-1/2
22-1/2	45	5-5/8
23	46	5-3/4
23-1/2	47	5-7/8
24	48	6
25	50	6-1/4
26	52	6-1/2
27	54	6-3/4
28	56	7
29	58	7-1/4
30	60	7-1/2
31	62	7-3/4
32	64	8
33	66	8-1/4
34	68	8-1/2
35	70	8-3/4
36	72	9

Note: All dimensions are shown in inches.

Note: The conduit will shorten 1/4 in. per inch of offset depth.

Note: With most hand benders, a straight up handle (vertical to the floor) indicates a 30 degree bend.

Table 13.7 45 Degree Angle of Offset Bends

Offset Depth in Inches	Distance Between Bends (Depth x 1.4)	Conduit Shrink (3/8 in. per Inch of Offset Depth)
4	5-5/8	1-1/2
4-1/2	6-1/4	1-11/16
5	7	1-7/8
5-1/2	7-3/4	2-1/16
6	8-3/8	2-1/4
6-1/2	9-1/8	2-7/16
7	9-3/4	2-5/8
7-1/2	10-1/2	2-13/16
8	11-1/4	3
8-1/2	11-7/8	3-3/16
9	12-5/8	3-3/8
9-1/2	13-1/4	3-9/16
10	14	3-3/4
10-1/2	14-3/4	3-15/16
11	15-3/8	4-1/8
11-1/2	16-1/8	4-5/16
12	16-3/4	4-1/2
12-1/2	17-1/2	4-11/16
13	18-1/4	4-7/8
13-1/2	18-7/8	5-1/16
14	19-5/8	5-1/4
14-1/2	20-1/4	5-7/16
15	21	5-5/8
15-1/2	21-3/4	5-13/16
16	22-3/8	6
16-1/2	23-1/8	6-3/16
17	23-3/4	6-3/8
17-1/2	24-1/2	6-9/16
18	25-1/4	6-3/4
18-1/2	25-7/8	6-15/16
19	26-5/8	7-1/8
19-1/2	27-1/4	7-5/16
20	28	7-1/2
20-1/2	28-3/4	7-11/16
21	29-3/8	7-7/8
21-1/2	30-1/8	8-1/16
22	30-3/4	8-1/4
22-1/2	31-1/2	8-7/16
23	32-1/4	8-5/8
23-1/2	32-7/8	8-13/16
24	33-5/8	9

Continued

CONDUIT BENDING

194

CONDUIT BENDING

Table 13.7   45 Degree Angle of Offset Bends   *(continued)*

Offset Depth in Inches	Distance Between Bends (Depth x 1.4)	Conduit Shrink (3/8 in. per Inch of Offset Depth)
25	35	9-3/8
26	36-3/8	9-3/4
27	37-3/4	10-1/8
28	39-1/4	10-1/2
29	40-5/8	10-7/8
30	42	11-1/4
31	43-3/8	11-5/8
32	44-3/4	12
33	46-1/4	12-3/8
34	47-5/8	12-3/4
35	49	13-1/8
36	50-3/8	13-1/2
37	51-3/4	13-7/8
38	53-1/4	14-1/4
39	54-5/8	14-5/8
40	56	15
41	57-3/8	15-3/8
42	58-3/4	15-3/4
43	60-1/4	16-1/8
44	61-5/8	16-1/2
45	63	16-7/8
46	64-3/8	17-1/4
47	65-3/4	17-5/8
48	67-1/4	18

Note: All dimensions are shown in inches.

Note: Distance between bends is rounded to the nearest eighth.

Note: The conduit will shorten 3/8 in. per inch of offset depth.

Table 13.8   60 Degree Angle of Offset Bends

Offset Depth in Inches	Distance Between Bends (Depth x 1.2)	Conduit Shrink (1/2 in. per Inch of Offset Depth)
5	6	2-1/2
5-1/2	6-5/8	2-3/4
6	7-1/4	3
6-1/2	7-3/4	3-1/4
7	8-3/8	3-1/2
7-1/2	9	3-3/4
8	9-5/8	4

Table continues below.

Table 13.8 60 Degree Angle of Offset Bends *(continued)*

Offset Depth in Inches	Distance Between Bends (Depth x 1.2)	Conduit Shrink (1/2 in. per Inch of Offset Depth)
8-1/2	10-1/4	4-1/4
9	10-3/4	4-1/2
9-1/2	11-3/8	4-3/4
10	12	5
10-1/2	12-5/8	5-1/4
11	13-1/4	5-1/2
11-1/2	13-3/4	5-3/4
12	14-3/8	6
12-1/2	15	6-1/4
13	15-5/8	6-1/2
13-1/2	16-1/4	6-3/4
14	16-3/4	7
14-1/2	17-3/8	7-1/4
15	18	7-1/2
15-1/2	18-5/8	7-3/4
16	19-1/4	8
16-1/2	19-3/4	8-1/4
17	20-3/8	8-1/2
17-1/2	21	8-3/4
18	21-5/8	9
18-1/2	22-1/4	9-1/4
19	22-3/4	9-1/2
19-1/2	23-3/8	9-3/4
20	24	10
20-1/2	24-5/8	10-1/4
21	25-1/4	10-1/2
21-1/2	25-3/4	10-3/4
22	26-3/8	11
22-1/2	27	11-1/4
23	27-5/8	11-1/2
23-1/2	28-1/4	11-3/4
24	28-3/4	12
25	30	12-1/2
26	31-1/4	13
27	32-3/8	13-1/2
28	33-5/8	14
29	34-3/4	14-1/2
30	36	15
31	37-1/4	15-1/2
32	38-3/8	16
33	39-5/8	16-1/2
34	40-3/4	17
35	42	17-1/2
36	43-1/4	18

Continued

196

CONDUIT BENDING

Table 13.8   60 Degree Angle of Offset Bends *(continued)*

Offset Depth in Inches	Distance Between Bends (Depth x 1.2)	Conduit Shrink (1/2 in. per Inch of Offset Depth)
37	44-3/8	18-1/2
38	45-5/8	19
39	46-3/4	19-1/2
40	48	20
41	49-1/4	20-1/2
42	50-3/8	21
43	51-5/8	21-1/2
44	52-3/4	22
45	54	22-1/2
46	55-1/4	23
47	56-3/8	23-1/2
48	57-5/8	24
49	58-3/4	24-1/2
50	60	25
51	61-1/4	25-1/2
52	62-3/8	26
53	63-5/8	26-1/2
54	64-3/4	27
55	66	27-1/2
56	67-1/4	28
57	68-3/8	28-1/2
58	69-5/8	29
59	70-3/4	29-1/2
60	72	30

Note: All dimensions are shown in inches.
Note: Distance between bends is rounded to the nearest eighth.
Note: The conduit will shorten 1/2 in. per inch of offset depth.

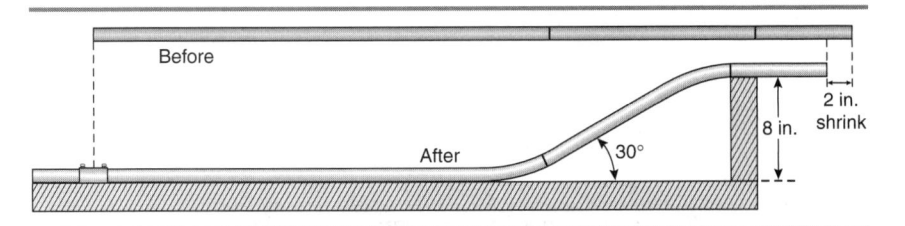

Figure 13.15   The conduit bend.

THREE-POINT (OR 3-BEND) SADDLES

Saddles are useful when the conduit needs to offset around an obstruction and then return to the same plane. While 3-point (or 3-bend) saddles can clear narrow obstacles, 4-point (or 4-bend) saddles can clear wide obstacles. Three-point saddles are useful when encountering a single pipe or conduit. Three-point saddles can be made with many combinations of angles, providing the angle of each outside bend is half the angle of the bend in the center. See Figure 13.16 for an example of a 3-point saddle.

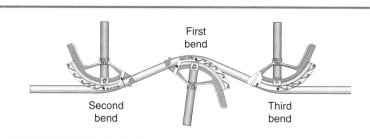

Figure 13.16   Three-point saddle.

The most common 3-point saddle is a 45 degree center bend and two 22-1/2 degree outer bends (see Figure 13.17). Using this combination, the multiplier to find the distance from the center bend to each of the outside bends is 2-1/2 times the height needed to clear the obstacle. Conduit shrink must also be considered

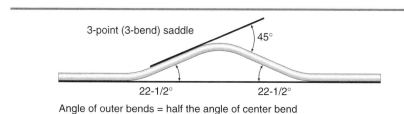

Figure 13.17   Three-point saddle with one 45 degree bend and two 22-1/2 degree bends.

CONDUIT BENDING

when bending a 3-point saddle. The conduit will shrink 3/16 in. per inch of saddle depth to the center marking. (Double this number for the total amount of conduit shrink.) Measure to the center of the obstruction; add to that measurement the amount of conduit shrink, and mark the conduit. Multiply the saddle depth by 2-1/2 and add the outside marks. Align the rim notch on the bender with the center mark and make the center bend. Since the rim notch is the exact center, it doesn't matter which direction the bender is facing. Other than the direction of the bender, the last two bends are the same.

The steps for creating a 3-point saddle bend are as follows:

Step 1: Determine the depth of the saddle.

Step 2: Measure to the center of the obstruction.

Step 3: Add the conduit shrink to the measurement and mark the conduit.

Step 4: Determine the distance from the center to the outer mark and mark the conduit on each side of the center marking.

Step 5: Align the rim notch on the bender with the center mark and bend a 45 degree angle.

Step 6: Align the arrow on the bender with one of the outside marks and bend a 22-1/2 degree angle.

Step 7: Reverse the bender, align the arrow with the other outside mark, and bend another 22-1/2 degree angle.

Example

A saddle is needed to clear a pipe. The distance to the center of the pipe is 20 in., and the depth needed for the saddle is 4 inches. Since the amount of conduit shrink to the center mark will be 3/4 inch (3/16 × 4), make the center mark at 20-3/4 in. (see Figure 13.18). The distance from the center to each outside mark is 10 in. (2-1/2 × 10), as shown in Figure 13.19. Align the rim notch on the bender with the center mark and bend a 45 degree angle (see Figure 13.20). Align the arrow on the bender with one of the outside marks and bend a 22-1/2 degree angle (see Figure 13.21). Reverse the bender and align the arrow with the other outside mark (see Figure 13.22). Bend another 22-1/2 degree angle and install the conduit (see Figure 13.23).

Table 13.9 contains useful measurements for making 3-point saddle bends.

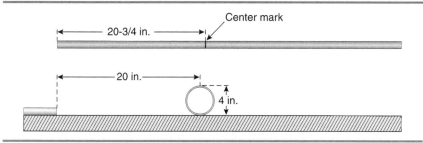

Figure 13.18 Marking the center.

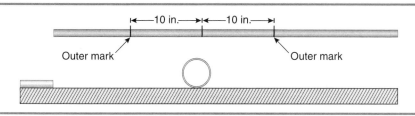

Figure 13.19 Distance from center.

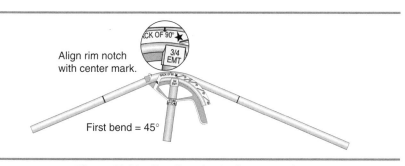

Figure 13.20 Bending the 45 degree angle.

CONDUIT BENDING

CONDUIT BENDING

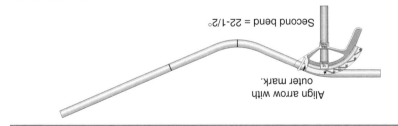

Figure 13.21 Bending the first 22-1/2 degree angle.

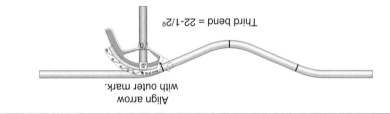

Figure 13.22 Aligning for and bending the second 22-1/2 degree angle.

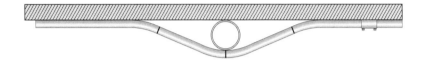

Figure 13.23 Installing the conduit.

Table 13.9 Three-Point Saddle Bends

Saddle Depth in Inches	Distance Between Bends (Depth x 2.5)	Conduit Shrink (3/16 in. per Inch of Saddle Depth)	Saddle Depth in Inches	Distance Between Bends (Depth x 2.5)	Conduit Shrink (3/16 in. per Inch of Saddle Depth)
1	2-1/2	3/16	12-1/2	31-1/4	2-3/8
1-1/2	3-3/4	5/16	13	32-1/2	2-7/16
2	5	3/8	13-1/2	33-3/4	2-9/16
2-1/2	6-1/4	1/2	14	35	2-5/8
3	7-1/2	9/16	14-1/2	36-1/4	2-3/4
3-1/2	8-3/4	11/16	15	37-1/2	2-13/16
4	10	3/4	15-1/2	38-3/4	2-15/16
4-1/2	11-1/4	7/8	16	40	3
5	12-1/2	15/16	16-1/2	41-1/4	3-1/8
5-1/2	13-3/4	1-1/16	17	42-1/2	3-3/16
6	15	1-1/8	17-1/2	43-3/4	3-5/16
6-1/2	16-1/4	1-1/4	18	45	3-3/8
7	17-1/2	1-5/16	18-1/2	46-1/4	3-1/2
7-1/2	18-3/4	1-7/16	19	47-1/2	3-9/16
8	20	1-1/2	19-1/2	48-3/4	3-11/16
8-1/2	21-1/4	1-5/8	20	50	3-3/4
9	22-1/2	1-11/16	20-1/2	51-1/4	3-7/8
9-1/2	23-3/4	1-13/16	21	52-1/2	3-15/16
10	25	1-7/8	21-1/2	53-3/4	4-1/16
10-1/2	26-1/4	2	22	55	4-1/8
11	27-1/2	2-1/16	22-1/2	56-1/4	4-1/4
11-1/2	28-3/4	2-3/16	23	57-1/2	4-5/16
12	30	2-1/4	23-1/2	58-3/4	4-7/16

Note: All dimensions are shown in inches.
Note: Conduit shrink is rounded to the nearest sixteenth.
Note: The conduit will shorten 3/16 in. per inch of saddle depth.

SEGMENT BENDS

Segment bends are multiple bends, spaced apart evenly, in a section of conduit (see Figure 13.24). The multiple bends combine to make one sweeping bend, such as a large radius ninety. The degree of angle for each bend should be the same. By incorporating smaller angles for each bend, the overall bend will appear as one bend. A 90 degree bend made with nine 10 degree bends will look better than one made with three 30 degree bends. If a 90 degree bend will be the overall bend when completed, the number of bends multiplied by the degree of each bend must equal 90 degrees. Some combinations for a large-radius ninety include eighteen 5 degree bends, fifteen 6 degree bends, nine 10 degree bends, six 15 degree bends, four 22-1/2 degree bends, three 30 degree bends, and two 45 degree bends. While many combinations are possible, the most common is nine 10 degree bends.

CONDUIT BENDING

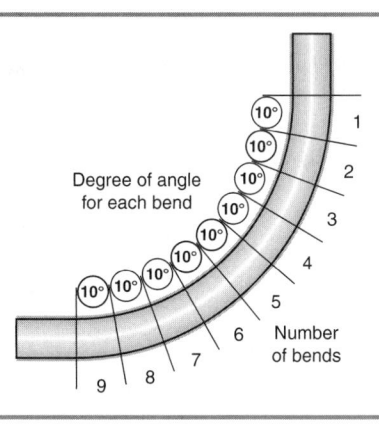

Figure 13.24 Segment bends.

A conduit can be bent to a particular radius by employing segment bends. Determine the desired radius for the centerline of the conduit. Choose the degree of angle, and calculate the number of bends. If the degree of angle selected for a ninety is 10 degrees, the conduit must be bent in nine segments. An overall ninety degree bend divided by ten degree bends requires nine bends. A multiplier is needed to find the distance between bends. To calculate the distance between bends, multiply the tangent of the angle (for the segment) by the centerline radius. The multiplier (or the tangent of the angle) for a 10 degree bend is 0.176. (Multipliers are shown in Table 13.10.) After marking the conduit, bend each segment by the degree of angle selected.

The steps for creating a segment bend are as follows:

Step 1: Determine the desired radius for the centerline of the conduit.

Step 2: Choose the degree of angle, and calculate the number of bends.

Step 3: Find the distance between bends and mark the conduit.

Table 13.10 Angles per Bend and Multipliers

Angle per Bend	Multiplier
5°	0.087
6°	0.105
10°	0.176
15°	0.268
18°	0.325
22-1/2°	0.414
30°	0.577

Centerline radius × multiplier = distance between bends

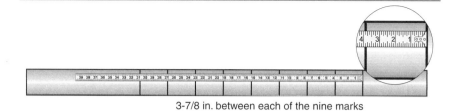

Figure 13.25 Distance between bends.

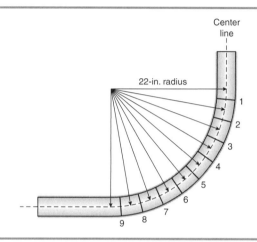

Figure 13.26 22-in. radius at the centerline.

Step 4: Align the bender with the first mark and bend the degree of angle for that segment.

Step 5: Bend each segment the same degree of angle.

Example

A 90 degree bend with a 22-in. radius (at the centerline) is needed in a 3-in. rigid metal conduit. The conduit will be bent in nine segments, and each segment will be a 10 degree bend. The calculated distance between bends for a ninety with nine 10 degree bends is 3.87 or 3-7/8 in. (22 × 0.176 = 3.87) (see Figure 13.25). Mark the conduit at the point of the first bend. Add eight more marks with 3-7/8 in. between each. Align the bender with the first mark and bend the first segment 10 degrees. Bend each of the remaining segments 10 degrees. Upon completion of the final bend, the ninety will have a 22-in. radius at the centerline (see Figure 13.26).

Table 13.11 shows the distances between bends and the degrees per segment for segment bends.

204

CONDUIT BENDING

Table 13.11 Distance Between Bends for Segment Bends

Radius	Degrees per Segment						
	5	6	10	15	18	22-1/2	30
5	7/16	1/2	7/8	1-5/16	1-5/8	2-1/16	2-7/8
6	1/2	5/8	1-1/16	1-5/8	1-15/16	2-1/2	3-7/16
7	5/8	3/4	1-1/4	1-7/8	2-1/4	2-7/8	4-1/16
8	11/16	13/16	1-7/16	2-1/8	2-5/8	3-5/16	4-5/8
9	13/16	15/16	1-9/16	2-7/16	2-15/16	3-3/4	5-3/16
10	7/8	1-1/16	1-3/4	2-11/16	3-1/4	4-1/8	5-3/4
11	15/16	1-1/8	1-15/16	2-15/16	3-9/16	4-9/16	6-3/8
12	1-1/16	1-1/4	2-1/8	3-3/16	3-7/8	5	6-15/16
13	1-1/8	1-3/8	2-5/16	3-1/2	4-1/4	5-3/8	7-1/2
14	1-1/4	1-1/2	2-7/16	3-3/4	4-9/16	5-13/16	8-1/16
15	1-5/16	1-9/16	2-5/8	4	4-7/8	6-3/16	8-11/16
16	1-3/8	1-11/16	2-13/16	4-5/16	5-3/16	6-5/8	9-1/4
17	1-1/2	1-13/16	3	4-9/16	5-1/2	7-1/16	9-13/16
18	1-9/16	1-7/8	3-3/16	4-13/16	5-7/8	7-7/16	10-3/8
19	1-11/16	2	3-3/8	5-1/16	6-3/16	7-7/8	11
20	1-3/4	2-1/8	3-1/2	5-3/8	6-1/2	8-5/16	11-9/16
21	1-13/16	2-3/16	3-11/16	5-5/8	6-13/16	8-11/16	12-1/8
22	1-15/16	2-5/16	3-7/8	5-7/8	7-1/8	9-1/8	12-11/16
23	2	2-7/16	4-1/16	6-3/16	7-1/2	9-1/2	13-1/4
24	2-1/8	2-1/2	4-1/4	6-7/16	7-13/16	9-15/16	13-7/8
25	2-3/16	2-5/8	4-7/16	6-11/16	8-1/8	10-3/8	14-7/16
26	2-1/4	2-3/4	4-9/16	6-15/16	8-7/16	10-3/4	15
27	2-3/8	2-13/16	4-3/4	7-1/4	8-3/4	11-3/16	15-9/16
28	2-7/16	2-15/16	4-15/16	7-1/2	9-1/8	11-5/8	16-3/16
29	2-9/16	3-1/16	5-1/8	7-3/4	9-7/16	12	16-3/4
30	2-5/8	3-1/8	5-5/16	8-1/16	9-3/4	12-7/16	17-5/16
31	2-11/16	3-1/4	5-7/16	8-5/16	10-1/16	12-13/16	17-7/8
32	2-13/16	3-3/8	5-5/8	8-9/16	10-3/8	13-1/4	18-1/2
33	2-7/8	3-7/16	5-13/16	8-13/16	10-3/4	13-11/16	19-1/16
34	3	3-9/16	6	9-1/8	11-1/16	14-1/16	19-5/8
35	3-1/16	3-11/16	6-3/16	9-3/8	11-3/8	14-1/2	20-3/16
36	3-1/8	3-13/16	6-3/8	9-5/8	11-11/16	14-15/16	20-13/16
37	3-1/4	3-7/8	6-1/2	9-15/16	12	15-5/16	21-3/8
38	3-5/16	4	6-11/16	10-3/16	12-3/8	15-3/4	21-15/16
39	3-7/16	4-1/8	6-7/8	10-7/16	12-11/16	16-1/8	22-1/2
40	3-1/2	4-3/16	7-1/16	10-11/16	13	16-9/16	23-1/8
41	3-9/16	4-5/16	7-1/4	11	13-5/16	17	23-11/16
42	3-11/16	4-7/16	7-3/8	11-1/4	13-5/8	17-3/8	24-1/4
43	3-3/4	4-1/2	7-9/16	11-1/2	14	17-13/16	24-13/16

Table continues below.

Table 13.11 Distance Between Bends for Segment Bends *(continued)*

	Degrees per Segment						
Radius	5	6	10	15	18	22-1/2	30
44	3-7/8	4-5/8	7-3/4	11-13/16	14-5/16	18-1/4	25-3/8
45	3-15/16	4-3/4	7-15/16	12-1/16	14-5/8	18-5/8	26
46	4	4-13/16	8 1/8	12-5/16	14-15/16	19-1/16	26-9/16
47	4-1/8	4-15/16	8-5/16	12-9/16	15-1/4	19-7/16	27-1/8
48	4-3/16	5-1/16	8-7/16	12-7/8	15-5/8	19-7/8	27-11/16
49	4-5/16	5-1/8	8-5/8	13-1/8	15-15/16	20-5/16	28-5/16
50	4-3/8	5-1/4	8-13/16	13-3/8	16-1/4	20-11/16	28-7/8
51	4-7/16	5-3/8	9	13-11/16	16-9/16	21-1/8	29-7/16
52	4-9/16	5-7/16	9-3/16	13-15/16	16-7/8	21-9/16	30
53	4-5/8	5-9/16	9-3/8	14-3/16	17-1/4	21-15/16	30-5/8
54	4-3/4	5-11/16	9-1/2	14-1/2	17-9/16	22-3/8	31-3/16
55	4-13/16	5-3/4	9-11/16	14-3/4	17-7/8	22-13/16	31-3/4
56	4-7/8	5-7/8	9-7/8	15	18-3/16	23-3/16	32-5/16
57	5	6	10-1/16	15-1/4	18-1/2	23-5/8	32-15/16
58	5-1/16	6-1/8	10-1/4	15-9/16	18-7/8	24	33-1/2
59	5-3/16	6-3/16	10-3/8	15-13/16	19-3/16	24-7/16	34-1/16
60	5-1/4	6-5/16	10-9/16	16-1/16	19-1/2	24-7/8	34-5/8
61	5-5/16	6-7/16	10-3/4	16-3/8	19-13/16	25-1/4	35-3/16
62	5-7/16	6-1/2	10-15/16	16-5/8	20-1/8	25-11/16	35-13/16
63	5-1/2	6-5/8	11-1/8	16-7/8	20-1/2	26-1/8	36-3/8
64	5-5/8	6-3/4	11-5/16	17-1/8	20-13/16	26-1/2	36-15/16
65	5-11/16	6-13/16	11-7/16	17-7/16	21-1/8	26-15/16	37-1/2
66	5-3/4	6-15/16	11-5/8	17-11/16	21-7/16	27-5/16	38-1/8
67	5-7/8	7-1/16	11-13/16	17-15/16	21-3/4	27-3/4	38-11/16
68	5-15/16	7-1/8	12	18-1/4	22-1/8	28-3/16	39-1/4
69	6-1/16	7-1/4	12-3/16	18-1/2	22-7/16	28-9/16	39-13/16
70	6-1/8	7-3/8	12-5/16	18-3/4	22-3/4	29	40-7/16
71	6-3/16	7-7/16	12-1/2	19	23-1/16	29-7/16	41
72	6-5/16	7-9/16	12-11/16	19-5/16	23-3/8	29-13/16	41-9/16

Note: The radius is for the centerline of the conduit.
Note: All dimensions are shown in inches.
Note: Distance between bends is rounded to the nearest sixteenth.

CONCENTRIC BENDS

Concentric bends share a common center point, but the centerline radius on each conduit is different. Fabricating conduit runs with concentric bends can be accomplished easily by understanding segment bends. Start with finding the radius for the centerline of the first or inside conduit. Choose the degree of angle and calculate the number of bends. Find the distance between bends and mark the first conduit. Next, determine the desired radius for the centerline of the next conduit. Find

CONDUIT BENDING

the distance between bends and mark this conduit. Although the degree of angle and the number of bends will remain the same with each conduit, the distance between bends will not. Continue determining the radius for each remaining conduit, find the distance between bends, and mark the conduits. Align the bender with each mark on each conduit and bend each to the degree of angle selected.

Example

Concentric bends are needed in three 3-in. rigid metal conduits. The overall angle of each conduit will be 90 degrees. A 22-in. radius (at the centerline) is needed for the first conduit. The outside diameter of a 3-in. rigid metal conduit is 3.5 in. A 2-in. space is needed between the conduits. Each conduit will be bent in nine segments, and each segment will be a 10 degree bend. The calculated distance between bends for the first ninety with nine 10 degree bends is 3.87, or 3-7/8 in. (22 × 0.176 = 3.872). The first conduit is marked with 3-7/8 in. between each mark (see Figure 13.27).

The radius of the second conduit can be found by adding the radius of the first conduit, half the diameter of the first conduit, the 2-in. space, and half the diameter of the second conduit (22 + 1-3/4 + 2 + 1-3/4 = 27-1/2-in. radius). The calculated distance between bends for the second conduit is 4.84 or 4-13/16 in. (27.5 × 0.176 = 4.84).

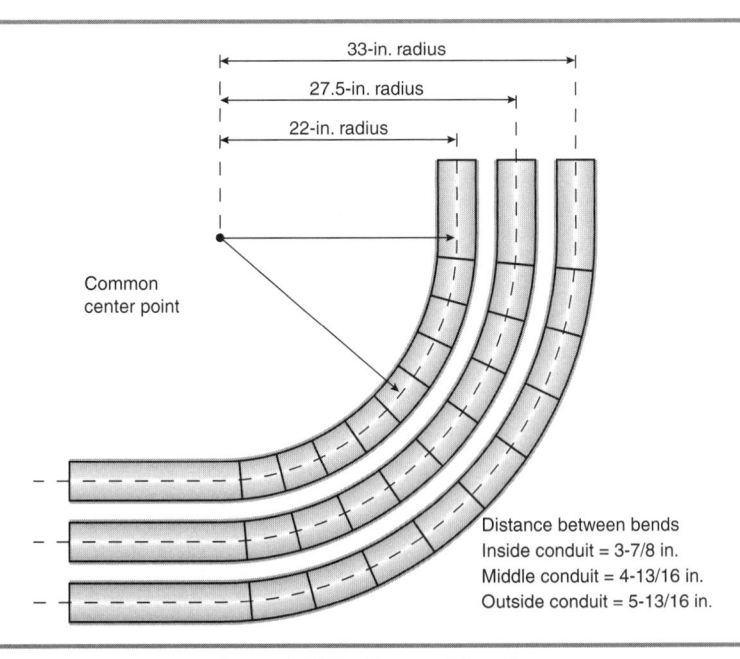

Figure 13.27   Segment bends.

The radius of the third conduit can be found by adding the radius of the second conduit, half the diameter of the second conduit, the 2-in. space, and half the diameter of the third conduit (27-1/2 + 1-3/4 + 2 + 1-3/4 = 33-in. radius). The calculated distance between bends for the third conduit is 5.81, or 5-13/16 in. (33 × 0.176 = 5.808).

NATURAL TRIGONOMETRIC FUNCTIONS

On rare occasions, you may find trigonometric functions useful in bending offsets. The trigonometric functions are as follows:

Sine (sin)

Cosine (cos)

Tangent (tan)

Cotangent (cot) = 1/tan

Secant (sec) = 1/cos

Cosecant (csc) = 1/sin

Table 13.12 lists the values of the different trigonometric functions for degree angles from 0 to 90.

Table 13.12 Natural Trigonometric Functions

Angle	Sine	Cosine	Tangent	Cotangent	Secant	Cosecant
0	0.0000	1.0000	0.0000	infinity	1.0000	infinity
1	0.0175	0.9998	0.0175	57.2900	1.0002	57.2987
2	0.0349	0.9994	0.0349	28.6363	1.0006	28.6537
3	0.0523	0.9986	0.0524	19.0811	1.0014	19.1073
4	0.0698	0.9976	0.0699	14.3007	1.0024	14.3356
5	0.0872	0.9962	0.0875	11.4301	1.0038	11.4737
6	0.1045	0.9945	0.1051	9.5144	1.0055	9.5668
7	0.1219	0.9925	0.1228	8.1443	1.0075	8.2055
8	0.1392	0.9903	0.1405	7.1154	1.0098	7.1853
9	0.1564	0.9877	0.1584	6.3138	1.0125	6.3925
10	0.1736	0.9848	0.1763	5.6713	1.0154	5.7588
11	0.1908	0.9816	0.1944	5.1446	1.0187	5.2408
12	0.2079	0.9781	0.2126	4.7046	1.0223	4.8097
13	0.2250	0.9744	0.2309	4.3315	1.0263	4.4454
14	0.2419	0.9703	0.2493	4.0108	1.0306	4.1336
15	0.2588	0.9659	0.2679	3.7321	1.0353	3.8637
16	0.2756	0.9613	0.2867	3.4874	1.0403	3.6280
17	0.2924	0.9563	0.3057	3.2709	1.0457	3.4203
18	0.3090	0.9511	0.3249	3.0777	1.0515	3.2361
19	0.3256	0.9455	0.3443	2.9042	1.0576	3.0716

Continued

208

CONDUIT BENDING

Table 13.12   Natural Trigonometric Functions *(continued)*

Angle	Sine	Cosine	Tangent	Cotangent	Secant	Cosecant
20	0.3420	0.9397	0.3640	2.7475	1.0642	2.9238
21	0.3584	0.9336	0.3839	2.6051	1.0711	2.7904
22	0.3746	0.9272	0.4040	2.4751	1.0785	2.6695
23	0.3907	0.9205	0.4245	2.3559	1.0864	2.5593
24	0.4067	0.9135	0.4452	2.2460	1.0946	2.4586
25	0.4226	0.9063	0.4663	2.1445	1.1034	2.3662
26	0.4384	0.8988	0.4877	2.0503	1.1126	2.2812
27	0.4540	0.8910	0.5095	1.9626	1.1223	2.2027
28	0.4695	0.8829	0.5317	1.8807	1.1326	2.1301
29	0.4848	0.8746	0.5543	1.8040	1.1434	2.0627
30	0.5000	0.8660	0.5774	1.7321	1.1547	2.0000
31	0.5150	0.8572	0.6009	1.6643	1.1666	1.9416
32	0.5299	0.8480	0.6249	1.6003	1.1792	1.8871
33	0.5446	0.8387	0.6494	1.5399	1.1924	1.8361
34	0.5592	0.8290	0.6745	1.4826	1.2062	1.7883
35	0.5736	0.8192	0.7002	1.4281	1.2208	1.7434
36	0.5878	0.8090	0.7265	1.3764	1.2361	1.7013
37	0.6018	0.7986	0.7536	1.3270	1.2521	1.6616
38	0.6157	0.7880	0.7813	1.2799	1.2690	1.6243
39	0.6293	0.7771	0.8098	1.2349	1.2868	1.5890
40	0.6428	0.7660	0.8391	1.1918	1.3054	1.5557
41	0.6561	0.7547	0.8693	1.1504	1.3250	1.5243
42	0.6691	0.7431	0.9004	1.1106	1.3456	1.4945
43	0.6820	0.7314	0.9325	1.0724	1.3673	1.4663
44	0.6947	0.7193	0.9657	1.0355	1.3902	1.4396
45	0.7071	0.7071	1.0000	1.0000	1.4142	1.4142
46	0.7193	0.6947	1.0355	0.9657	1.4396	1.3902
47	0.7314	0.6820	1.0724	0.9325	1.4663	1.3673
48	0.7431	0.6691	1.1106	0.9004	1.4945	1.3456
49	0.7547	0.6561	1.1504	0.8693	1.5243	1.3250
50	0.7660	0.6428	1.1918	0.8391	1.5557	1.3054
51	0.7771	0.6293	1.2349	0.8098	1.5890	1.2868
52	0.7880	0.6157	1.2799	0.7813	1.6243	1.2690
53	0.7986	0.6018	1.3270	0.7536	1.6616	1.2521
54	0.8090	0.5878	1.3764	0.7265	1.7013	1.2361
55	0.8192	0.5736	1.4281	0.7002	1.7434	1.2208
56	0.8290	0.5592	1.4826	0.6745	1.7883	1.2062
57	0.8387	0.5446	1.5399	0.6494	1.8361	1.1924
58	0.8480	0.5299	1.6003	0.6249	1.8871	1.1792
59	0.8572	0.5150	1.6643	0.6009	1.9416	1.1666
60	0.8660	0.5000	1.7321	0.5774	2.0000	1.1547
61	0.8746	0.4848	1.8040	0.5543	2.0627	1.1434
62	0.8829	0.4695	1.8807	0.5317	2.1301	1.1326
63	0.8910	0.4540	1.9626	0.5095	2.2027	1.1223

Table continues below.

Table 13.12 Natural Trigonometric Functions *(continued)*

Angle	Sine	Cosine	Tangent	Cotangent	Secant	Cosecant
64	0.8988	0.4384	2.0503	0.4877	2.2812	1.1126
65	0.9063	0.4226	2.1445	0.4663	2.3662	1.1034
66	0.9135	0.4067	2.2460	0.4452	2.4586	1.0946
67	0.9205	0.3907	2.3559	0.4245	2.5593	1.0864
68	0.9272	0.3746	2.4751	0.4040	2.6695	1.0785
69	0.9336	0.3584	2.6051	0.3839	2.7904	1.0711
70	0.9397	0.3420	2.7475	0.3640	2.9238	1.0642
71	0.9455	0.3256	2.9042	0.3443	3.0716	1.0576
72	0.9511	0.3090	3.0777	0.3249	3.2361	1.0515
73	0.9563	0.2924	3.2709	0.3057	3.4203	1.0457
74	0.9613	0.2756	3.4874	0.2867	3.6280	1.0403
75	0.9659	0.2588	3.7321	0.2679	3.8637	1.0353
76	0.9703	0.2419	4.0108	0.2493	4.1336	1.0306
77	0.9744	0.2250	4.3315	0.2309	4.4454	1.0263
78	0.9781	0.2079	4.7046	0.2126	4.8097	1.0223
79	0.9816	0.1908	5.1446	0.1944	5.2408	1.0187
80	0.9848	0.1736	5.6713	0.1763	5.7588	1.0154
81	0.9877	0.1564	6.3138	0.1584	6.3925	1.0125
82	0.9903	0.1392	7.1154	0.1405	7.1853	1.0098
83	0.9925	0.1219	8.1443	0.1228	8.2055	1.0075
84	0.9945	0.1045	9.5144	0.1051	9.5668	1.0055
85	0.9962	0.0872	11.4301	0.0875	11.4737	1.0038
86	0.9976	0.0698	14.3007	0.0699	14.3356	1.0024
87	0.9986	0.0523	19.0811	0.0524	19.1073	1.0014
88	0.9994	0.0349	28.6363	0.0349	28.6537	1.0006
89	0.9998	0.0175	57.2900	0.0175	57.2987	1.0002
90	1.0000	0.0000	infinity	0.0000	infinity	1.0000

CONDUIT BENDING

CHAPTER 14
CONDUCTORS AND GROUNDING

Chapter 14 covers conductors in vertical raceways, minimum size conductors, wire bending, conductor ampacities, and grounding. Most of this chapter is extracted from the *NEC* for the reader's quick and easy reference.

CONDUCTORS

Conductors in Vertical Raceways

See Exhibit 14.1.

EXHIBIT 14.1

NEC Table 300.19(A)   Spacings for Conductor Supports

		Conductors			
	Support of Conductors in Vertical	Aluminum or Copper-Clad Aluminum		Copper	
Size of Wire	Raceways	m	ft	m	ft
18 AWG through 8 AWG	Not greater than	30	100	30	100
6 AWG through 1/0 AWG	Not greater than	60	200	30	100
2/0 AWG through 4/0 AWG	Not greater than	55	180	25	80
Over 4/0 AWG through 350 kcmil	Not greater than	41	135	18	60
Over 350 kcmil through 500 kcmil	Not greater than	36	120	15	50
Over 500 kcmil through 750 kcmil	Not greater than	28	95	12	40
Over 750 kcmil	Not greater than	26	85	11	35

Minimum Size Conductors

See Exhibit 14.2.

CONDUCTORS AND GROUNDING

EXHIBIT 14.2

NEC Table 310.5   Minimum Size of Conductors

| Conductor Voltage Rating (Volts) | Minimum Conductor Size (AWG) | |
	Copper	Aluminum or Copper-Clad Aluminum
0–2000	14	12
2001–8000	8	8
8001–15,000	2	2
15,001–28,000	1	1
28,001–35,000	1/0	1/0

Wire Bending

312.6(A) Width of Wiring Gutters. Conductors shall not be deflected within a cabinet or cutout box unless a gutter having a width in accordance with Table 312.6(A) [shown here as Exhibit 14.3] is provided. Conductors in parallel in accordance with 310.4 shall be judged on the basis of the number of conductors in parallel.

EXHIBIT 14.3

NEC Table 312.6(A)   Minimum Wire-Bending Space
at Terminals and Minimum Width of Wiring Gutters

| Wire Size (AWG or kcmil) | Wires per Terminal | | | | | | | | | |
| | 1 | | 2 | | 3 | | 4 | | 5 | |
	mm	in.	mm	in.	mm	in.	mm	in.	mm	in.
14–10	Not specified		—							
8–6	38.1	1½	—							
4–3	50.8	2	—							
2	63.5	2½	—							
1	76.2	3	—							
1/0–2/0	88.9	3½	127	5	178	7	—	—	—	—
3/0–4/0	102	4	152	6	203	8	—	—	—	—
250	114	4½	152	6	203	8	254	10	—	—
300–350	127	5	203	8	254	10	305	12	—	—
400–500	152	6	203	8	254	10	305	12	356	14
600–700	203	8	254	10	305	12	356	14	406	16
750–900	203	8	305	12	356	14	406	16	457	18
1000–1250	254	10	—	—	—	—	—	—	—	—
1500–2000	305	12	—	—	—	—	—	—	—	—

Note: Bending space at terminals shall be measured in a straight line from the end of the lug or wire connector (in the direction that the wire leaves the terminal) to the wall, barrier, or obstruction.

(B) Wire-Bending Space at Terminals. Wire-bending space at each terminal shall be provided in accordance with 312.6(B)(1) or (B)(2).

(1) Conductors Not Entering or Leaving Opposite Wall. Table 312.6(A) shall apply where the conductor does not enter or leave the enclosure through the wall opposite its terminal.

376.23 Insulated Conductors. Insulated conductors installed in a metallic wireway shall comply with 376.23(A) and (B).

(A) Deflected Insulated Conductors. Where insulated conductors are deflected within a metallic wireway, either at the ends or where conduits, fittings, or other raceways or cables enter or leave the metallic wireway, or where the direction of the metallic wireway is deflected greater than 30 degrees, dimensions corresponding to one wire per terminal in Table 312.6(A) shall apply.

378.23(A) Deflected Insulated Conductors. Where insulated conductors are deflected within a metallic wireway, either at the ends or where conduits, fittings, or other raceways or cables enter or leave the metallic wireway, or where the direction of the metallic wireway is deflected greater than 30 degrees, dimensions corresponding to one wire per terminal in Table 312.6(A) shall apply.

312.6(B)(2) Conductors Entering or Leaving Opposite Wall. Table 312.6(B) [shown here as Exhibit 14.4] shall apply where the conductor does enter or leave the enclosure through the wall opposite its terminal.

Exception No. 1: Where the distance between the wall and its terminal is in accordance with Table 312.6(A), a conductor shall be permitted to enter or leave an enclosure through the wall opposite its terminal, provided the conductor enters or leaves the enclosure where the gutter joins an adjacent gutter that has a width that conforms to Table 312.6(B) for the conductor.

Exception No. 2: A conductor not larger than 350 kcmil shall be permitted to enter or leave an enclosure containing only a meter or leave an enclosure containing only a meter socket(s) through the wall opposite its terminal, provided the distance between the terminal and the opposite wall is not less than that specified in Table 312.6(A) and the terminal is a lay-in type where the terminal is either of the following:

(a) Directed toward the opening in the enclosure and within a 45 degree angle of directly facing the enclosure wall

(b) Directly facing the enclosure wall and offset not greater than 50 percent of the bending space specified in Table 312.6(A)

FPN: *Offset* is the distance measured along the enclosure wall from the axis of the centerline of the terminal to a line passing through the center of the opening in the enclosure.

CONDUCTORS AND GROUNDING

EXHIBIT 14.4

NEC Table 312.6(B) Minimum Wire-Bending Space at Terminals

Wire Size (AWG or kcmil)		Wires per Terminal							
		1		2		3		4 or More	
All Other Conductors	Compact Stranded AA-8000 Aluminum Alloy Conductors (See Note 3.)	mm	in.	mm	in.	mm	in.	mm	in.
14–10	12–8	Not specified		—	—	—	—	—	—
8	6	38.1	1½	—	—	—	—	—	—
6	4	50.8	2	—	—	—	—	—	—
4	2	76.2	3	—	—	—	—	—	—
3	1	76.2	3	—	—	—	—	—	—
2	1/0	88.9	3½	—	—	—	—	—	—
1	2/0	114	4½	—	—	—	—	—	—
1/0	3/0	140	5½	140	5½	178	7	—	—
2/0	4/0	152	6	152	6	190	7½	—	—
3/0	250	165ᵃ	6½ᵃ	165ᵃ	6½ᵃ	203	8	—	—
4/0	300	178ᵇ	7ᵇ	190ᶜ	7½ᶜ	216ᵃ	8½ᵃ	—	—
250	350	216ᵈ	8½ᵈ	229ᵈ	8½ᵈ	254ᵇ	9ᵇ	254	10
300	400	254ᵉ	10ᵉ	254ᵈ	10ᵈ	279ᵇ	11ᵇ	305	12
350	500	305ᵉ	12ᵉ	305ᵉ	12ᵉ	330ᵉ	13ᵉ	356ᵈ	14ᵈ
400	600	330ᵉ	13ᵉ	330ᵉ	13ᵉ	356ᵉ	14ᵉ	381ᵉ	15ᵉ
500	700–750	356ᵉ	14ᵉ	356ᵉ	14ᵉ	381ᵉ	15ᵉ	406ᵉ	16ᵉ
600	800–900	381ᵉ	15ᵉ	406ᵉ	16ᵉ	457ᵉ	18ᵉ	483ᵉ	19ᵉ
700	1000	406ᵉ	16ᵉ	457ᵉ	18ᵉ	508ᵉ	20ᵉ	559ᵉ	22ᵉ
750	—	432ᵉ	17ᵉ	483ᵉ	19ᵉ	559ᵉ	22ᵉ	610ᵉ	24ᵉ
800	—	457	18	508	20	559	22	610	24
900	—	483	19	559	22	610	24	610	24
1000	—	508	20	—	—	—	—	—	—
1250	—	559	22	—	—	—	—	—	—
1500	—	610	24	—	—	—	—	—	—
1750	—	610	24	—	—	—	—	—	—
2000	—	610	24	—	—	—	—	—	—

1. Bending space at terminals shall be measured in a straight line from the end of the lug or wire connector in a direction perpendicular to the enclosure wall.

2. For removable and lay-in wire terminals intended for only one wire, bending space shall be permitted to be reduced by the following number of millimeters (inches):

ᵃ 12.7 mm (½ in.)
ᵇ 25.4 mm (1 in.)
ᶜ 38.1 mm (1½ in.)
ᵈ 50.8 mm (2 in.)
ᵉ 76.2 mm (3 in.)

3. This column shall be permitted to determine the required wire-bending space for compact stranded aluminum conductors in sizes up to 1000 kcmil and manufactured using AA-8000 series electrical grade aluminum alloy conductor material in accordance with 310.14.

Conductor Ampacities

310.15 Ampacities for Conductors Rated 0–2000 Volts.

310.15(B) Tables. Ampacities for conductors rated 0 to 2000 volts shall be as specified in the Allowable Ampacity Tables 310.16 through 310.19, and Ampacity Tables 310.20 and 310.21 as modified by (B)(1) through (B)(6).

310.15(B)(2) Adjustment Factors.

(a) *More Than Three Current-Carrying Conductors in a Raceway or Cable.* Where the number of current-carrying conductors in a raceway or cable exceeds three, or where single conductors or multiconductor cables are stacked or bundled longer than 600 mm (24 in.) without maintaining spacing and are not installed in raceways, the allowable ampacity of each conductor shall be reduced as shown in Table 310.15(B)(2)(a) [shown here as Exhibit 14.5]. Each current-carrying conductor of a paralleled set of conductors shall be counted as a current-carrying conductor.

Exception No. 1: Where conductors of different systems, as provided in 300.3, are installed in a common raceway or cable, the derating factors shown in Table 310.15(B)(2)(a) shall apply only to the number of power and lighting conductors (Articles 210, 215, 220, and 230).

Exception No. 2: For conductors installed in cable trays, the provisions of 392.11 shall apply.

Exception No. 3: Derating factors shall not apply to conductors in nipples having a length not exceeding 600 mm (24 in.).

Exception No. 4: Derating factors shall not apply to underground conductors entering or leaving an outdoor trench if those conductors have physical protection in the form of rigid metal conduit, intermediate metal conduit, or rigid nonmetallic conduit having a length not exceeding 3.05 m (10 ft) and if the number of conductors does not exceed four.

EXHIBIT 14.5

NEC Table 310.15(B)(2)(a) Adjustment Factors for More Than Three Current-Carrying Conductors in a Raceway or Cable

Number of Current-Carrying Conductors	Percent of Values in Tables 310.16 through 310.19 as Adjusted for Ambient Temperature if Necessary
4–6	80
7–9	70
10–20	50
21–30	45
31–40	40
41 and above	35

CONDUCTORS AND GROUNDING

CONDUCTORS AND GROUNDING

Exception No. 5: Adjustment factors shall not apply to Type AC cable or to Type MC cable without an overall outer jacket under the following conditions:

(1) Each cable has not more than three current-carrying conductors.
(2) The conductors are 12 AWG copper.
(3) Not more than 20 current-carrying conductors are bundled, stacked, or supported on "bridle rings."

A 60 percent adjustment factor shall be applied where the current-carrying conductors in these cables that are stacked or bundled longer than 600 mm (24 in.) without maintaining spacing exceeds 20.

(b) *More Than One Conduit, Tube, or Raceway.* Spacing between conduits, tubing, or raceways shall be maintained.

Various conductor ampacities are shown in Tables 14.1 through 14.6.

Table 14.1   Adjusted Ampacities for 60°C (140°F) Conductors

Size AWG or kcmil	Copper						
	Ampacity Adjustment Factor						
	100%	80%	70%	50%	45%	40%	35%
14	20	16	14	10	9	8	7
12	25	20	17	12	11	10	8
10	30	24	21	15	13	12	10
8	40	32	28	20	18	16	14
6	55	44	38	27	24	22	19
4	70	56	49	35	31	28	24
3	85	68	59	42	38	34	29
2	95	76	66	47	42	38	33
1	110	88	77	55	49	44	38
1/0	125	100	87	62	56	50	43
2/0	145	116	101	72	65	58	50
3/0	165	132	115	82	74	66	57
4/0	195	156	136	97	87	78	68
250	215	172	150	107	96	86	75
300	240	192	168	120	108	96	84
350	260	208	182	130	117	104	91
400	280	224	196	140	126	112	98
500	320	256	224	160	144	128	112
600	355	284	248	177	159	142	124
700	385	308	269	192	173	154	134
750	400	320	280	200	180	160	140

Note: Ampacities are taken from *NEC* Table 310.16.
Note: Fractions in this table have been rounded down.
Note: Unless specifically permitted in 240.4(E) or 240.4(G), the overcurrent protection must not exceed 15 amperes for 14 AWG, 20 amperes for 12 AWG, and 30 amperes for 10 AWG copper.

Table 14.2 Adjusted Ampacities for 75°C (167°F) Conductors

Size AWG or kcmil	Copper					
	Ampacity Adjustment Factor					
	100%	80%	70%	50%	45%	40%	35%
14	20	16	14	10	9	8	7
12	25	20	17	12	11	10	8
10	35	28	24	17	15	14	12
8	50	40	35	25	22	20	17
6	65	52	45	32	29	26	22
4	85	68	59	42	38	34	29
3	100	80	70	50	45	40	35
2	115	92	80	57	51	46	40
1	130	104	91	65	58	52	45
1/0	150	120	105	75	67	60	52
2/0	175	140	122	87	78	70	61
3/0	200	160	140	100	90	80	70
4/0	230	184	161	115	103	92	80
250	255	204	178	127	114	102	89
300	285	228	199	142	128	114	99
350	310	248	217	155	139	124	108
400	335	268	234	167	150	134	117
500	380	304	266	190	171	152	133
600	420	336	294	210	189	168	147
700	460	368	322	230	207	184	161
750	475	380	332	237	213	190	166

Note: Ampacities are taken from *NEC* Table 310.16.
Note: Fractions in this table have been rounded down.
Note: Unless specifically permitted in 240.4(E) or 240.4(G), the overcurrent protection must not exceed 15 amperes for 14 AWG, 20 amperes for 12 AWG, and 30 amperes for 10 AWG copper.

Table 14.3 Adjusted Ampacities for 90°C (194°F) Conductors

Size AWG or kcmil	Copper					
	Ampacity Adjustment Factor					
	100%	80%	70%	50%	45%	40%	35%
14	25	20	17	12	11	10	8
12	30	24	21	15	13	12	10
10	40	32	28	20	18	16	14

Continued

CONDUCTORS AND GROUNDING

CONDUCTORS AND GROUNDING

Table 14.3   Adjusted Ampacities for 90°C (194°F) Conductors   *(continued)*

Size AWG or kcmil	Copper						
	Ampacity Adjustment Factor						
	100%	80%	70%	50%	45%	40%	35%
8	55	44	38	27	24	22	19
6	75	60	52	37	33	30	26
4	95	76	66	47	42	38	33
3	110	88	77	55	49	44	38
2	130	104	91	65	58	52	45
1	150	120	105	75	67	60	52
1/0	170	136	119	85	76	68	59
2/0	195	156	136	97	87	78	68
3/0	225	180	157	112	101	90	78
4/0	260	208	182	130	117	104	91
250	290	232	203	145	130	116	101
300	320	256	224	160	144	128	112
350	350	280	245	175	157	140	122
400	380	304	266	190	171	152	133
500	430	344	301	215	193	172	150
600	475	380	332	237	213	190	166
700	520	416	364	260	234	208	182
750	535	428	374	267	240	214	187

Note: Ampacities are taken from *NEC* Table 310.16.
Note: Fractions in this table have been rounded down.
Note: Unless specifically permitted in 240.4(E) or 240.4(G), the overcurrent protection must not exceed 15 amperes for 14 AWG, 20 amperes for 12 AWG, and 30 amperes for 10 AWG copper.

Table 14.4   Adjusted Ampacities for 60°C (140°F) Conductors

Size AWG or kcmil	Aluminum or Copper-Clad Aluminum						
	Ampacity Adjustment Factor						
	100%	80%	70%	50%	45%	40%	35%
14	—	—	—	—	—	—	—
12	20	16	14	10	9	8	7
10	25	20	17	12	11	10	8
8	30	24	21	15	13	12	10
6	40	32	28	20	18	16	14
4	55	44	38	27	24	22	19
3	65	52	45	32	29	26	22
2	75	60	52	37	33	30	26
1	85	68	59	42	38	34	29

Table continues below.

Table 14.4 Adjusted Ampacities for 60°C (140°F) Conductors *(continued)*

Size AWG or kcmil	Aluminum or Copper-Clad Aluminum Ampacity Adjustment Factor						
	100%	80%	70%	50%	45%	40%	35%
1/0	100	80	70	50	45	40	35
2/0	115	92	80	57	51	46	40
3/0	130	104	91	65	58	52	45
4/0	150	120	105	75	67	60	52
250	170	136	119	85	76	68	59
300	190	152	133	95	85	76	66
350	210	168	147	105	94	84	73
400	225	180	157	112	101	90	78
500	260	208	182	130	117	104	91
600	285	228	199	142	128	114	99
700	310	248	217	155	139	124	108
750	320	256	224	160	144	128	112

Note: Ampacities are taken from *NEC* Table 310.16.
Note: Fractions in this table have been rounded down.
Note: Unless specifically permitted in 240.4(E) or 240.4(G), the overcurrent protection must not exceed 15 amperes for 12 AWG, and 25 amperes for 10 AWG aluminum and copper-clad aluminum.

Table 14.5 Adjusted Ampacities for 75°C (167°F) Conductors

Size AWG or kcmil	Aluminum or Copper-Clad Aluminum Ampacity Adjustment Factor						
	100%	80%	70%	50%	45%	40%	35%
14	—	—	—	—	—	—	—
12	20	16	14	10	9	8	7
10	30	24	21	15	13	12	10
8	40	32	28	20	18	16	14
6	50	40	35	25	22	20	17
4	65	52	45	32	29	26	22
3	75	60	52	37	33	30	26
2	90	72	63	45	40	36	31
1	100	80	70	50	45	40	35
1/0	120	96	84	60	54	48	42
2/0	135	108	94	67	60	54	47
3/0	155	124	108	77	69	62	54
4/0	180	144	126	90	81	72	63

Continued

CONDUCTORS AND GROUNDING

220

CONDUCTORS AND GROUNDING

Table 14.5   Adjusted Ampacities for 75°C (167°F) Conductors   *(continued)*

Size AWG or kcmil	Aluminum or Copper-Clad Aluminum						
	Ampacity Adjustment Factor						
	100%	80%	70%	50%	45%	40%	35%
250	205	164	143	102	92	82	71
300	230	184	161	115	103	92	80
350	250	200	175	125	112	100	87
400	270	216	189	135	121	108	94
500	310	248	217	155	139	124	108
600	340	272	238	170	153	136	119
700	375	300	262	187	168	150	131
750	385	308	269	192	173	154	134

Note: Ampacities are taken from *NEC* Table 310.16.
Note: Fractions in this table have been rounded down.
Note: Unless specifically permitted in 240.4(E) or 240.4(G), the overcurrent protection must not exceed 15 amperes for 12 AWG, and 25 amperes for 10 AWG aluminum and copper-clad aluminum.

Table 14.6   Adjusted Ampacities for 90°C (194°F) Conductors

Size AWG or kcmil	Aluminum or Copper-Clad Aluminum						
	Ampacity Adjustment Factor						
	100%	80%	70%	50%	45%	40%	35%
14	—	—	—	—	—	—	—
12	25	20	17	12	11	10	8
10	35	28	24	17	15	14	12
8	45	36	31	22	20	18	15
6	60	48	42	30	27	24	21
4	75	60	52	37	33	30	26
3	85	68	59	42	38	34	29
2	100	80	70	50	45	40	35
1	115	92	80	57	51	46	40
1/0	135	108	94	67	60	54	47
2/0	150	120	105	75	67	60	52
3/0	175	140	122	87	78	70	61
4/0	205	164	143	102	92	82	71
250	230	184	161	115	103	92	80
300	255	204	178	127	114	102	89
350	280	224	196	140	126	112	98
400	305	244	213	152	137	122	106
500	350	280	245	175	157	140	122

Continued

Table continues below.

Table 14.6 Adjusted Ampacities for 90°C (194°F) Conductors *(continued)*

Size AWG or kcmil	Aluminum or Copper-Clad Aluminum						
	Ampacity Adjustment Factor						
	100%	80%	70%	50%	45%	40%	35%
600	385	308	269	192	173	154	134
700	420	336	294	210	189	168	147
750	435	348	304	217	195	174	152

Note: Ampacities are taken from *NEC* Table 310.16.
Note: Fractions in this table have been rounded down.
Note: Unless specifically permitted in 240.4(E) or 240.4(G), the overcurrent protection must not exceed 15 amperes for 12 AWG, and 25 amperes for 10 AWG aluminum and copper-clad aluminum.

310.15(B)(4) Neutral Conductor.

(a) A neutral conductor that carries only the unbalanced current from other conductors of the same circuit shall not be required to be counted when applying the provisions of 310.15(B)(2)(a).

(b) In a 3-wire circuit consisting of two phase wires and the neutral of a 4-wire, 3-phase, wye-connected system, a common conductor carries approximately the same current as the line-to-neutral load currents of the other conductors and shall be counted when applying the provisions of 310.15(B)(2)(a).

(c) On a 4-wire, 3-phase wye circuit where the major portion of the load consists of nonlinear loads, harmonic currents are present in the neutral conductor; the neutral shall therefore be considered a current-carrying conductor.

310.15(B)(5) Grounding or Bonding Conductor. A grounding or bonding conductor shall not be counted when applying the provisions of 310.15(B)(2)(a).

310.15(B)(6) 120/240-Volt, 3-Wire, Single-Phase Dwelling Services and Feeders. For individual dwelling units of one family, two-family, and multifamily dwellings, conductors, as listed in Table 310.15(B)(6) [shown here as Exhibit 14.6], shall be permitted as 120/240-volt, 3-wire, single-phase service-entrance conductors, service lateral conductors, and feeder conductors that serve as the main power feeder to each dwelling unit and are installed in raceway or cable with or without an equipment grounding conductor. For application of this section, the main power feeder shall be the feeder(s) between the main disconnect and the lighting and appliance branch-circuit panelboard(s). The feeder conductors to a dwelling unit shall not be required to have an allowable ampacity rating greater than their service-entrance conductors. The grounded conductor shall be permitted to be smaller than the ungrounded conductors, provided the requirements of 215.2, 220.61, and 230.42 are met.

See Exhibit 14.7 for *NEC* Table 310.16.

CONDUCTORS AND GROUNDING

EXHIBIT 14.6

NEC Table 310.15(B)(6)   Conductor Types and Sizes for 120/240-Volt, 3-Wire, Single-Phase Dwelling Services and Feeders. Conductor Types RHH, RHW, RHW-2, THHN, THHW, THW, THW-2, THWN, THWN-W, XHHW, XHHW-2, SE, USE, USE-2

Conductor (AWG or kcmil)		
Copper	Aluminum or Copper-Clad Aluminum	Service or Feeder Rating (Amperes)
4	2	100
3	1	110
2	1/0	125
1	2/0	150
1/0	3/0	175
2/0	4/0	200
3/0	250	225
4/0	300	250
250	350	300
350	500	350
400	600	400

EXHIBIT 14.7

NEC Table 310.16 Allowable Ampacities of Insulated Conductors Rated 0 Through 2000 Volts, 60°C Through 90°C (140°F Through 194°F), Not More Than Three Current-Carrying Conductors in Raceway, Cable, or Earth (Directly Buried), Based on Ambient Temperature of 30°C (86°F)

	Temperature Rating of Conductor (See Table 310.13.)						
	60°C (140°F)	75°C (167°F)	90°C (194°F)	60°C (140°F)	75°C (167°F)	90°C (194°F)	
	Types TW, UF	Types RHW, THHW, THW, THWN, XHHW, USE, ZW	Types TBS, SA, SIS, FEP, FEPB, MI, RHH, RHW-2, THHN, THHW, THW-2, THWN-2, USE-2, XHH, XHHW, XHHW-2, ZW-2	Types TW, UF	Types RHW, THHW, THW, THWN, XHHW, USE	Types TBS, SA, SIS, THHN, THHW, THW-2, THWN-2, RHH, RHW-2, USE-2, XHH, XHHW, XHHW-2, ZW-2	
Size AWG or kcmil		COPPER			ALUMINUM OR COPPER-CLAD ALUMINUM		Size AWG or kcmil
18	—	—	14	—	—	—	—
16	—	—	18	—	—	—	—
14*	20	20	25	—	—	—	—
12*	25	25	30	20	20	25	12*
10*	30	35	40	25	30	35	10*
8	40	50	55	30	40	45	8
6	55	65	75	40	50	60	6
4	70	85	95	55	65	75	4
3	85	100	110	65	75	85	3
2	95	115	130	75	90	100	2
1	110	130	150	85	100	115	1
1/0	125	150	170	100	120	135	1/0
2/0	145	175	195	115	135	150	2/0
3/0	165	200	225	130	155	175	3/0
4/0	195	230	260	150	180	205	4/0
250	215	255	290	170	205	230	250
300	240	285	320	190	230	255	300
350	260	310	350	210	250	280	350
400	280	335	380	225	270	305	400
500	320	380	430	260	310	350	500
600	355	420	475	285	340	385	600
700	385	460	520	310	375	420	700
750	400	475	535	320	385	435	750
800	410	490	555	330	395	450	800
900	435	520	585	355	425	480	900
1000	455	545	615	375	445	500	1000
1250	495	590	665	405	485	545	1250
1500	520	625	705	435	520	585	1500
1750	545	650	735	455	545	615	1750
2000	560	665	750	470	560	630	2000
CORRECTION FACTORS							
Ambient Temp. (°C)	For ambient temperatures other than 30°C (86°F), multiply the allowable ampacities shown above by the appropriate factor shown below.						Ambient Temp. (°F)
21–25	1.08	1.05	1.04	1.08	1.05	1.04	70–77
26–30	1.00	1.00	1.00	1.00	1.00	1.00	78–86
31–35	0.91	0.94	0.96	0.91	0.94	0.96	87–95
36–40	0.82	0.88	0.91	0.82	0.88	0.91	96–104
41–45	0.71	0.82	0.87	0.71	0.82	0.87	105–113
46–50	0.58	0.75	0.82	0.58	0.75	0.82	114–122
51–55	0.41	0.67	0.76	0.41	0.67	0.76	123–131
56–60	—	0.58	0.71	—	0.58	0.71	132–140
61–70	—	0.33	0.58	—	0.33	0.58	141–158
71–80	—	—	0.41	—	—	0.41	159–176

*See 240.4(D).

CONDUCTORS AND GROUNDING

CONDUCTORS AND GROUNDING

402.5 Allowable Ampacities for Fixture Wires. The allowable ampacity of fixture wire shall be as specified in Table 402.5 [shown here as Exhibit 14.8].

No conductor shall be used under such conditions that its operating temperature exceeds the temperature specified in Table 402.3 for the type of insulation involved.

EXHIBIT 14.8

NEC **Table 402.5   Allowable Ampacity for Fixture Wires**

Size (AWG)	Allowable Ampacity
18	6
16	8
14	17
12	23
10	28

Conductor Properties

See Exhibits 14.9 and 14.10.

EXHIBIT 14.9

NEC Table 8   Conductor Properties

Size (AWG or kcmil)	Area mm²	Circular mlls	Stranding Quantity	Stranding Diameter mm	Stranding Diameter in.	Overall Diameter mm	Overall Diameter in.	Overall Area mm²	Overall Area in.²	Copper Uncoated ohm/km	Copper Uncoated ohm/kFT	Copper Coated ohm/km	Copper Coated ohm/kFT	Aluminum ohm/km	Aluminum ohm/kFT
18	0.823	1620	1	—	—	1.02	0.040	0.823	0.001	25.5	7.77	26.5	8.08	42.0	12.8
18	0.823	1620	7	0.39	0.015	1.16	0.046	1.06	0.002	26.1	7.95	27.7	8.45	42.8	13.1
16	1.31	2580	1	—	—	1.29	0.051	1.31	0.002	16.0	4.89	16.7	5.08	26.4	8.05
16	1.31	2580	7	0.49	0.019	1.46	0.058	1.68	0.003	16.4	4.99	17.3	5.29	26.9	8.21
14	2.08	4110	1	—	—	1.63	0.064	2.08	0.003	10.1	3.07	10.4	3.19	16.6	5.06
14	2.08	4110	7	0.62	0.024	1.85	0.073	2.68	0.004	10.3	3.14	10.7	3.26	16.9	5.17
12	3.31	6530	1	—	—	2.05	0.081	3.31	0.005	6.34	1.93	6.57	2.01	10.45	3.18
12	3.31	6530	7	0.78	0.030	2.32	0.092	4.25	0.006	6.50	1.98	6.73	2.05	10.69	3.25
10	5.261	10380	1	—	—	2.588	0.102	5.26	0.008	3.984	1.21	4.148	1.26	6.561	2.00
10	5.261	10380	7	0.98	0.038	2.95	0.116	6.76	0.011	4.070	1.24	4.226	1.29	6.679	2.04
8	8.367	16510	1	—	—	3.264	0.128	8.37	0.013	2.506	0.764	2.579	0.786	4.125	1.26
8	8.367	16510	7	1.23	0.049	3.71	0.146	10.76	0.017	2.551	0.778	2.653	0.809	4.204	1.28
6	13.30	26240	7	1.56	0.061	4.67	0.184	17.09	0.027	1.608	0.491	1.671	0.510	2.652	0.808
4	21.15	41740	7	1.96	0.077	5.89	0.232	27.19	0.042	1.010	0.308	1.053	0.321	1.666	0.508
3	26.67	52620	7	2.20	0.087	6.60	0.260	34.28	0.053	0.802	0.245	0.833	0.254	1.320	0.403
2	33.62	66360	7	2.47	0.097	7.42	0.292	43.23	0.067	0.634	0.194	0.661	0.201	1.045	0.319
1	42.41	83690	19	1.69	0.066	8.43	0.332	55.80	0.087	0.505	0.154	0.524	0.160	0.829	0.253
1/0	53.49	105600	19	1.89	0.074	9.45	0.372	70.41	0.109	0.399	0.122	0.415	0.127	0.660	0.201
2/0	67.43	133100	19	2.13	0.084	10.62	0.418	88.74	0.137	0.3170	0.0967	0.329	0.101	0.523	0.159
3/0	85.01	167800	19	2.39	0.094	11.94	0.470	111.9	0.173	0.2512	0.0766	0.2610	0.0797	0.413	0.126
4/0	107.2	211600	19	2.68	0.106	13.41	0.528	141.1	0.219	0.1996	0.0608	0.2050	0.0626	0.328	0.100
250	127	—	37	2.09	0.082	14.61	0.575	168	0.260	0.1687	0.0515	0.1753	0.0535	0.2778	0.0847
300	152	—	37	2.29	0.090	16.00	0.630	201	0.312	0.1409	0.0429	0.1463	0.0446	0.2318	0.0707
350	177	—	37	2.47	0.097	17.30	0.681	235	0.364	0.1205	0.0367	0.1252	0.0382	0.1984	0.0605
400	203	—	37	2.64	0.104	18.49	0.728	268	0.416	0.1053	0.0321	0.1084	0.0331	0.1737	0.0529
500	253	—	37	2.95	0.116	20.65	0.813	336	0.519	0.0845	0.0258	0.0869	0.0265	0.1391	0.0424
600	304	—	61	2.52	0.099	22.68	0.893	404	0.626	0.0704	0.0214	0.0732	0.0223	0.1159	0.0353
700	355	—	61	2.72	0.107	24.49	0.964	471	0.730	0.0603	0.0184	0.0622	0.0189	0.0994	0.0303
750	380	—	61	2.82	0.111	25.35	0.998	505	0.782	0.0563	0.0171	0.0579	0.0176	0.0927	0.0282
800	405	—	61	2.91	0.114	26.16	1.030	538	0.834	0.0528	0.0161	0.0544	0.0166	0.0868	0.0265
900	456	—	61	3.09	0.122	27.79	1.094	606	0.940	0.0470	0.0143	0.0481	0.0147	0.0770	0.0235
1000	507	—	61	3.25	0.128	29.26	1.152	673	1.042	0.0423	0.0129	0.0434	0.0132	0.0695	0.0212
1250	633	—	91	2.98	0.117	32.74	1.289	842	1.305	0.0338	0.0103	0.0347	0.0106	0.0554	0.0169
1500	760	—	91	3.26	0.128	35.86	1.412	1011	1.566	0.02814	0.00858	0.02814	0.00883	0.0464	0.0141
1750	887	—	127	2.98	0.117	38.76	1.526	1180	1.829	0.02410	0.00735	0.02410	0.00756	0.0397	0.0121
2000	1013	—	127	3.19	0.126	41.45	1.632	1349	2.092	0.02109	0.00643	0.02109	0.00662	0.0348	0.0106

Notes:

1. These resistance values are valid **only** for the parameters as given. Using conductors having coated strands, different stranding type, and, especially, other temperatures changes the resistance.

2. Formula for temperature change: $R_2 = R_{11} [1 + \alpha (T_2 - 75)]$ where $\alpha_{cu} = 0.00323$, $\alpha_{AL} = 0.00330$ at 75°C.

3. Conductors with compact and compressed stranding have about 9 percent and 3 percent, respectively, smaller bare conductor diameters than those shown. See Table 5A for actual compact cable dimensions.

4. The IACS conductivities used: bare copper = 100%, aluminum = 61%.

5. Class B stranding is listed as well as solid for some sizes. Its overall diameter and area is that of its circumscribing circle.

FPN: The construction information is per NEMA WC8-1992 or ANSI/UL 1581-1998. The resistance is calculated per National Bureau of Standards Handbook 100, dated 1966, and Handbook 109, dated 1972.

EXHIBIT 14.10

NEC Table 9 Alternating-Current Resistance and Reactance for 600-Volt Cables, 3-Phase, 60Hz, 75°C (167°F)—Three Single Conductors in Conduit

Size (AWG or kcmil)	X_L (Reactance) for All Wires		PVC, Aluminum Conduits	Alternating-Current Resistance for Uncoated Copper Wires			Alternating-Current Resistance for Aluminum Wires			Effective Z at 0.85 PF for Uncoated Copper Wires			Effective Z at 0.85 PF for Aluminum Wires			Size (AWG or kcmil)
	Steel Conduit			PVC Conduit	Aluminum Conduit	Steel Conduit	PVC Conduit	Aluminum Conduit	Steel Conduit	PVC Conduit	Aluminum Conduit	Steel Conduit	PVC Conduit	Aluminum Conduit	Steel Conduit	
										Ohms to Neutral per Kilometer						
										Ohms to Neutral per 1000 Feet						
14	0.190 0.058	0.240 0.073		10.2 3.1	10.2 3.1	10.2 3.1	— —	— —	— —	8.9 2.7	8.9 2.7	8.9 2.7	— —	— —	— —	14
12	0.177 0.054	0.223 0.068		6.6 2.0	6.6 2.0	6.6 2.0	10.5 3.2	10.5 3.2	10.5 3.2	5.6 1.7	5.6 1.7	5.6 1.7	9.2 2.8	9.2 2.8	9.2 2.8	12
10	0.164 0.050	0.207 0.063		3.9 1.2	3.9 1.2	3.9 1.2	6.6 2.0	6.6 2.0	6.6 2.0	3.6 1.1	3.6 1.1	3.6 1.1	5.9 1.8	5.9 1.8	5.9 1.8	10
8	0.171 0.052	0.213 0.065		2.56 0.78	2.56 0.78	2.56 0.78	4.3 1.3	4.3 1.3	4.3 1.3	2.26 0.69	2.26 0.69	2.30 0.70	3.6 1.1	3.6 1.1	3.6 1.1	8
6	0.167 0.051	0.210 0.064		1.61 0.49	1.61 0.49	1.61 0.49	2.66 0.81	2.66 0.81	2.66 0.81	1.44 0.44	1.48 0.45	1.48 0.45	2.33 0.71	2.36 0.72	2.36 0.72	6
4	0.157 0.048	0.197 0.060		1.02 0.31	1.02 0.31	1.02 0.31	1.67 0.51	1.67 0.51	1.67 0.51	0.95 0.29	0.95 0.29	0.98 0.30	1.51 0.46	1.51 0.46	1.51 0.46	4
3	0.154 0.047	0.194 0.059		0.82 0.25	0.82 0.25	0.82 0.25	1.31 0.40	1.35 0.41	1.31 0.40	0.75 0.23	0.79 0.24	0.79 0.24	1.21 0.37	1.21 0.37	1.21 0.37	3

Continued

EXHIBIT 14.10 (continued)

NEC Table 9 Alternating-Current Resistance and Reactance for 600-Volt Cables, 3-Phase, 60Hz, 75°C (167°F)—Three Single Conductors in Conduit

Ohms to Neutral per Kilometer
Ohms to Neutral per 1000 Feet

Size (AWG or kcmil)	X_L (Reactance) for All Wires		Alternating-Current Resistance for Uncoated Copper Wires			Alternating-Current Resistance for Aluminum Wires			Effective Z at 0.85 PF for Uncoated Copper Wires			Effective Z at 0.85 PF for Aluminum Wires			Size (AWG or kcmil)
	PVC, Aluminum Conduits	Steel Conduit	PVC Conduit	Aluminum Conduit	Steel Conduit	PVC Conduit	Aluminum Conduit	Steel Conduit	PVC Conduit	Aluminum Conduit	Steel Conduit	PVC Conduit	Aluminum Conduit	Steel Conduit	
2	0.148 / 0.045	0.187 / 0.057	0.62 / 0.19	0.66 / 0.20	0.66 / 0.20	1.05 / 0.32	1.05 / 0.32	1.05 / 0.32	0.62 / 0.19	0.66 / 0.20	0.66 / 0.20	0.98 / 0.30	0.98 / 0.30	0.98 / 0.30	2
1	0.151 / 0.046	0.187 / 0.057	0.49 / 0.15	0.52 / 0.16	0.52 / 0.16	0.82 / 0.25	0.85 / 0.26	0.82 / 0.25	0.52 / 0.16	0.52 / 0.16	0.52 / 0.16	0.79 / 0.24	0.79 / 0.24	0.82 / 0.25	1
1/0	0.144 / 0.044	0.180 / 0.055	0.39 / 0.12	0.43 / 0.13	0.39 / 0.12	0.66 / 0.20	0.69 / 0.21	0.66 / 0.20	0.43 / 0.13	0.43 / 0.13	0.43 / 0.13	0.62 / 0.19	0.66 / 0.20	0.66 / 0.20	1/0
2/0	0.141 / 0.043	0.177 / 0.054	0.33 / 0.10	0.33 / 0.10	0.33 / 0.10	0.52 / 0.16	0.52 / 0.16	0.52 / 0.16	0.36 / 0.11	0.36 / 0.11	0.36 / 0.11	0.52 / 0.16	0.52 / 0.16	0.52 / 0.16	2/0
3/0	0.138 / 0.042	0.171 / 0.052	0.253 / 0.077	0.269 / 0.082	0.259 / 0.079	0.43 / 0.13	0.43 / 0.13	0.43 / 0.13	0.289 / 0.088	0.302 / 0.092	0.308 / 0.094	0.43 / 0.13	0.43 / 0.13	0.46 / 0.14	3/0
4/0	0.135 / 0.041	0.167 / 0.051	0.203 / 0.062	0.220 / 0.067	0.207 / 0.063	0.33 / 0.10	0.36 / 0.11	0.33 / 0.10	0.243 / 0.074	0.256 / 0.078	0.262 / 0.080	0.36 / 0.11	0.36 / 0.11	0.36 / 0.11	4/0
250	0.135 / 0.041	0.171 / 0.052	0.171 / 0.052	0.187 / 0.057	0.177 / 0.054	0.279 / 0.085	0.295 / 0.090	0.282 / 0.086	0.217 / 0.066	0.230 / 0.070	0.240 / 0.073	0.308 / 0.094	0.322 / 0.098	0.33 / 0.10	250

Continued

EXHIBIT 14.10 (continued)

NEC Table 9 Alternating-Current Resistance and Reactance for 600-Volt Cables, 3-Phase, 60Hz, 75°C (167°F)—Three Single Conductors in Conduit

Ohms to Neutral per Kilometer / Ohms to Neutral per 1000 Feet

Size (AWG or kcmil)	X_L (Reactance) for All Wires			Alternating-Current Resistance for Uncoated Copper Wires			Alternating-Current Resistance for Aluminum Wires			Effective Z at 0.85 PF for Uncoated Copper Wires			Effective Z at 0.85 PF for Aluminum Wires			Size (AWG or kcmil)
	PVC, Aluminum Conduits	Steel Conduit	PVC Conduit	Aluminum Conduit	Steel Conduit	PVC Conduit	Aluminum Conduit	Steel Conduit	PVC Conduit	Aluminum Conduit	Steel Conduit	PVC Conduit	Aluminum Conduit	Steel Conduit		
300	0.135 0.041	0.167 0.051	0.144 0.044	0.161 0.049	0.148 0.045	0.233 0.071	0.249 0.076	0.236 0.072	0.194 0.059	0.207 0.063	0.213 0.065	0.269 0.082	0.282 0.086	0.289 0.088	300
350	0.131 0.040	0.164 0.050	0.125 0.038	0.141 0.043	0.128 0.039	0.200 0.061	0.217 0.066	0.207 0.063	0.174 0.053	0.190 0.058	0.197 0.060	0.240 0.073	0.253 0.077	0.262 0.080	350
400	0.131 0.040	0.161 0.049	0.108 0.033	0.125 0.038	0.115 0.035	0.177 0.054	0.194 0.059	0.180 0.055	0.161 0.049	0.174 0.053	0.184 0.056	0.217 0.066	0.233 0.071	0.240 0.073	400
500	0.128 0.039	0.157 0.048	0.089 0.027	0.105 0.032	0.095 0.029	0.141 0.043	0.157 0.048	0.148 0.045	0.141 0.043	0.157 0.048	0.164 0.050	0.187 0.057	0.200 0.061	0.210 0.064	500
600	0.128 0.039	0.157 0.048	0.075 0.023	0.092 0.028	0.082 0.025	0.118 0.036	0.135 0.041	0.125 0.038	0.131 0.040	0.144 0.044	0.154 0.047	0.167 0.051	0.180 0.055	0.190 0.058	600
750	0.125 0.038	0.157 0.048	0.062 0.019	0.079 0.024	0.069 0.021	0.095 0.029	0.112 0.034	0.102 0.031	0.118 0.036	0.131 0.040	0.141 0.043	0.148 0.045	0.161 0.049	0.171 0.052	750
1000	0.121 0.037	0.151 0.046	0.049 0.015	0.062 0.019	0.059 0.018	0.075 0.023	0.089 0.027	0.082 0.025	0.105 0.032	0.118 0.036	0.131 0.040	0.128 0.039	0.138 0.042	0.151 0.046	1000

Continued

EXHIBIT 14.10 *(continued)*

NEC **Table 9 Alternating-Current Resistance and Reactance for 600-Volt Cables, 3-Phase, 60Hz, 75°C (167°F)—Three Single Conductors in Conduit**

Notes:

1. These values are based on the following constants: UL-Type RHH wires with Class B stranding, in cradled configuration. Wire conductivities are 100 percent IACS copper and 61 percent IACS aluminum, and aluminum conduit is 45 percent IACS. Capacitive reactance is ignored, since it is negligible at these voltages. These resistance values are valid only at 75°C (167°F) and for the parameters as given, but are representative for 600-volt wire types operating at 60 Hz.

2. *Effective Z* is defined as $R \cos(\theta) + X \sin(\theta)$, where θ is the power factor angle of the circuit. Multiplying current by effective impedance gives a good approximation for line-to-neutral voltage drop. Effective impedance values shown in this table are valid only at 0.85 power factor. For another circuit power factor (PF), effective impedance (Ze) can be calculated from R and X_L values given in this table as follows: $Ze = R \times PF + X_L \sin[\arccos(PF)]$.

CONDUCTORS AND GROUNDING

CONDUCTORS AND GROUNDING

GROUNDING

Grounding Electrode Conductors

250.66 Size of Alternating-Current Grounding Electrode Conductor. The size of the grounding electrode conductor of a grounded or ungrounded ac system shall not be less than given in Table 250.66 [shown here as Exhibit 14.11], except as permitted in 250.66(A) through (C).

EXHIBIT 14.11

NEC Table 250.66   Grounding Electrode Conductor
for Alternating-Current Systems

Size of Largest Ungrounded Service-Entrance Conductor or Equivalent Area for Parallel Conductors[a] (AWG/kcmil)		Size of Grounding Electrode Conductor (AWG/kcmil)	
Copper	Aluminum or Copper-Clad Aluminum	Copper	Aluminum or Copper-Clad Aluminum[b]
2 or smaller	1/0 or smaller	8	6
1 or 1/0	2/0 or 3/0	6	4
2/0 or 3/0	4/0 or 250	4	2
Over 3/0 through 350	Over 250 through 500	2	1/0
Over 350 through 600	Over 500 through 900	1/0	3/0
Over 600 through 1100	Over 900 through 1750	2/0	4/0
Over 1100	Over 1750	3/0	250

Notes:

1. Where multiple sets of service-entrance conductors are used as permitted in 230.40, Exception No. 2, the equivalent size of the largest service-entrance conductor shall be determined by the largest sum of the areas of the corresponding conductors of each set.

2. Where there are no service-entrance conductors, the grounding electrode conductor size shall be determined by the equivalent size of the largest service-entrance conductor required for the load to be served.

[a]This table also applies to the derived conductors of separately derived ac systems.

(A) Connections to Rod, Pipe, or Plate Electrodes. Where the grounding electrode conductor is connected to rod, pipe, or plate electrodes as permitted in 250.52(A)(5) or (A)(6), that portion of the conductor that is the sole connection to the grounding electrode shall not be required to be larger than 6 AWG copper wire or 4 AWG aluminum wire.

(B) Connections to Concrete-Encased Electrodes. Where the grounding electrode conductor is connected to a concrete-encased electrode as permitted in 250.52(A)(3), that portion of the conductor that is the sole connection to the grounding electrode shall not be required to be larger than 4 AWG copper wire.

(C) Connections to Ground Rings. Where the grounding electrode conductor is connected to a ground ring as permitted in 250.52(A)(4), that portion of the conductor that is the sole connection to the grounding electrode shall not be required to be larger than the conductor used for the ground ring.

Equipment Grounding Conductors

250.122 Size of Equipment Grounding Conductors.

(A) General. Copper, aluminum, or copper-clad aluminum equipment grounding conductors of the wire type shall not be smaller than shown in Table 250.122 but shall not be required to be larger than the circuit conductors supplying the equipment. Where a raceway or a cable armor or sheath is used as the equipment grounding conductor, as provided in 250.118 and 250.134(A), it shall comply with 250.4(A)(5) or (B)(4).

(B) Increased in Size. Where ungrounded conductors are increased in size, equipment grounding conductors, where installed, shall be increased in size proportionately according to circular mil area of the ungrounded conductors.

(C) Multiple Circuits. Where a single equipment grounding conductor is run with multiple circuits in the same raceway or cable, it shall be sized for the largest overcurrent device protecting conductors in the raceway or cable.

(D) Motor Circuits. Where the overcurrent device consists of an instantaneous trip circuit breaker or a motor short-circuit protector, as allowed in 430.52, the equipment grounding conductor size shall be permitted to be based on the rating of the motor overload protective device but shall not be less than the size shown in Table 250.122 [shown here as Exhibit 14.12].

(E) Flexible Cord and Fixture Wire. The equipment grounding conductor in a flexible cord with the largest circuit conductor 10 AWG or smaller, and the equipment grounding conductor used with fixture wires of any size in accordance with 240.5, shall not be smaller than 18 AWG copper and shall not be smaller than the circuit conductors. The equipment grounding conductor in a flexible cord with a circuit conductor larger than 10 AWG shall be sized in accordance with Table 250.122 [shown here as Exhibit 14.12].

CONDUCTORS AND GROUNDING

EXHIBIT 14.12

NEC Table 250.122   Minimum Size Equipment Grounding
Conductors for Grounding Raceway and Equipment

Rating or Setting of Automatic Overcurrent Device in Circuit Ahead of Equipment, Conduit, etc., Not Exceeding (Amperes)	Size (AWG or kcmil)	
	Copper	Aluminum or Copper-Clad Aluminum*
15	14	12
20	12	10
30	10	8
40	10	8
60	10	8
100	8	6
200	6	4
300	4	2
400	3	1
500	2	1/0
600	1	2/0
800	1/0	3/0
1000	2/0	4/0
1200	3/0	250
1600	4/0	350
2000	250	400
2500	350	600
3000	400	600
4000	500	800
5000	700	1200
6000	800	1200

Note: Where necessary to comply with 250.4(A)(5) or (B)(4), the equipment grounding conductor shall be sized larger than given in this table.

*See installation requirements in 250.120.

(F) Conductors in Parallel. Where conductors are run in parallel in multiple raceways or cables as permitted in 310.4, the equipment grounding conductors, where used, shall be run in parallel in each raceway or cable. One of the methods in 250.122(F)(1) or (F)(2) shall be used to ensure the equipment grounding conductors are protected.

(1) Based on Rating of Overcurrent Protective Device. Each parallel equipment grounding conductor shall be sized on the basis of the ampere rating of the overcurrent device protecting the circuit conductors on the raceway or cable in accordance with Table 250.122.

(2) Ground-Fault Protection of Equipment Installed. Where ground-fault protection of equipment is installed, each parallel equipment grounding conductor in a multi-conductor cable shall be permitted to be sized in accordance with Table 250.122 on the basis of the trip rating of the ground-fault protection where the following conditions are met:

(1) Conditions of maintenance and supervision ensure that only qualified persons will service the installation.
(2) The ground-fault protection equipment is set to trip at not more than the ampacity of a single ungrounded conductor of one of the cables in parallel.
(3) The ground-fault protection is listed for the purpose of protecting the equipment grounding conductor.

(G) Feeder Taps. Equipment grounding conductors run with feeder taps shall not be smaller than shown in Table 250.122 based on the rating of the overcurrent device ahead of the feeder but shall not be required to be larger than the tap conductors.

CHAPTER 15
VOLTAGE DROP

Chapter 15 covers single-phase and three-phase voltage drop formulas. Formulas include: (1) finding the amount of voltage dropped in a circuit, (2) determining the minimum conductor size, (3) finding the maximum distance, and (4) finding the maximum amount of current in amperes the circuit can carry. Alternative voltage drop formulas using Ohm's Law are also covered.

DETERMINING VOLTAGE DROP

Because of resistance or impedance in conductors, some voltage will drop through the conductors. The amount of voltage depends on several factors, including: the conductor's size, the conductor's length, and the current. The amount of voltage drop may be significant enough to cause damage to equipment. Fine print notes (FPNs) in 210.19(A)(1) and 215.2(A)(3) provide a recommendation for voltage drop in branch circuits and feeders.

The recommended voltage drop is a percentage of the source voltage. The *Code* recommends a maximum voltage drop of 3 percent for branch circuits. To find the allowable voltage drop, multiply the source voltage by 3 percent. The voltage drop on a combination of both feeders and branch circuits is recommended not to exceed 5 percent. Unless the voltage drop on the branch circuit is 3 percent, the feeder is not limited to only 2 percent of the source voltage.

Different methods are available to calculate voltage drop. This book will discuss voltage drop formulas using K values and formulas using conductor resistance.

VOLTAGE DROP FORMULAS

Formula Definitions

V_D = volts dropped in a circuit

2 = the multiplying factor for single-phase voltage

1.732 = the multiplying factor for three-phase voltage

Cmil = the circular mill area of the conductor

K = a factor obtained by multiplying the conductor's circular mill area by one foot of the conductor's resistance (note: the Ohms in Tables 8 and 9 are per 1000 feet)
  approximate K value for copper = 12.9
  approximate K value for aluminum = 21.2

L = the length or distance from the voltage source to the load (one way only)

I = the actual current of the load

236

VOLTAGE DROP

Single-Phase Voltage Drop Formulas

To find the volts dropped: $V_D = \dfrac{2\,K\,L\,I}{Cmil}$ or $V_D = \dfrac{2 \times K \times L \times I}{Cmil}$

To find the conductor size: $Cmil = \dfrac{2\,K\,L\,I}{V_D}$ or $Cmil = \dfrac{2 \times K \times L \times I}{V_D}$

To find the distance (one way): $L = \dfrac{Cmil\,V_D}{2\,K\,I}$ or $L = \dfrac{Cmil \times V_D}{2 \times K \times I}$

To find the current in amperes: $I = \dfrac{Cmil\,V_D}{2\,K\,L}$ or $I = \dfrac{Cmil \times V_D}{2 \times K \times L}$

Three-Phase Voltage Drop Formulas

(Use the single-phase formulas, but replace the "2" with "1.732.")

To find the volts dropped: $V_D = \dfrac{1.732\,K\,L\,I}{Cmil}$ or $V_D = \dfrac{1.732 \times K \times L \times I}{Cmil}$

To find the conductor size: $Cmil = \dfrac{1.732\,K\,L\,I}{V_D}$ or $Cmil = \dfrac{1.732 \times K \times L \times I}{V_D}$

To find the distance (one way): $L = \dfrac{Cmil\,V_D}{1.732\,K\,I}$ or $L = \dfrac{Cmil \times V_D}{1.732 \times K \times I}$

To find the current in amperes: $I = \dfrac{Cmil\,V_D}{1.732\,K\,L}$ or $I = \dfrac{Cmil \times V_D}{1.732 \times K \times L}$

Note: Instead of replacing the "2" with "1.732," some formulas include "× 0.866" on the same line containing the "2" (2 × 0.866 = 1.732).

Examples of Voltage Drop Formulas

Generally, only approximate values are needed for voltage drop calculations. Unless precise values are needed, approximate K values are used with voltage drop formulas. Where performing precise voltage drop computations, insert exact K values. The following examples are calculated with approximate K values.

Example 1, Voltage Drop

What is the voltage drop of a 208-volt, single-phase circuit that has a load of 20 amperes? The circuit is fed with 8 AWG, THW, copper conductors and is 190 feet from the load.

Solution

Using the single-phase voltage drop formula for volts dropped, and substituting the known values gives:

$$V_D = \frac{2\,KLI}{Cmil} = \frac{2 \times 12.9 \times 190 \times 20}{16510} = 5.9 \text{ volts}$$

Example 2, Voltage Drop

What is the voltage drop of a 230-volt, single-phase circuit that has a load of 40 amperes? The circuit is fed with 6 AWG, THW, aluminum conductors and is 100 feet from the load.

Solution

Select the single-phase voltage drop formula. Note that this example is fed with aluminum conductors.

$$V_D = \frac{2\,KLI}{Cmil} = \frac{2 \times 21.2 \times 100 \times 40}{26240} = 6.5 \text{ volts}$$

Example 3, Voltage Drop

What is the voltage drop of a 208-volt, three-phase circuit that has a load of 60 amperes? The circuit is fed with 4 AWG, THW, copper conductors and is 180 feet from the load.

Solution

Use the three-phase voltage drop formula to find volts dropped in this circuit.

$$V_D = \frac{1.732\,KLI}{Cmil} = \frac{1.732 \times 12.9 \times 180 \times 60}{41740} = 5.8 \text{ volts}$$

Example 4, Conductor Size

What size branch-circuit copper conductors are needed for a load of 10 amperes? The single-phase voltage is 120 volts, and the load is 100 feet from the source.

Solution

Select the single-phase formula to find the conductor size.

$$Cmil = \frac{2\,KLI}{V_D} = \frac{2 \times 12.9 \times 100 \times 10}{3.6} = 7167 \text{ Cmil} = 10 \text{ AWG copper}$$

Example 5, Conductor Size

What size branch-circuit aluminum conductors are needed for a load of 50 amperes? The single-phase voltage is 240 volts, and the load is 100 feet from the source.

VOLTAGE DROP

Solution

Find the minimum size conductors with the single-phase formula for conductor size. Note that this circuit will be fed with aluminum conductors.

$$Cmil = \frac{2\,K\,L\,I}{V_D} = \frac{2 \times 21.2 \times 100 \times 50}{7.2} = 29{,}444 \text{ Cmil} = 4 \text{ AWG aluminum}$$

Example 6, Conductor Size

What size branch-circuit copper conductors are needed for a load of 35 amperes? The three-phase voltage is 208 volts, and the load is 150 feet from the source.

Solution

The three-phase formula for conductor size is needed to solve this example.

$$Cmil = \frac{1.732\,K\,L\,I}{V_D} = \frac{1.732 \times 12.9 \times 150 \times 35}{6.24} = 18{,}798 \text{ Cmil} = 6 \text{ AWG copper}$$

Example 7, Distance

What is the maximum distance permitted for 2 AWG, THW, copper conductors that have a load of 115 amperes, and a single-phase voltage of 240 volts?

Solution

Use the single-phase formula that solves for the maximum one-way distance.

$$L = \frac{Cmil\,V_D}{2\,K\,I} = \frac{66360 \times 7.2}{2 \times 12.9 \times 115} = 161 \text{ feet}$$

Example 8, Distance

What is the maximum distance permitted for 8 AWG, THW, stranded, aluminum conductors that have a load of 40 amperes, and a single-phase voltage of 208 volts?

Solution

Select the single-phase formula for distance (one way.) Note that this example is fed with aluminum conductors.

$$L = \frac{Cmil\,V_D}{2\,K\,I} = \frac{16510 \times 6.24}{2 \times 21.2 \times 40} = 61 \text{ feet}$$

Example 9, Distance

What is the maximum distance permitted for 12 AWG, THW, copper conductors that have a load of 12 amperes, and a three-phase voltage of 208 volts?

Solution

Choose the formula for finding the maximum one-way distance in a three-phase circuit.

$$L = \frac{Cmil\ V_D}{1.732\ K\ I} = \frac{6530 \times 6.24}{1.732 \times 12.9 \times 12} = 152 \text{ feet}$$

Example 10, Current (in amperes)

What is the maximum load permitted on 10 AWG, THW, copper conductors supplying a 120-volt load 100 feet from the source?

Solution

Use the single-phase formula for finding the maximum current in amperes.

$$I = \frac{Cmil\ V_D}{2\ K\ L} = \frac{10380 \times 3.6}{2 \times 12.9 \times 80} = 18 \text{ amperes}$$

Example 11, Current (in amperes)

What is the maximum load permitted on 250 kcmil, THW, aluminum conductors supplying a 240-volt, single-phase load 250 feet from the source?

Solution

Select the single-phase formula for current. Note that this example is fed with aluminum conductors.

$$I = \frac{Cmil\ V_D}{2\ K\ L} = \frac{250000 \times 7.2}{2 \times 21.2 \times 250} = 170 \text{ amperes}$$

Example 12, Current (in amperes)

What is the maximum load permitted on 500 kcmil, THW, copper conductors supplying a 208-volt, three-phase load 550 feet from the source?

Solution

Use the formula for finding the maximum current in a three-phase circuit.

$$I = \frac{Cmil\ V_D}{1.732\ K\ L} = \frac{500000 \times 6.24}{1.732 \times 12.9 \times 550} = 254 \text{ amperes}$$

VOLTAGE DROP

CALCULATING EXACT *K* VALUES

If the conductor size is known, the exact *K* value can be calculated. To find the exact *K*, multiply the conductor's circular mill area by the resistance for one foot of the conductor. Conductor resistances are shown in Tables 8 and 9 of Chapter 9 in the *NEC*. Tables 8 and 9 can also be found in Chapter 14 of this pocket guide. Note, the resistance values in Tables 8 and 9 are per 1000 feet.

To find the exact *K* value, multiply the conductor's circular mill area by the ohm/kFT and divide by 1000, as shown in the following formula.

$$K = \frac{Cmil \times R}{1000}$$

Example

What is the exact *K* value for 6 AWG copper at 75°C (167°F)?

Solution

$$K = \frac{Cmil \times R}{1000} = \frac{26,240 \times 0.491}{1000} = 12.88384$$

Tables 15.1 through 15.5 provide *K* values for different types of conductors.

Table 15.1   Direct-Current Resistance and *K* Values for Uncoated Copper Conductors

Size (AWG or kcmil)	Circular mils	Ohm/kFT at 75°C (167°F)	*K* Value at 60°C (140°F)	*K* Value at 75°C (167°F)	*K* Value at 90°C (194°F)
18 Solid	1620	7.77	11.97754	12.5874	13.19726
18 Stranded	1620	7.95	12.25501	12.879	13.50299
16 Solid	2580	4.89	12.00495	12.6162	13.22745
16 Stranded	2580	4.99	12.25045	12.8742	13.49795
14 Solid	4110	3.07	12.00637	12.6177	13.22903
14 Stranded	4110	3.14	12.28013	12.9054	13.53067
12 Solid	6530	1.93	11.99229	12.6029	13.21351
12 Stranded	6530	1.98	12.30297	12.9294	13.55583
10 Solid	10380	1.21	11.95128	12.5598	13.16832
10 Stranded	10380	1.24	12.24759	12.8712	13.49481
8 Solid	16510	0.764	12.00251	12.61364	13.22477
8 Stranded	16510	0.778	12.22245	12.84478	13.46711
6	26240	0.491	12.25962	12.88384	13.50806
4	41740	0.308	12.23305	12.85592	13.47879

Table continues below.

Table 15.1 Direct-Current Resistance and
K Values for Uncoated Copper Conductors *(continued)*

Size (AWG or kcmil)	Circular mils	Ohm/kFT at 75°C (167°F)	K Value at 60°C (140°F)	K Value at 75°C (167°F)	K Value at 90°C (194°F)
3	52620	0.245	12.26729	12.8919	13.51651
2	66360	0.194	12.2501	12.87384	13.49758
1	83690	0.154	12.26382	12.88826	13.5127
1/0	105600	0.122	12.25901	12.8832	13.50739
2/0	133100	0.0967	12.24718	12.87077	13.49436
3/0	167800	0.0766	12.23073	12.85348	13.47623
4/0	211600	0.0608	12.24196	12.86528	13.4886
250	250000	0.0515	12.25121	12.875	13.49879
300	300000	0.0429	12.24645	12.87	13.49355
350	350000	0.0367	12.22266	12.845	13.46734
400	400000	0.0321	12.2179	12.84	13.4621
500	500000	0.0258	12.275	12.9	13.52501
600	600000	0.0214	12.2179	12.84	13.4621
700	700000	0.0184	12.25596	12.88	13.50404
750	750000	0.0171	12.20363	12.825	13.44637
800	800000	0.0161	12.25596	12.88	13.50404
900	900000	0.0143	12.24645	12.87	13.49355
1000	1000000	0.0129	12.275	12.9	13.52501
1250	1250000	0.0103	12.25121	12.875	13.49879
1500	1500000	0.00858	12.24645	12.87	13.49355
1750	1750000	0.00735	12.23931	12.8625	13.48569
2000	2000000	0.00643	12.23693	12.86	13.48307

Note: K values are derived from Chapter 9, Table 8 of the *NEC*®.

Table 15.2 Direct-Current Resistance and
K Values for Coated Copper Conductors

Size (AWG or kcmil)	Circular mils	Ohm/kFT at 75°C (167°F)	K Value at 60°C (140°F)	K Value at 75°C (167°F)	K Value at 90°C (194°F)
18 Solid	1620	8.08	12.45541	13.0896	13.72379
18 Stranded	1620	8.45	13.02577	13.689	14.35223
16 Solid	2580	5.08	12.47139	13.1064	13.74141
16 Stranded	2580	5.29	12.98694	13.6482	14.30946
14 Solid	4110	3.19	12.47568	13.1109	13.74612
14 Stranded	4110	3.26	12.74944	13.3986	14.04776
12 Solid	6530	2.01	12.48938	13.1253	13.76122
12 Stranded	6530	2.05	12.73792	13.3865	14.03508
10 Solid	10380	1.26	12.44513	13.0788	13.71247
10 Stranded	10380	1.29	12.74144	13.3902	14.03896

Continued

242

VOLTAGE DROP

Table 15.2 Direct-Current Resistance and K Values for Coated Copper Conductors *(continued)*

Size (AWG or kcmil)	Circular mils	Ohm/kFT at 75°C (167°F)	K Value at 60°C (140°F)	K Value at 75°C (167°F)	K Value at 90°C (194°F)
8 Solid	16510	0.786	12.34813	12.97686	13.60559
8 Stranded	16510	0.809	12.70946	13.35659	14.00372
6	26240	0.51	12.73402	13.3824	14.03078
4	41740	0.321	12.74938	13.39854	14.0477
3	52620	0.254	12.71792	13.36548	14.01304
2	66360	0.201	12.69212	13.33836	13.9846
1	83690	0.16	12.74164	13.3904	14.03916
1/0	105600	0.127	12.76143	13.4112	14.06097
2/0	133100	0.101	12.79178	13.4431	14.09442
3/0	167800	0.0797	12.72571	13.37366	14.02161
4/0	211600	0.0626	12.60438	13.24616	13.88794
250	250000	0.0535	12.72698	13.375	14.02302
300	300000	0.0446	12.73174	13.38	14.02826
350	350000	0.0382	12.72222	13.37	14.01778
400	400000	0.0331	12.59852	13.24	13.88148
500	500000	0.0265	12.60804	13.25	13.89196
600	600000	0.0223	12.73174	13.38	14.02826
700	700000	0.0189	12.58901	13.23	13.87099
750	750000	0.0176	12.56046	13.2	13.83954
800	800000	0.0166	12.63658	13.28	13.92342
900	900000	0.0147	12.58901	13.23	13.87099
1000	1000000	0.0132	12.56046	13.2	13.83954
1250	1250000	0.0106	12.60804	13.25	13.89196
1500	1500000	0.00883	12.60328	13.245	13.88672
1750	1750000	0.00756	12.58901	13.23	13.87099
2000	2000000	0.00662	12.59852	13.24	13.88148

Note: K values are derived from Chapter 9, Table 8 of the *NEC*®.

Table 15.3 Direct-Current Resistance and K Values for Aluminum Conductors

Size (AWG or kcmil)	Circular mils	Ohm/kFT at 75°C (167°F)	K Value at 60°C (140°F)	K Value at 75°C (167°F)	K Value at 90°C (194°F)
18 Solid	1620	12.8	19.70957	20.736	21.76243
18 Stranded	1620	13.1	20.17151	21.222	22.27249
16 Solid	2580	8.05	19.74093	20.769	21.79707
16 Stranded	2580	8.21	20.1333	21.1818	22.2303
14 Solid	4110	5.06	19.76717	20.7966	21.82603
14 Stranded	4110	5.17	20.19689	21.2487	22.30051
12 Solid	6530	3.18	19.73751	20.7654	21.79329

Table continues below.

Table 15.3 Direct-Current Resistance and K Values
for Aluminum Conductors *(continued)*

Size (AWG or kcmil)	Circular mils	Ohm/kFT at 75°C (167°F)	K Value at 60°C (140°F)	K Value at 5°C (167°F)	K Value at 90°C (194°F)
12 Stranded	6530	3.25	20.17199	21.2225	22.27301
10 Solid	10380	2.0	19.73238	20.76	21.78762
10 Stranded	10380	2.04	20.12703	21.1752	22.22337
8 Solid	16510	1.26	19.77287	20.8026	21.83233
8 Stranded	16510	1.28	20.08673	21.1328	22.17887
6	26240	0.808	20.15242	21.20192	22.25142
4	41740	0.508	20.15433	21.20392	22.25351
3	52620	0.403	20.15617	21.20586	22.25555
2	66360	0.319	20.12098	21.16884	22.2167
1	83690	0.253	20.12548	21.17357	22.22166
1/0	105600	0.201	20.17493	21.2256	22.27627
2/0	133100	0.159	20.11534	21.1629	22.21046
3/0	167800	0.126	20.09623	21.1428	22.18937
4/0	211600	0.1	20.11258	21.16	22.20742
250	250000	0.0847	20.12684	21.175	22.22316
300	300000	0.0707	20.16011	21.21	22.2599
350	350000	0.0605	20.12684	21.175	22.22316
400	400000	0.0529	20.11258	21.16	22.20742
500	500000	0.0424	20.1506	21.2	22.2494
600	600000	0.0353	20.13159	21.18	22.22841
700	700000	0.0303	20.16011	21.21	22.25990
750	750000	0.0282	20.10308	21.15	22.19693
800	800000	0.0265	20.1506	21.2	22.2494
900	900000	0.0235	20.10308	21.15	22.19693
1000	1000000	0.0212	20.1506	21.2	22.2494
1250	1250000	0.0169	20.07931	21.125	22.17069
1500	1500000	0.0141	20.10308	21.15	22.19693
1750	1750000	0.0121	20.12684	21.175	22.22316
2000	2000000	0.0106	20.1506	21.2	22.2494

Note: *K* values are derived from Chapter 9, Table 8 of the *NEC*®.

Table 15.4 Alternating-Current Resistance and K Values for
Uncoated Copper Conductors at 75°C (167°F)

Size (AWG or kcmil)	PVC Conduit		Aluminum Conduit		Steel Conduit	
	Ohms/kFT	K Values	Ohms/kFT	K Values	Ohms/kFT	K Values
14	3.1	12.741	3.1	12.741	3.1	12.741
12	2.0	13.06	2.0	13.06	2.0	13.06
10	1.2	12.456	1.2	12.456	1.2	12.456
8	0.78	12.8778	0.78	12.8778	0.78	12.8778
6	0.49	12.8576	0.49	12.8576	0.49	12.8576

Continued

VOLTAGE DROP

244

VOLTAGE DROP

Table 15.4 Alternating-Current Resistance and K Values for Uncoated Copper Conductors at 75°C (167°F) (continued)

Size (AWG or kcmil)	PVC Conduit		Aluminum Conduit		Steel Conduit	
	Ohms/kFT	K Values	Ohms/kFT	K Values	Ohms/kFT	K Values
4	0.31	12.9394	0.31	12.9394	0.31	12.9394
3	0.25	13.155	0.25	13.155	0.25	13.155
2	0.19	12.6084	0.2	13.272	0.2	13.272
1	0.15	12.5535	0.16	13.3904	0.16	13.3904
1/0	0.12	12.672	0.13	13.728	0.12	12.672
2/0	0.1	13.31	0.1	13.31	0.1	13.31
3/0	0.077	12.9206	0.082	13.7596	0.079	13.2562
4/0	0.062	13.1192	0.067	14.1772	0.063	13.3308
250	0.052	13.0	0.057	14.25	0.054	13.5
300	0.044	13.2	0.049	14.7	0.045	13.5
350	0.038	13.3	0.043	15.05	0.039	13.65
400	0.033	13.2	0.038	15.2	0.035	14.0
500	0.027	13.5	0.032	16.0	0.029	14.5
600	0.023	13.8	0.028	16.8	0.025	15.0
750	0.019	14.25	0.024	18.0	0.021	15.75
1000	0.015	15.0	0.019	19.0	0.018	18.0

Note: K values are derived from Chapter 9, Table 9 of the NEC®.

Table 15.5 Alternating-Current Resistance and K Values for Aluminum Conductors at 75°C (167°F)

Size (AWG or kcmil)	PVC Conduit		Aluminum Conduit		Steel Conduit	
	Ohms/kFT	K Values	Ohms/kFT	K Values	Ohms/kFT	K Values
14	—	—	—	—	—	—
12	3.2	20.896	3.2	20.896	3.2	20.896
10	2.0	20.76	2.0	20.76	2.0	20.76
8	1.3	21.463	1.3	21.463	1.3	21.463
6	0.81	21.2544	0.81	21.2544	0.81	21.2544
4	0.51	21.2874	0.51	21.2874	0.51	21.2874
3	0.4	21.048	0.41	21.5742	0.4	21.048
2	0.32	21.2352	0.32	21.2352	0.32	21.2352
1	0.25	20.9225	0.26	21.7594	0.25	20.9225
1/0	0.2	21.12	0.21	22.176	0.2	21.12
2/0	0.16	21.296	0.16	21.296	0.16	21.296
3/0	0.13	21.814	0.13	21.814	0.13	21.814
4/0	0.1	21.16	0.11	23.276	0.1	21.16
250	0.085	21.25	0.09	22.5	0.086	21.5
300	0.071	21.3	0.076	22.8	0.072	21.6

Table continues below.

Table 15.5 Alternating-Current Resistance and *K* Values for Aluminum Conductors at 75°C (167°F) *(continued)*

Size (AWG or kcmil)	PVC Conduit Ohms/kFT	PVC Conduit *K* Values	Aluminum Conduit Ohms/kFT	Aluminum Conduit *K* Values	Steel Conduit Ohms/kFT	Steel Conduit *K* Values
350	0.061	21.35	0.066	23.1	0.063	22.05
400	0.054	21.6	0.059	23.6	0.055	22.0
500	0.043	21.5	0.048	24	0.045	22.5
600	0.036	21.6	0.041	24.6	0.038	22.8
750	0.029	21.75	0.034	25.5	0.031	23.25
1000	0.023	23.0	0.027	27	0.025	25.0

Note: *K* values are derived from Chapter 9, Table 9 of the *NEC*®.

ALTERNATIVE VOLTAGE DROP FORMULAS USING OHM'S LAW

The following formulas use the conductor's resistance instead of the conductor's *K* value to determine voltage drop. The "R" represents the conductor's resistance for 1000 feet. Insert the ohm/kFT values from Tables 8 and 9 of Chapter 9 for the conductor size.

Single-Phase Voltage Drop Formulas

(To convert to three-phase formulas, replace the "2" with "1.732.")

To find the volts dropped: $V_D = \dfrac{2\,R\,L\,I}{1000}$ or $V_D = \dfrac{2 \times R \times L \times I}{1000}$

To find the conductor size: $R = \dfrac{V_D\,1000}{2\,L\,I}$ or $R = \dfrac{V_D \times 1000}{2 \times L \times I}$

To find the distance (one way): $L = \dfrac{V_D\,1000}{2\,R\,I}$ or $L = \dfrac{V_D \times 1000}{2 \times R \times I}$

To find the current in amperes: $I = \dfrac{V_D\,1000}{2\,R\,L}$ or $I = \dfrac{V_D \times 1000}{2 \times R \times L}$

(Note, in the above formulas, the "R" is per 1000 feet.)

VOLTAGE DROP

VOLTAGE DROP

EXAMPLES OF ALTERNATIVE VOLTAGE DROP FORMULAS USING OHM'S LAW

Example 1, Voltage Drop

What is the voltage drop of a 208-volt, single-phase circuit that has a load of 20 amperes? The circuit is fed with 8 AWG, THW, stranded copper conductors and is 190 feet from the load.

Solution

Use the alternative formula that solves for single-phase voltage dropped in a circuit.

$$V_D = \frac{2\,R\,L\,I}{1000} = \frac{2 \times 0.778 \times 190 \times 20}{1000} = 5.9 \text{ volts}$$

Example 2, Conductor Size

What size branch-circuit copper conductors are needed for a load of 10 amperes? The single-phase voltage is 120 volts and the load is 100 feet from the source.

Solution

Select the formula for finding the conductor size in a single-phase circuit.

$$R = \frac{V_D\,1000}{2\,L\,I} = \frac{3.6 \times 1000}{2 \times 100 \times 10} = 1.8 \text{ ohm/kFT (maximum)} = 10 \text{ AWG copper}$$

Example 3, Distance

What is the maximum distance permitted for 8 AWG, THW, stranded, aluminum conductors that have a load of 40 amperes, and a single-phase voltage of 208 volts?

Solution

Use the single-phase formula for distance (one way). Note that this example is fed with aluminum conductors.

$$L = \frac{V_D\,1000}{2\,R\,I} = \frac{6.24 \times 1000}{2 \times 1.28 \times 40} = 61 \text{ feet}$$

Example 4, Current (in amperes)

What is the maximum load permitted on 500 kcmil, THW, copper conductors supplying a 208-volt, three-phase load 550 feet from the source?

Solution

Use the alternative formula for finding the maximum current in amperes. Since this is a three-phase circuit, replace the "2" with "1.732."

(Because the circuit is three phase, replace the 2 with 1.732.)

$$I = \frac{V_D\ 1000}{1.732\ R\ L} = \frac{6.24 \times 1000}{1.732 \times 0.0258 \times 550} = 254 \text{ amperes}$$

CHAPTER 16

RECEPTACLES

Chapter 16 covers receptacles and branch circuit requirements. It consists of extracted *NEC*® material from Article 210 and Article 406.

BRANCH-CIRCUIT REQUIREMENTS

See Exhibits 16.1 to 16.3.

EXHIBIT 16.1

NEC Table 210.21(B)(2) Maximum Cord-and-Plug-Connected Load to Receptacle

Circuit Rating (Amperes)	Receptacle Rating (Amperes)	Maximum Load (Amperes)
15 or 20	15	12
20	20	16
30	30	24

EXHIBIT 16.2

NEC Table 210.21(B)(3) Receptacle Ratings for Various Size Circuits

Circuit Rating (Amperes)	Receptacle Rating (Amperes)
15	Not over 15
20	15 or 20
30	30
40	40 or 50
50	50

RECEPTACLES

EXHIBIT 16.3

NEC Table 210.24   Summary of Branch-Circuit Requirements

Circuit Rating	15 A	20 A	30 A	40 A	50 A
Conductors (min. size):					
Circuit wires[1]	14	12	10	8	6
Taps	14	14	14	12	12
Fixture wires and cords — see 240.5					
Overcurrent Protection	**15 A**	**20 A**	**30 A**	**40 A**	**50 A**
Outlet devices:					
Lampholders permitted	Any type	Any type	Heavy duty	Heavy duty	Heavy duty
Receptacle rating[2]	15 max. A	15 or 20 A	30 A	40 or 50 A	50 A
Maximum Load	**15 A**	**20 A**	**30 A**	**40 A**	**50 A**
Permissible load	See 210.23(A)	See 210.23(A)	See 210.23(B)	See 210.23(C)	See 210.23(C)

[1]These gauges are for copper conductors.

[2]For receptacle rating of cord-connected electric-discharge luminaires (lighting fixtures), see 410.30(C).

RECEPTACLE AND PLUG CONFIGURATIONS

Figures 16.1 and 16.2 are NEMA configuration charts for plugs and receptacles. While Figure 16.1 shows general-purpose nonlocking plugs and receptacles, Figure 16.2 shows specific-purpose locking plugs and receptacles. These charts are useful in quickly selecting the right plug and/or receptacle.

DESCRIPTION		NEMA NUMBER	15 AMPERE		20 AMPERE		30 AMPERE		50 AMPERE		60 AMPERE	
			RECEPTACLE	PLUG	RECEPTACLE	PLUG	RECEPTACLE	PLUG	RECEPTACLE	PLUG	RECEPTACLE	PLUG
2–POLE 2–WIRE	125V	1	1–15R	1–15P POLARIZED / NON POLAR.		2–20P MATES WITH 5–20R		1–30P MATES WITH 5–30R				
2–POLE 2–WIRE	250V	2		2–15P MATES WITH 6–15R	2–20R	2–20P	2–30R	2–30P				
2–POLE 2–WIRE	277V AC	3										
2–POLE 2–WIRE	600V	4										
2–POLE 3–WIRE GROUNDING	125V	5	5–15R	5–15P	5–20R	5–20P	5–30R	5–30P	5–50R	5–50P		
2–POLE 3–WIRE GROUNDING	125V	5ALT			5ALT–20R							
2–POLE 3–WIRE GROUNDING	250V	6	6–15R	6–15P	6–20R	6–20P	6–30R	6–30P	6–50R	6–50P		
2–POLE 3–WIRE GROUNDING	250V	6ALT			6ALT–20R							
2–POLE 3–WIRE GROUNDING	277V AC	7	7–15R	7–15P	7–20R	7–20P	7–30R	7–30P	7–50R	7–50P		
2–POLE 3–WIRE GROUNDING	347V AC	24	24–15R	24–15P	24–20R	24–20P	24–30R	24–30P	24–50R	24–50P		
2–POLE 3–WIRE GROUNDING	480V AC	8										
2–POLE 3–WIRE GROUNDING	600V AC	9										
3–POLE 3–WIRE	125 / 250V	10			10–20R	10–20P	10–30R	10–30P	10–50R	10–50P		
3–POLE 3–WIRE	3 Ø 250V	11	11–15R	11–15P	11–20R	11–20P	11–30R	11–30P	11–50R	11–50P		
3–POLE 3–WIRE	3 Ø 480V	12										
3–POLE 3–WIRE	3 Ø 600V	13										
3–POLE 4–WIRE GROUNDING	125 / 250V	14	14–15R	14–15P	14–20R	14–20P	14–30R	14–30P	14–50R	14–50P	14–60R	14–60P
3–POLE 4–WIRE GROUNDING	3 Ø 250V	15	15–15R	15–15P	15–20R	15–20P	15–30R	15–30P	15–50R	15–50P	15–60R	15–60P
3–POLE 4–WIRE GROUNDING	3 Ø 480V	16										
3–POLE 4–WIRE GROUNDING	3 Ø 600V	17										
4–POLE 4–WIRE	3 Ø Y 120 / 280V	18	18–15R	18–15P	18–20R	18–20P	18–30R	18–30P	18–50R	18–50P	18–60R	18–60P
4–POLE 4–WIRE	3 Ø Y 277 / 480V	19										
4–POLE 4–WIRE	3 Ø Y 347 / 600V	20										
4–POLE 5–WIRE GROUNDING	3 Ø Y 120 / 208V	21										
4–POLE 5–WIRE GROUNDING	3 Ø Y 277 / 480V	22										
4–POLE 5–WIRE GROUNDING	3 Ø Y 347 / 600V	23										

Note: Blank spaces reserved for future configurations.

Figure 16.1 Configuration chart for general-purpose locking plugs and receptacles.

Reproduced from *Wiring, Devices—Dimensional Requirements,* NEMA WD 6–1997.

RECEPTACLES

DESCRIPTION	NEMA NUMBER	15 AMPERE RECEPTACLE	15 AMPERE PLUG	20 AMPERE RECEPTACLE	20 AMPERE PLUG	30 AMPERE RECEPTACLE	30 AMPERE PLUG	50 AMPERE RECEPTACLE	50 AMPERE PLUG	60 AMPERE RECEPTACLE	60 AMPERE PLUG
2-POLE 2-WIRE 125V	1	L1-15R	L1-15P								
250V	2			L2-20R	L2-20P						
277V AC	3										
600V	4										
2-POLE 3-WIRE GROUNDING 125V	5	L5-15R	L5-15P	L5-20R	L5-20P	L5-30R	L5-30P	L5-50R	L5-50P	L5-60R	L5-60P
250V	6	L6-15R	L6-15P	L6-20R	L6-20P	L6-30R	L6-30P	L6-50R	L6-50P	L6-60R	L6-60P
277V AC	7	L7-15R	L7-15P	L7-20R	L7-20P	L7-30R	L7-30P	L7-50R	L7-50P	L7-60R	L7-60P
347V AC	24			L24-20R	L24-20P						
480V AC	8			L8-20R	L8-20P	L8-30R	L8-30P	L8-50R	L8-50P	L8-60R	L8-60P
600V AC	9			L9-20R	L9-20P	L9-30R	L9-30P	L9-50R	L9-50P	L9-60R	L9-60P
3-POLE 3-WIRE 125/250V	10			L10-20R	L10-20P	L10-30R	L10-30P				
3 Ø 250V	11	11-15R	11-15P	L11-20R	L11-20P	L11-30R	L11-30P				
3 Ø 480V	12			L12-20R	L12-20P	L12-30R	L12-30P				
3 Ø 600V	13					L13-30R	L13-30P				
3-POLE 4-WIRE GROUNDING 125/250V	14			L14-20R	L14-20P	L14-30R	L14-30P	L14-50R	L14-50P	L14-60R	L14-60P
3 Ø 250V	15			L15-20R	L15-20P	L15-30R	L15-30P	L15-50R	L15-50P	L15-60R	L15-60P
3 Ø 480V	16			L16-20R	L16-20P	L16-30R	L16-30P	L16-50R	L16-50P	L16-60R	L16-60P
3 Ø 600V	17					L17-30R	L17-30P	L17-50R	L17-50P	L17-60R	L17-60P
4-POLE 4-WIRE 3 Ø Y 120/208V	18			L18-20R	L18-20P	L18-30R	L18-30P				
3 Ø Y 277/480V	19			L19-20R	L19-20P	L19-30R	L19-30P				
3 Ø Y 347/600V	20			L20-20R	L20-20P	L20-30R	L20-30P				
4-POLE 5-WIRE GROUNDING 3 Ø Y 120/208V	21			L21-20R	L21-20P	L21-30R	L21-30P	L21-50R	L21-50P	L21-60R	L21-60P
3 Ø Y 277/480V	22			L22-20R	L22-20P	L22-30R	L22-30P	L22-50R	L22-50P	L22-60R	L22-60P
3 Ø Y 347/600V	23			L23-20R	L23-20P	L23-30R	L23-30P	L23-50R	L23-50P	L23-60R	L23-60P

Note: Blank spaces reserved for future configurations.

Figure 16.2   Configuration chart for specific-purpose locking plugs and receptacles.

Reproduced from *Wiring, Devices—Dimensional Requirements*, NEMA WD 6–1997.

CHAPTER 17
SWITCHES AND LIGHTING

Chapter 17 covers switches and lighting, including wiring diagrams for single-pole, 3-way, and 4-way switches.

SWITCHES

Note the white conductor cannot be used as the return conductor from the switch to the switched outlet. The white conductor must be reidentified to indicate its use. In these applications, the conductor with white insulation must be permanently re-identified by painting or other effective means at its terminations and at each location where the conductor is visible and accessible. [See *NEC* 200.7(C)(2).]

Single-Pole Switches

Figure 17.1 shows a single-pole switch with the power supply feeding the lighting outlet.

Figure 17.2 shows a single-pole switch with the power supply feeding the wall switch.

Two 3-Way Switches

Figures 17.3 and 17.4 show two 3-way switches with the power supply feeding the lighting outlet. Figures 17.5 through 17.7 each show two 3-way switches with the power supply feeding one wall switch. Each of these diagrams show different configurations for wiring two 3-way switches.

One 4-Way and Two 3-Way Switches

Figures 17.8 and 17.9 show one 4-way and two 3-way switches feeding the lighting outlet. Figures 17.10 through 17.12 show one 4-way and two 3-way switches feeding one wall switch. Each of these diagrams shows different configurations for wiring one 4-way and two 3-way switches.

More than Three Wall Switches

Additional wall switches can be added to any of the previous configurations. Always start with two three-way switches, one receiving power from the supply and the other feeding power to the lighting outlet. Now, any number of four-way switches can be added. Simply install each additional four-way switch between the two three-way switches. One set of travelers will feed into the four-way switch and the other set of travelers will feed out, as shown in Figure 17.13.

254

SWITCHES AND LIGHTING

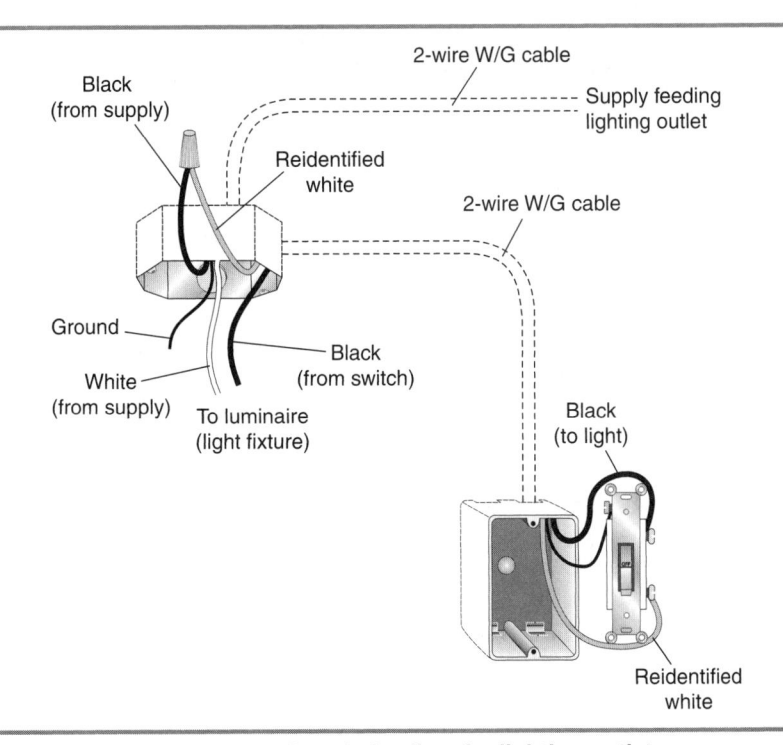

Figure 17.1 Supply feeding the lighting outlet.

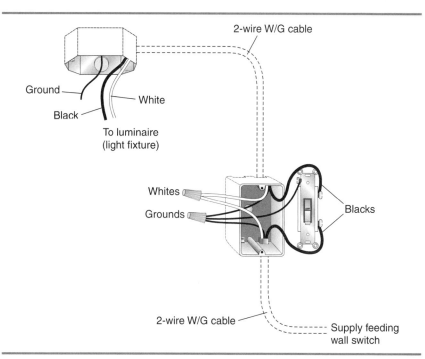

Figure 17.2 Supply feeding the wall switch.

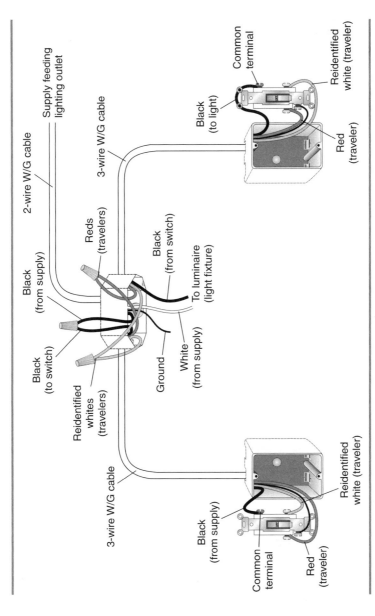

Figure 17.3 Supply feeding the lighting outlet with wiring running from the lighting outlet to each switch.

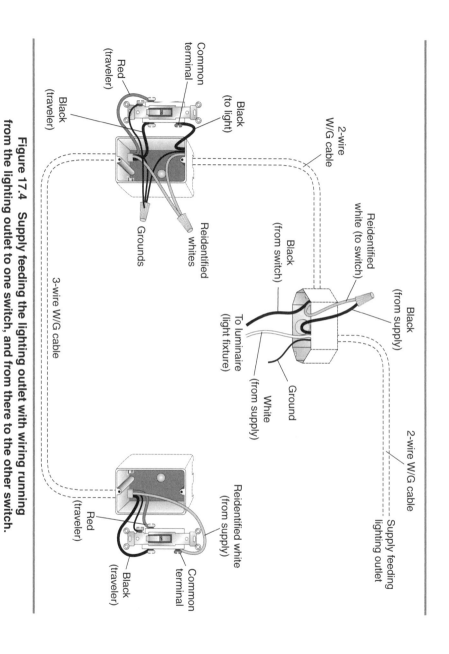

Figure 17.4 Supply feeding the lighting outlet with wiring running from the lighting outlet to one switch, and from there to the other switch.

SWITCHES AND LIGHTING

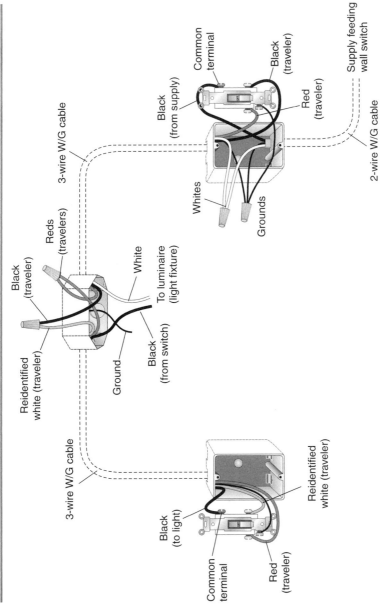

Figure 17.5 Supply feeding one wall switch with wiring running from the switch supplied with power to the lighting outlet, and from there to the other switch.

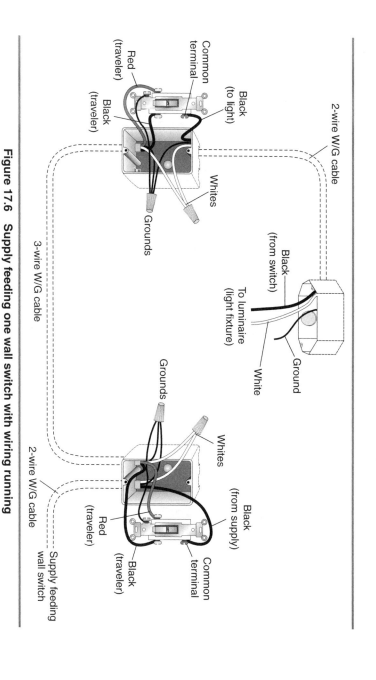

Figure 17.6 Supply feeding one wall switch with wiring running from the switch supplied with power to the other switch, and then to the lighting outlet.

SWITCHES AND LIGHTING

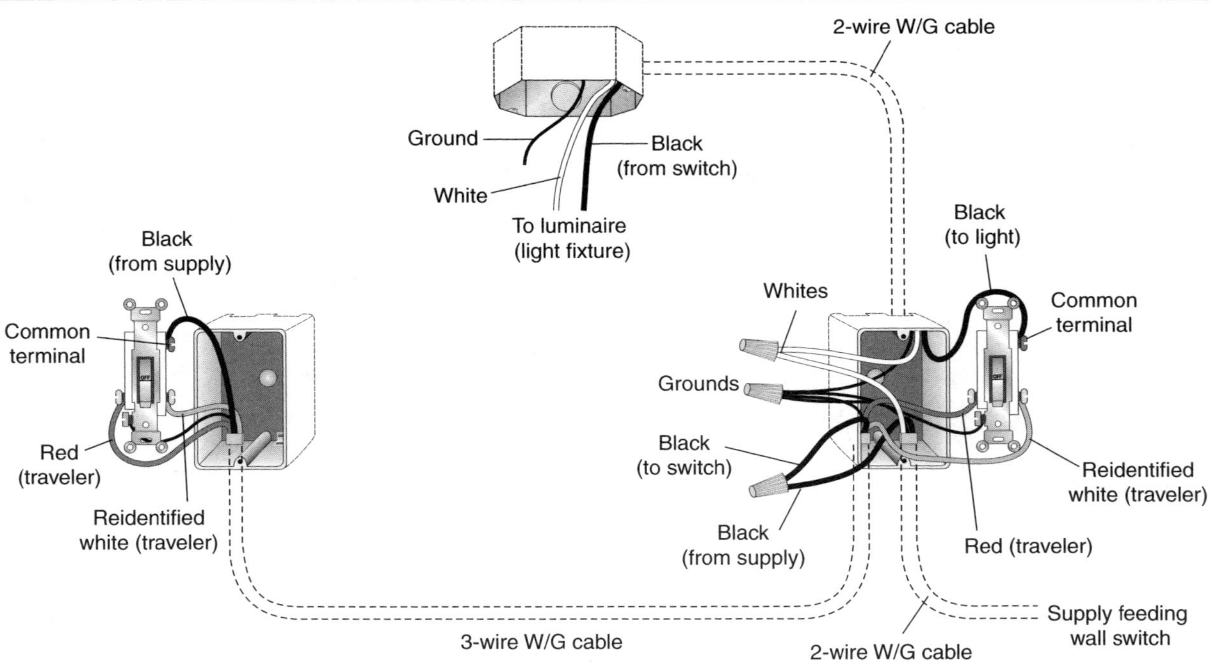

Figure 17.7 Supply feeding one wall switch with wiring running from the switch supplied with power to the other switch and also to the lighting outlet.
Note: Do not exceed box fill calculations as specified in 314.16. (See Chapter 11.)

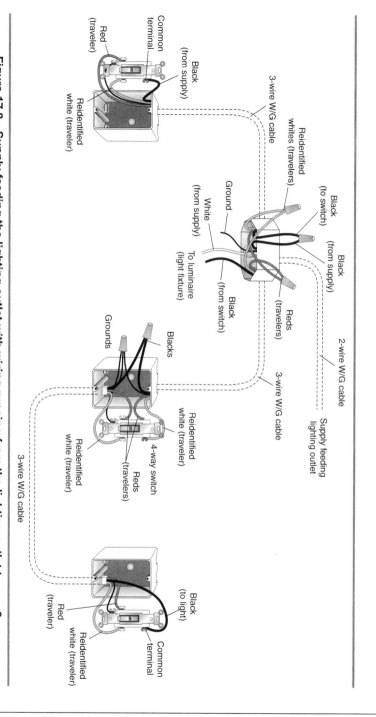

Figure 17.8 Supply feeding the lighting outlet with wiring running from the lighting outlet to one 3-way switch and also the 4-way switch; wiring also runs from the 4-way switch to the other 3-way switch.

SWITCHES AND LIGHTING

261

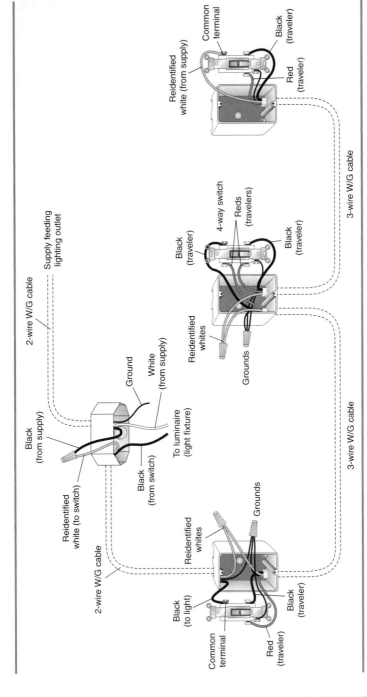

Figure 17.9 Supply feeding the lighting outlet with wiring running from the lighting outlet to one 3-way switch... then to the 4-way switch... and then to the other 3-way switch.

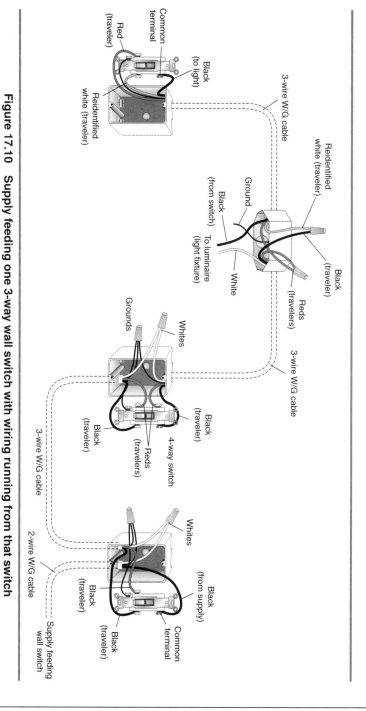

Figure 17.10 Supply feeding one 3-way wall switch with wiring running from that switch to the 4-way switch... to the lighting outlet... and then to the other 3-way switch.

SWITCHES AND LIGHTING

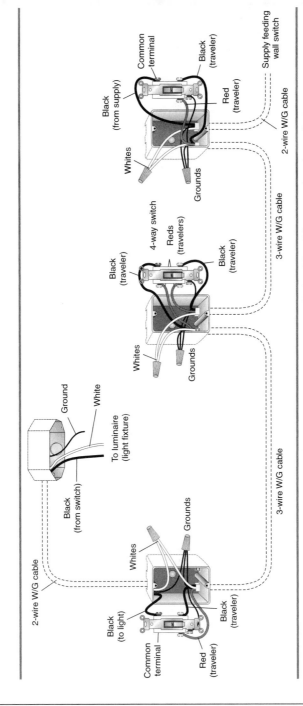

Figure 17.11 Supply feeding one 3-way wall switch with wiring running from that switch to the 4-way switch... to the other 3-way switch... and then to the lighting outlet.

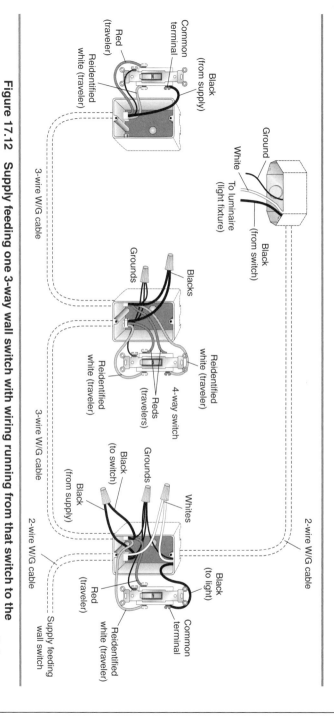

Figure 17.12 Supply feeding one 3-way wall switch with wiring running from that switch to the 4-way switch and also to the lighting outlet; wiring also runs from the 4-way switch to the other 3-way switch.

Note: Do not exceed box fill calculations as specified in 314.16. (See Chapter 12.)

SWITCHES AND LIGHTING

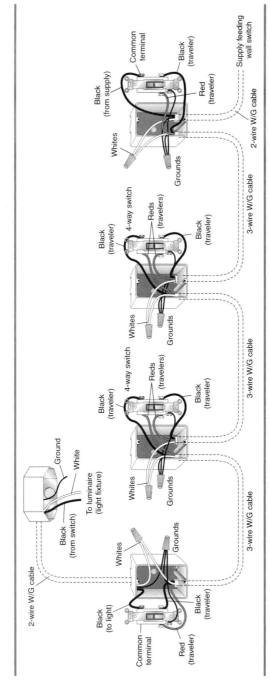

Figure 17.13 Installing 4-way switches between two 3-way switches.

LIGHTING

Luminaires (light fixtures) in clothes closets are covered in *NEC*, Section 410.8. The following paragraphs are extracted from Section 410.8.

410.8 Luminaires (Fixtures) in Clothes Closets.

(A) Definition.

Storage Space. The volume bounded by the sides and back closet walls and planes extending from the closet floor vertically to a height of 1.8 m (6 ft) or to the highest clothes-hanging rod and parallel to the walls at a horizontal distance of 600 mm (24 in.) from the sides and back of the closet walls, respectively, and continuing vertically to the closet ceiling parallel to the walls at a horizontal distance of 300 mm (12 in.) or the width of the shelf, whichever is greater; for a closet that permits access to both sides of a hanging rod, this space includes the volume below the highest rod extending 300 mm (12 in.) on either side of the rod on a plane horizontal to the floor extending the entire length of the rod.

FPN: See Figure 410.8 [shown here as Exhibit 17.1].

EXHIBIT 17.1
NEC **Figure 410.8   Closet Storage Space**

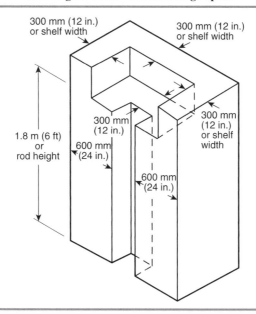

SWITCHES AND LIGHTING

(B) Luminaire (Fixture) Types Permitted. Listed luminaires (fixtures) of the following types shall be permitted to be installed in a closet:

(1) A surface-mounted or recessed incandescent luminaire (fixture) with a completely enclosed lamp

(2) A surface-mounted or recessed fluorescent luminaire (fixture)

(C) Luminaire (Fixture) Types Not Permitted. Incandescent luminaires (fixtures) with open or partially enclosed lamps and pendant luminaires (fixtures) or lampholders shall not be permitted.

(D) Location. Luminaires (fixtures) in clothes closets shall be permitted to be installed as follows:

(1) Surface-mounted incandescent luminaires (fixtures) installed on the wall above the door or on the ceiling, provided there is a minimum clearance of 300 mm (12 in.) between the luminaire (fixture) and the nearest point of a storage space

(2) Surface-mounted fluorescent luminaires (fixtures) installed on the wall above the door or on the ceiling, provided there is a minimum clearance of 150 mm (6 in.) between the luminaire (fixture) and the nearest point of a storage space

(3) Recessed incandescent luminaires (fixtures) with a completely enclosed lamp installed in the wall or the ceiling, provided there is a minimum clearance of 150 mm (6 in.) between the luminaire (fixture) and the nearest point of a storage space

(4) Recessed fluorescent luminaires (fixtures) installed in the wall or the ceiling, provided there is a minimum clearance of 150 mm (6 in.) between the luminaire (fixture) and the nearest point of a storage space

CHAPTER 18
MOTORS

Chapter 18 covers motors, including *NEC* tables and other relevant tables, wiring diagrams, NEMA starters, and motor terminal markings and connections.

NEC ARTICLE 430 TABLES

The following material is extracted from the *NEC*. It provides the reader with a collection of tables from Article 430 that are most often referenced (see Exhibits 18.1 to 18.10). See Chapter 11 for NEMA enclosures.

EXHIBIT 18.1

NEC Table 430.7(B)   Locked-Rotor Indicating Code Letters

Code Letter	Kilovolt-Amperes per Horsepower with Locked Rotor
A	0–3.14
B	3.15–3.54
C	3.55–3.99
D	4.0–4.49
E	4.5–4.99
F	5.0–5.59
G	5.6–6.29
H	6.3–7.09
J	7.1–7.99
K	8.0–8.99
L	9.0–9.99
M	10.0–11.19
N	11.2–12.49
P	12.5–13.99
R	14.0–15.99
S	16.0–17.99
T	18.0–19.99
U	20.0–22.39
V	22.4 and up

MOTORS

EXHIBIT 18.2

NEC Table 430.37   Overload Units

Kind of Motor	Supply System	Number and Location of Overload Units, Such as Trip Coils or Relays
1-phase ac or dc	2-wire, 1-phase ac or dc ungrounded	1 in either conductor
1-phase ac or dc	2-wire, 1-phase ac or dc, one conductor grounded	1 in ungrounded conductor
1-phase ac or dc	3-wire, 1-phase ac or dc, grounded neutral	1 in either ungrounded conductor
1-phase ac	Any 3-phase	1 in ungrounded conductor
2-phase ac	3-wire, 2-phase ac, ungrounded	2, one in each phase
2-phase ac	3-wire, 2-phase ac, one conductor grounded	2 in ungrounded conductors
2-phase ac	4-wire, 2-phase ac, grounded or ungrounded	2, one per phase in ungrounded conductors
2-phase ac	Grounded neutral or 5-wire, 2-phase ac, ungrounded	2, one per phase in any ungrounded phase wire
3-phase ac	Any 3-phase	3, one in each phase*

Exception: An overload unit in each phase shall not be required where overload protection is provided by other approved means.

EXHIBIT 18.3

NEC Table 430.52 Maximum Rating or Setting of Motor Branch-Circuit Short-Circuit and Ground-Fault Protective Devices

	Percentage of Full-Load Current			
Type of Motor	Nontime Delay Fuse[1]	Dual Element (Time-Delay) Fuse[1]	Instantaneous Trip Breaker	Inverse Time Breaker[2]
Single-phase motors	300	175	800	250
AC polyphase motors other than wound-rotor				
Squirrel cage — other than Design B energy-efficient	300	175	800	250
Design B energy-efficient	300	175	1100	250
Synchronous[3]	300	175	800	250
Wound rotor	150	150	800	150
Direct current (constant voltage)	150	150	250	150

Note: For certain exceptions to the values specified, see 430.54.

[1] The values in the Nontime Delay Fuse column apply to Time-Delay Class CC fuses.

[2] The values given in the last column also cover the ratings of nonadjustable inverse time types of circuit breakers that may be modified as in 430.52(C), Exception No. 1 and No. 2.

[3] Synchronous motors of the low-torque, low-speed type (usually 450 rpm or lower), such as are used to drive reciprocating compressors, pumps, and so forth, that start unloaded, do not require a fuse rating or circuit-breaker setting in excess of 200 percent of full-load current.

MOTORS

MOTORS

EXHIBIT 18.4

NEC Table 430.91   Motor Controller Enclosure Selection

For Outdoor Use

Provides a Degree of Protection Against the Following Environmental Conditions	Enclosure Type Number[1]									
	3	3R	3S	3X	3RX	3SX	4	4X	6	6P
Incidental contact with the enclosed equipment	X	X	X	X	X	X	X	X	X	X
Rain, snow, and sleet	X	X	X	X	X	X	X	X	X	X
Sleet[2]	—	—	X	—	—	X	—	—	—	—
Windblown dust	X	—	X	X	—	X	X	X	X	X
Hosedown	—	—	—	—	—	—	X	X	X	X
Corrosive agents	—	—	—	X	X	X	—	X	—	X
Temporary submersion	—	—	—	—	—	—	—	—	X	X
Prolonged submersion	—	—	—	—	—	—	—	—	—	X

For Indoor Use

Provides a Degree of Protection Against the Following Environmental Conditions	Enclosure Type Number[1]									
	1	2	4	4X	5	6	6P	12	12K	13
Incidental contact with the enclosed equipment	X	X	X	X	X	X	X	X	X	X
Falling dirt	X	X	X	X	X	X	X	X	X	X
Falling liquids and light splashing	—	X	X	X	X	X	X	X	X	X
Circulating dust, lint, fibers, and flyings	—	—	X	X	—	X	X	X	X	X
Settling airborne dust, lint, fibers, and flyings	—	—	X	X	X	X	X	X	X	X
Hosedown and splashing water	—	—	X	X	—	X	X	—	—	—
Oil and coolant seepage	—	—	—	—	—	—	—	X	X	X
Oil or coolant spraying and splashing	—	—	—	—	—	—	—	—	—	X
Corrosive agents	—	—	—	X	—	—	X	—	—	—
Temporary submersion	—	—	—	—	—	X	X	—	—	—
Prolonged submersion	—	—	—	—	—	—	X	—	—	—

[1]Enclosure type number shall be marked on the motor controller enclosure.

[2]Mechanism shall be operable when ice covered.

FPN: The term *raintight* is typically used in conjunction with Enclosure Types 3, 3S, 3SX, 3X, 4, 4X, 6, 6P. The term *rainproof* is typically used in conjunction with Enclosure Types 3R, 3RX. The term *watertight* is typically used in conjunction with Enclosure Types 4, 4X, 6, 6P. The term *driptight* is typically used in conjunction with Enclosure Types 2, 5, 12, 12K, 13. The term *dusttight* is typically used in conjunction with Enclosure Types 3, 3S, 3SX, 3X, 5, 12, 12K, 13.

EXHIBIT 18.5

NEC Table 430.247   Full-Load Current in Amperes, Direct-Current Motors

The following values of full-load currents* are for motors running at base speed.

rsepower	Armature Voltage Rating*					
	90 Volts	120 Volts	180 Volts	240 Volts	500 Volts	550 Volts
¼	4.0	3.1	2.0	1.6	—	—
⅓	5.2	4.1	2.6	2.0	—	—
½	6.8	5.4	3.4	2.7	—	—
¾	9.6	7.6	4.8	3.8	—	—
1	12.2	9.5	6.1	4.7	—	—
1½	—	13.2	8.3	6.6	—	—
2	—	17	10.8	8.5	—	—
3	—	25	16	12.2	—	—
5	—	40	27	20	—	—
7½	—	58	—	29	13.6	12.2
10	—	76	—	38	18	16
15	—	—	—	55	27	24
20	—	—	—	72	34	31
25	—	—	—	89	43	38
30	—	—	—	106	51	46
40	—	—	—	140	67	61
50	—	—	—	173	83	75
60	—	—	—	206	99	90
75	—	—	—	255	123	111
100	—	—	—	341	164	148
125	—	—	—	425	205	185
150	—	—	—	506	246	222

*These are average dc quantities.

MOTORS

MOTORS

EXHIBIT 18.6

NEC Table 430.248   Full-Load Currents in Amperes, Single-Phase Alternating-Current Motors

The following values of full-load currents are for motors running at usual speeds and motors with normal torque characteristics.

The voltages listed are rated motor voltages. The currents listed shall be permitted for system voltage ranges of 110 to 120 and 220 to 240 volts.

Horsepower	115 Volts	200 Volts	208 Volts	230 Volts
⅙	4.4	2.5	2.4	2.2
¼	5.8	3.3	3.2	2.9
⅓	7.2	4.1	4.0	3.6
½	9.8	5.6	5.4	4.9
¾	13.8	7.9	7.6	6.9
1	16	9.2	8.8	8.0
1½	20	11.5	11.0	10
2	24	13.8	13.2	12
3	34	19.6	18.7	17
5	56	32.2	30.8	28
7½	80	46.0	44.0	40
10	100	57.5	55.0	50

EXHIBIT 18.7

NEC Table 430.249   Full-Load Current, Two-Phase Alternating-Current Motors (4-Wire)

The following values of full-load currents are for motors running at speeds usual for belted motors and motors with normal torque characteristics.

Current in the common conductor of a 2-phase, 3-wire system will be 1.41 times the value given. The voltages listed are listed are rated motor voltages. The currents listed shall be permitted for system voltage ranges of 110 to 120, 220 to 240, 440 to 480, and 550 to 600 volts.

Horsepower	Induction-Type Squirrel Cage and Wound Rotor (Amperes)				
	115 Volts	230 Volts	460 Volts	575 Volts	2300 Volts
½	4.0	2.0	1.0	0.8	—
¾	4.8	2.4	1.2	1.0	—
1	6.4	3.2	1.6	1.3	—
1½	9.0	4.5	2.3	1.8	—
2	11.8	5.9	3.0	2.4	—
3	—	8.3	4.2	3.3	—
5	—	13.2	6.6	5.3	—
7½	—	19	9.0	8.0	—

Table continues below.

EXHIBIT 18.7 (continued)

NEC Table 430.249 Full-Load Current, Two-Phase Alternating-Current Motors (4-Wire)

Horsepower	Induction-Type Squirrel Cage and Wound Rotor (Amperes)				
	115 Volts	230 Volts	460 Volts	575 Volts	2300 Volts
10	—	24	12	10	—
15	—	36	18	14	—
20	—	47	23	19	—
25	—	59	29	24	—
30	—	69	35	28	—
40	—	90	45	36	—
50	—	113	56	45	—
60	—	133	67	53	14
75	—	166	83	66	18
100	—	218	109	87	23
125	—	270	135	108	28
150	—	312	156	125	32
200	—	416	208	167	43

EXHIBIT 18.8

NEC Table 430.250 Full-Load Current, Three-Phase Alternating-Current Motors

The following values of full-load currents are typical for motors running at speeds usual for belted motors and motors with normal torque characteristics.

The voltages listed are rated motor voltages. The currents listed shall be permitted for system voltage ranges of 110 to 120, 220 to 240, 440 to 480, and 550 to 600 volts.

Horsepower	Induction-Type Squirrel Cage and Wound Rotor (Amperes)							Synchronous-Type Unity Power Factor* (Amperes)			
	115 Volts	200 Volts	208 Volts	230 Volts	460 Volts	575 Volts	2300 Volts	230 Volts	460 Volts	575 Volts	2300 Volts
½	4.4	2.5	2.4	2.2	1.1	0.9	—	—	—	—	—
¾	6.4	3.7	3.5	3.2	1.6	1.3	—	—	—	—	—
1	8.4	4.8	4.6	4.2	2.1	1.7	—	—	—	—	—
1½	12.0	6.9	6.6	6.0	3.0	2.4	—	—	—	—	—
2	13.6	7.8	7.5	6.8	3.4	2.7	—	—	—	—	—
3	—	11.0	10.6	9.6	4.8	3.9	—	—	—	—	—
5	—	17.5	16.7	15.2	7.6	6.1	—	—	—	—	—
7½	—	25.3	24.2	22	11	9	—	—	—	—	—
10	—	32.2	30.8	28	14	11	—	—	—	—	—
15	—	48.3	46.2	42	21	17	—	—	—	—	—
20	—	62.1	59.4	54	27	22	—	—	—	—	—
25	—	78.2	74.8	68	34	27	—	53	26	21	—
30	—	92	88	80	40	32	—	63	32	26	—
40	—	120	114	104	52	41	—	83	41	33	—
50	—	150	143	130	65	52	—	104	52	42	—
60	—	177	169	154	77	62	16	123	61	49	12
75	—	221	211	192	96	77	20	155	78	62	15
100	—	285	273	248	124	99	26	202	101	81	20
125	—	359	343	312	156	125	31	253	126	101	25
150	—	414	396	360	180	144	37	302	151	121	30
200	—	552	528	480	240	192	49	400	201	161	40
250	—	—	—	—	302	242	60	—	—	—	—
300	—	—	—	—	361	289	72	—	—	—	—
350	—	—	—	—	414	336	83	—	—	—	—
400	—	—	—	—	477	382	95	—	—	—	—
450	—	—	—	—	515	412	103	—	—	—	—
500	—	—	—	—	590	472	118	—	—	—	—

*For 90 and 80 percent power factor, the figures shall be multiplied by 1.1 and 1.25, respectively.

276

MOTORS

EXHIBIT 18.9

NEC Table 430.251(A)   Conversion Table of Single-Phase Locked-Rotor
Currents for Selection of Disconnecting Means and Controllers
as Determined from Horsepower and Voltage Rating

For use only with 430.110, 440.12, 440.41, and 455.8(C).

Rated Horsepower	Maximum Locked-Rotor Current in Amperes, Single Phase		
	115 Volts	208 Volts	230 Volts
½	58.8	32.5	29.4
¾	82.8	45.8	41.4
1	96	53	48
1	120	66	60
2	144	80	72
3	204	113	102
5	336	186	168
7½	480	265	240
10	600	332	300

EXHIBIT 18.10

NEC Table 430.251(B) Conversion Table of Polyphase Design B, C, and D Maximum Locked-Rotor Currents for Selection of Disconnecting Means and Controllers as Determined from Horsepower and Voltage Rating and Design Letter

For use only with 430.110, 440.12, 440.41, and 455.8(C).

	Maximum Motor Locked-Rotor Current in Amperes, Two- and Three-Phase, Design B, C, and D*					
	115 Volts	200 Volts	208 Volts	230 Volts	460 Volts	575 Volts
Rated Horsepower	B, C, D	B, C, D	B, C, D	B, C, D	B, C, D	B, C, D
½	40	23	22.1	20	10	8
¾	50	28.8	27.6	25	12.5	10
1	60	34.5	33	30	15	12
1½	80	46	44	40	20	16
2	100	57.5	55	50	25	20
3	—	73.6	71	64	32	25.6
5	—	105.8	102	92	46	36.8
7½	—	146	140	127	63.5	50.8
10	—	186.3	179	162	81	64.8
15	—	267	257	232	116	93
20	—	334	321	290	145	116
25	—	420	404	365	183	146
30	—	500	481	435	218	174
40	—	667	641	580	290	232
50	—	834	802	725	363	290
60	—	1001	962	870	435	348
75	—	1248	1200	1085	543	434
100	—	1668	1603	1450	725	580
125	—	2087	2007	1815	908	726
150	—	2496	2400	2170	1085	868
200	—	3335	3207	2900	1450	1160
250	—	—	—	—	1825	1460
300	—	—	—	—	2200	1760
350	—	—	—	—	2550	2040
400	—	—	—	—	2900	2320
450	—	—	—	—	3250	2600
500	—	—	—	—	3625	2900

*Design A motors are not limited to a maximum starting current or locked rotor current.

MOTORS

MOTOR CONTROL ABBREVIATIONS AND SYMBOLS

Tables 18.1 and 18.2 show motor control abbreviations and symbols.

Table 18.1   Motor Control Abbreviations

Abbrev.	Term	Abbrev.	Term
AUX	auxiliary	NC	normally closed
CB	circuit breaker	NO	normally open
CR	control relay	OL	overloads
DP	double pole	PL	pilot light
DPDT	double pole, double throw	PRI	primary
DPST	double pole, single throw	R	reverse
F	forward or fast	REV	reverse
FOR	forward	SCR	silicon controlled rectifier
FS	float switch	SEC	secondary
FTS	foot switch	SP	single pole
FWD	forward	SPDT	single pole, double throw
GND	ground	SPST	single pole, single throw
HOA	hand-off-automatic	T1	motor terminal one
HV	high voltage	T2	motor terminal two
L1	line one	T3	motor terminal three
L2	line two	TC	timed closed
L3	line three	TD	time delay
LS	limit switch	TO	timed open
LV	low voltage	TR	timing relay
M	motor or motor starter		

(See Chapter 2 for additional abbreviations.)

Table 18.2 Motor Control Symbols

Switches

Symbol	Description
	Disconnect
	Fused Disconnect
	Circuit Breaker
	Circuit Breaker with Thermal Overloads
	Circuit Breaker with Magnetic Overloads
	Limit Switch—Normally Open
	Limit Switch—Normally Closed
	Limit Switch—Normally Open, Held Closed
	Limit Switch—Normally Closed, Held Open
	Flow Switch—Normally Open
	Flow Switch—Normally Closed
	Pressure or Vacuum Switch—Normally Open (Closes on Pressure Rise)

Continued

MOTORS

MOTORS

Table 18.2   Motor Control Symbols *(continued)*

Switches

	Pressure or Vacuum Switch—Normally Closed (Opens on Pressure Rise)
	Liquid Level Float Switch—Normally Open
	Liquid Level Float Switch—Normally Closed
	Foot Switch—Normally Open
	Foot Switch—Normally Closed
	Temperature Actuated Switch—Normally Open
	Temperature Actuated Switch—Normally Closed
	Speed (Plugging)
	Anti-Plug

Push Buttons—Momentary Contact

	Normally Open
	Normally Closed
	Normally Open and Normally Closed (Double Circuit)

Table continues below.

Table 18.2 Motor Control Symbols *(continued)*

Symbol	Description
Push Buttons—Momentary Contact *(continued)*	
	Mushroom Head
	Wobble Stick
	Illuminated
Push Buttons—Maintained Contact	
	Two Single Circuits
	One Double Circuit
Instant Operating Contacts	
	Normally Open
	Normally Closed
	Normally Open with Blowout
	Normally Closed with Blowout
Timed Contacts	
	Normally Open, Timed Closing Time Delay Starts when Energized
	Normally Closed, Timed Opening Time Delay Starts when Energized
	Normally Open, Timed Opening Time Delay Starts when De-energized
	Normally Closed, Timed Closing Time Delay Starts when De-energized

Continued

282

MOTORS

Table 18.2   Motor Control Symbols *(continued)*

Single-Pole, Single-Throw Contacts	
	Normally Open, Single Break
	Normally Open, Double Break
	Normally Closed, Single Break
	Normally Closed, Double Break

Double-Pole, Single-Throw Contacts	
	Two Normally Open, Single Break
	Two Normally Open, Double Break
	Two Normally Closed, Single Break
	Two Normally Closed, Double Break

Single-Pole, Double-Throw Contacts	
	Single Break
	Double Break

Double-Pole, Double-Throw Contacts	
	Single Break
	Double Break

Table continues below.

Table 18.2 Motor Control Symbols (continued)

Symbol	Description
Pilot Lights	
—(R)—	Non Push-to-Test (Letter Indicates Color)
(G)	Push-to-Test (Letter Indicates Color)
Overload Relays	
	Thermal
	Magnetic
Resistors	
—[RES]—	Fixed
—[H]—	Heating Element
—[RES]—	Adjustable, by Fixed Taps
—[RH]—	Rheostat, Potentiometer or Adjustable Taps
Inductors	
—⌒⌒⌒—	Iron Core
—⌒⌒⌒—	Air Core
Capacitors	
—)(—	Fixed
—)/(—	Adjustable
Transformers	
—⌒⌒⌒—	Auto

Continued

MOTORS

Table 18.2 Motor Control Symbols *(continued)*

	Transformers
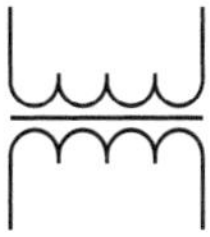	Iron Core
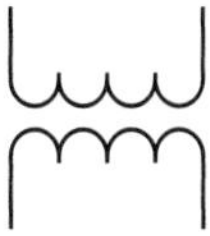	Air Core
	Current
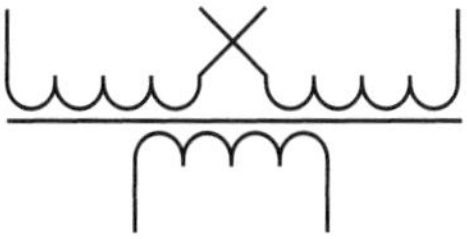	Dual Voltage
	AC Motors
	Single-Phase
	Three-Phase, Squirrel Cage
	Two-Phase, Four-Wire
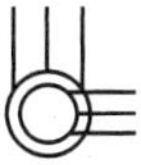	Wound Rotor
	DC Motors
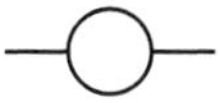	Armature
	Shunt Field (four loops)
	Series Field (three loops)
	Commutating or Compensating Field (two loops)

Table continues below.

Table 18.2 Motor Control Symbols *(continued)*

Symbol	Description
Wiring	
┼	Not Connected
┼ (with dot)	Connected
│ (thick)	Power
│ (thin)	Control
Connections	
- - - - -	Mechanical Connection
- - + - -	Mechanical Interlock Connection
Miscellaneous	
—⊏▬⊐—	Fuse (Power or Control)
⫞⎮⎮⫠ (+ −)	Battery
—(M)—	Starter Coil
O	Wiring Terminal
⏚	Ground
○ on □	Bell
◁	Buzzer

Continued

MOTORS

MOTORS

Table 18.2   Motor Control Symbols *(continued)*

Miscellaneous

	Horn, Alarm, Siren, etc.
	Annunciator
VM / AM	Meter (Letters Indicate Type)
	Meter Shunt
	Thermocouple

WIRING DIAGRAMS

Figures 18.1 through 18.10 show some common wiring diagrams.

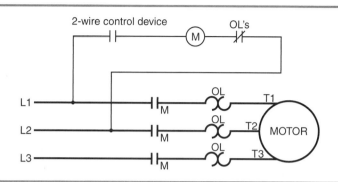

Figure 18.1   Basic diagram of two-wire control circuit.

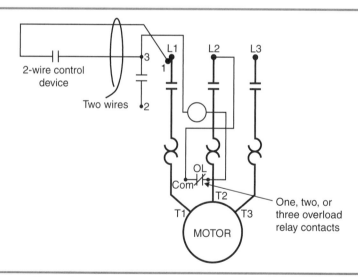

Figure 18.2   Wiring diagram of starter (two-wire control).

288

MOTORS

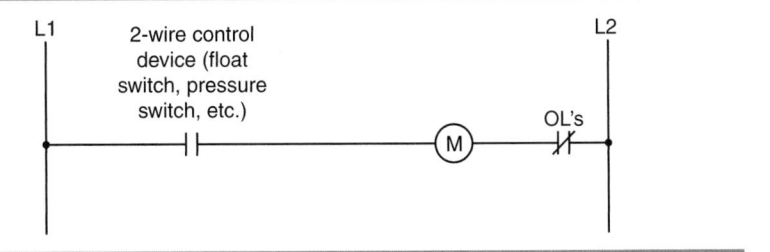

Figure 18.3   Control circuit only.

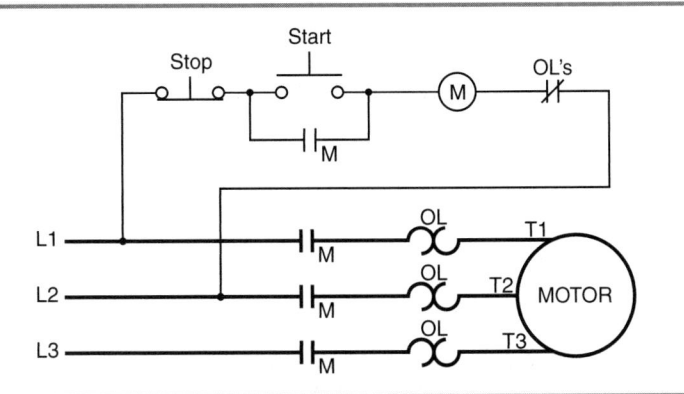

Figure 18.4   Basic three-wire control circuit.

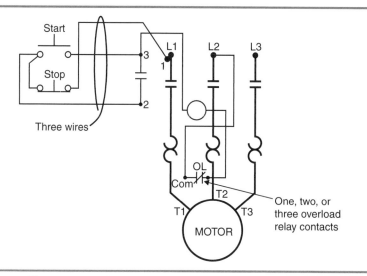

Figure 18.5 Wiring diagram of starter (three-wire control).

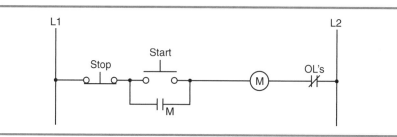

Figure 18.6 Control circuit only.

290

MOTORS

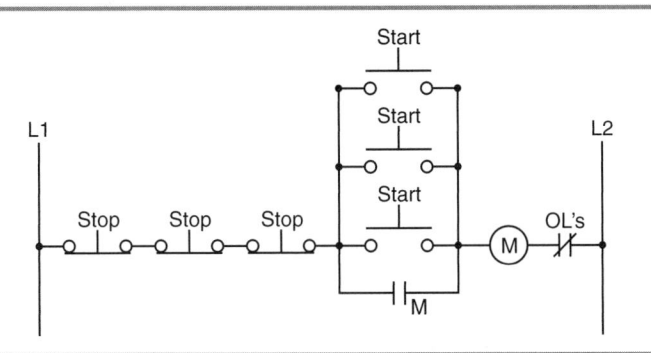

Figure 18.7   Multiple start and stop stations.

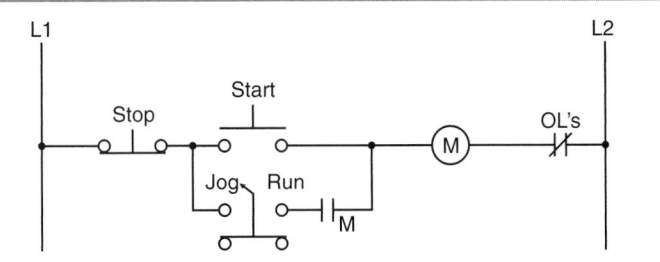

Figure 18.8   Start push button with jog selector switch.

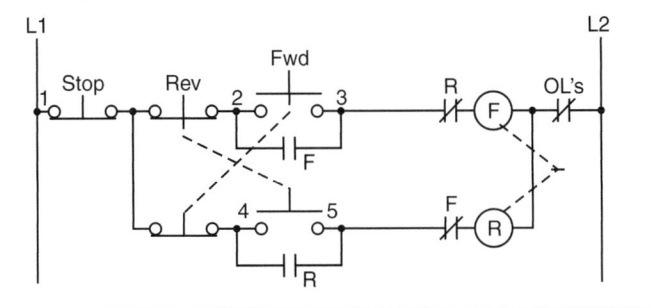

Figure 18.9   Reversing starter.

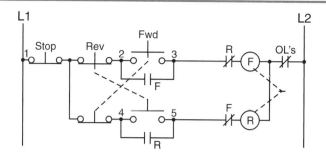

Figure 18.10 Reversing starter with limit switches.

NEMA STARTERS

Table 18.3 shows electrical ratings for NEMA starters and Table 18.4 shows NEMA starter sizes for various three-phase motors.

Table 18.3 Electrical Ratings for NEMA Starters

NEMA Size	Volts (VAC)	Single Phase	Three Phase	Continuous Current Rating (Amperes)
00	115	1/3	—	9
	200	—	1-1/2	
	230	1	1-1/2	
	460/575	—	2	
0	115	1	—	18
	200	—	3	
	230	2	3	
	460/575	—	5	
1	115	2	—	27
	200	—	7-1/2	
	230	3	7-1/2	
	460/575	—	10	
1P	115	3	—	36
	230	5	—	
2	115	3	—	45
	200	—	10	
	230	7-1/2	15	
	460/575	—	25	
3	115	7-1/2	—	90
	200	—	25	
	230	15	30	
	460/575	—	50	

Continued

MOTORS

MOTORS

Table 18.3 Electrical Ratings for NEMA Starters *(continued)*

NEMA Size	Volts (VAC)	Single Phase	Three Phase	Continuous Current Rating (Amperes)
4	200	—	40	
	230	—	50	135
	460/575	—	100	
5	200	—	75	
	230	—	100	270
	460/575	—	200	
6	200	—	150	
	230	—	200	540
	460/575	—	400	
7	230	—	300	
	460/575	—	600	90
8	230	—	450	
	460/575	—	400	1215
9	460/575	—	900	
	230	—	800	2250

Table 18.4 Maximum Horsepower

Three-Phase Motors

NEMA Size	Full Voltage Starting			Auto Transformer Starting			Part Winding Starting			Wye Delta Starting		
	200 VAC	230 VAC	460/575 VAC	200 VAC	230 VAC	460/575 VAC	200 VAC	230 VAC	460/575 VAC	200 VAC	230 VAC	460/575 VAC
00	1-1/2	1-1/2	2	—	—	—	—	—	—	—	—	—
0	3	3	5	—	—	—	—	—	—	—	—	—
1	7-1/2	7-1/2	10	7-1/2	7-1/2	10	10	10	15	10	10	15
2	10	15	25	10	15	25	20	25	40	20	25	40
3	25	30	50	25	30	50	40	50	75	40	50	75
4	40	50	100	40	50	100	75	75	150	60	75	150
5	75	100	200	75	100	200	150	150	350	150	150	300
6	150	200	400	150	200	400	—	300	600	300	350	700
7	—	300	600	—	300	600	—	450	900	500	500	1000
8	—	450	900	—	450	900	—	700	1400	750	800	1500
9	—	800	1600	—	800	1600	—	1300	2600	1500	1500	3000

TERMINAL MARKINGS AND CONNECTIONS

Note: In three-phase motors, the rotation direction can be reversed by swapping any two lines feeding the motor.

Note: Refer to each motor's nameplate for high- and low-voltage connections.

Wye-Connected Three-Phase Motors

Figures 18.11 through 18.13 show wye-connected three-phase motors.

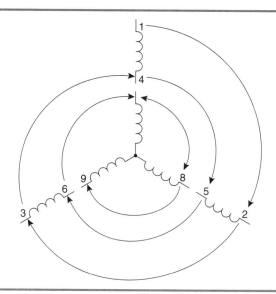

Figure 18.11   Standard numbering for a wye-connected motor.

MOTORS

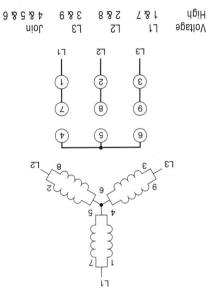

Figure 18.12 High-voltage connections for wye-connected motors.

Voltage	L1	L2	L3	Join
High	1	2	3	4&7, 5&8, 6&9

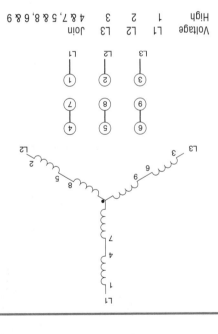

Figure 18.13 Low-voltage connections for wye-connected motors.

Voltage	L1	L2	L3	Join
High	1&7	2&8	3&9	4&5&6

Delta-Connected Three-Phase Motors

Figures 18.14 through 18.16 show delta-connected three-phase motors.

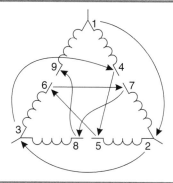

Figure 18.14   Standard numbering for a delta-connected motor.

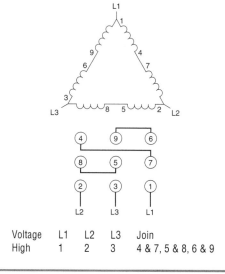

Voltage	L1	L2	L3	Join
High	1	2	3	4 & 7, 5 & 8, 6 & 9

Figure 18.15   High-voltage connections for delta-connected motors.

MOTORS

MOTORS

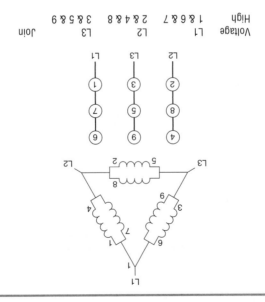

Voltage	L1	L2	L3	Join
High	1 & 6 & 7	2 & 4 & 8	3 & 5 & 9	

Figure 18.16 Low-voltage connections for delta-connected motors.

CHAPTER 19
TRANSFORMERS

Chapter 19 covers transformer ratings, common transformer connections, transformer formulas, and buck-boost transformer diagrams.

TRANSFORMER RATINGS

Exhibits 19.1 and 19.2 [*NEC* Tables 450.3(A) and 450.3(B)] are extracted from *NEC,* Article 450. They provide the reader with a quick reference for sizing overcurrent protection for transformers. Transformers over 600 volts, nominal must not exceed the ratings or settings of overcurrent protection specified in Exhibit 19.1. The maximum rating or setting of overcurrent protection for transformers 600 volts, nominal is specified in Exhibit 19.2.

EXHIBIT 19.1

NEC Table 450.3(A)   Maximum Rating or Setting of Overcurrent Protection for Transformers over 600 Volts (as a Percentage of Transformer-Rated Current)

		Primary Protection Over 600 Volts		Secondary Protection (See Note 2.)		
				Over 600 Volts		600 Volts or Less
Location Limitations	Transformer Rated Impedance	Circuit Breaker (See Note 4.)	Fuse Rating	Circuit Breaker (See Note 4.)	Fuse Rating	Circuit Breaker or Fuse Rating
Any location	Not more than 6%	600% (See Note 1.)	300% (See Note 1.)	300% (See Note 1.)	250% (See Note 1.)	125% (See Note 1.)
	More than 6% and not more than 10%	400% (See Note 1.)	300% (See Note 1.)	250% (See Note 1.)	225% (See Note 1.)	125% (See Note 1.)
Supervised locations only (See Note 3.)	Any	300% (See Note 1.)	250% (See Note 1.)	Not required	Not required	Not required
	Not more than 6%	600%	300%	300% (See Note 5.)	250% (See Note 5.)	250% (See Note 5.)
	More than 6% and not more than 10%	400%	300%	250% (See Note 5.)	225% (See Note 5.)	250% (See Note 5.)

Table continues below.

TRANSFORMERS

EXHIBIT 19.1 *(continued)*

NEC Table 450.3(A)   Maximum Rating or Setting of Overcurrent Protection for Transformers Over 600 Volts (as a Percentage of Transformer-Rated Current)

Notes:

1. Where the required fuse rating or circuit breaker setting does not correspond to a standard rating or setting, a higher rating or setting that does not exceed the next higher standard rating or setting shall be permitted.

2. Where secondary overcurrent protection is required, the secondary overcurrent device shall be permitted to consist of not more than six circuit breakers or six sets of fuses grouped in one location. Where multiple overcurrent devices are utilized, the total of all the device ratings shall not exceed the allowed value of a single overcurrent device. If both circuit breakers and fuses are used as the overcurrent device, the total of the device ratings shall not exceed that allowed for fuses.

3. A supervised location is a location where conditions of maintenance and supervision ensure that only qualified persons monitor and service the transformer installation.

4. Electrically actuated fuses that may be set to open at a specific current shall be set in accordance with settings for circuit breakers.

5. A transformer equipped with a coordinated thermal overload protection by the manufacturer shall be permitted to have separate secondary protection omitted.

EXHIBIT 19.2

NEC Table 450.3(B)   Maximum Rating or Setting of Overcurrent Protection for Transformers 600 Volts and Less (as a Percentage of Transformer-Rated Current)

Protection Method	Primary Protection			Secondary Protection (See Note 2.)	
	Currents of 9 Amperes or More	Currents Less Than 9 Amperes	Currents Less Than 2 Amperes	Currents of 9 Amperes or More	Currents Less Than 9 Amperes
Primary only protection	125% (See Note 1.)	167%	300%	Not required	Not required
Primary and secondary protection	250% (See Note 3.)	250% (See Note 3.)	250% (See Note 3.)	125% (See Note 1.)	167%

Notes:

1. Where 125 percent of this current does not correspond to a standard rating of a fuse or nonadjustable circuit breaker, a higher rating that does not exceed the next higher standard rating shall be permitted.

2. Where secondary overcurrent protection is required, the secondary overcurrent device shall be permitted to consist of not more than six circuit breakers or six sets of fuses grouped in one location. Where multiple overcurrent devices are utilized, the total of all the device ratings shall not exceed the allowed value of a single overcurrent device. If both breakers and fuses are utilized as the overcurrent device, the total of the device ratings shall not exceed that allowed for fuses.

3. A transformer equipped with coordinated thermal overload protection by the manufacturer and arranged to interrupt the primary current shall be permitted to have primary overcurrent protection rated or set at a current value that is not more than six times the rated current of the transformer for transformers having not more than 6 percent impedance and not more than four times the rated current of the transformer for transformers having more than 6 percent but not more than 10 percent impedance.

Table 19.1 lists single-phase transformer ratings and Table 19.2 lists 3-phase transformer ratings.

Table 19.1 Single-Phase Transformers

kVA Ratings	Full Load Current—Amperes							
	120 V	208 V	240 V	277 V	480 V	600 V	2400 V	4160 V
0.5	4.2	2.4	2.1	1.8	1.0	0.8	0.2	0.1
0.75	6.3	3.6	3.1	2.7	1.6	1.3	0.3	0.2
1	8.3	4.8	4.2	3.6	2.1	1.7	0.4	0.2
1.5	12.5	7.2	6.3	5.4	3.1	2.5	0.6	0.4
2	16.7	9.6	8.3	7.2	4.2	3.3	0.8	0.5
3	25.0	14.4	12.5	10.8	6.3	5.0	1.3	0.7
5	42.7	24.0	20.8	18.1	10.4	8.3	2.1	1.2
7.5	62.5	36.1	31.3	27.1	15.6	12.5	3.1	1.8
10	83.3	48.1	41.7	36.1	20.8	16.7	4.2	2.4
15	125.0	72.1	62.5	54.2	31.3	25.0	6.3	3.6
20	166.7	96.2	83.3	72.2	41.7	33.3	8.3	4.8
25	208.3	120.2	104.2	90.3	52.1	41.7	10.4	6.0
30	250.0	144.2	125.0	108.3	62.5	50.0	12.5	7.2
37.5	312.5	180.3	156.3	135.4	78.1	62.5	15.6	9.0
50	417.7	240.4	208.3	180.5	104.2	83.3	20.8	12.0
75	625.0	360.6	312.5	270.8	156.3	125.0	31.3	18.0
100	833.3	480.8	416.7	361.0	208.3	166.7	41.7	24.0
167	1391.7	802.9	695.8	602.9	347.9	278.3	69.6	40.1
250	2083.3	1201.9	1014.7	902.5	520.8	416.7	104.2	60.1
333	2775.0	1601.0	1387.5	1202.2	693.8	555.0	138.8	80.0
500	4166.7	2403.8	2083.3	1805.1	1041.7	833.3	208.3	120.2

kVA = (FLA × Line Voltage) / 1000
Amperes = (kVA × 1000) / Line Voltage

Table 19.2 Three-Phase Transformers

kVA Ratings	Full Load Current—Amperes							
	208 V	230 V	240 V	460 V	480 V	600 V	2400 V	4160 V
3	8.3	7.5	7.2	3.8	3.6	2.9	0.7	0.4
6	16.7	15.1	14.4	7.5	7.2	5.8	1.4	0.8
9	25.0	22.6	21.7	11.3	10.8	8.7	2.2	1.2
15	41.7	37.7	36.1	18.8	18.0	14.4	3.6	2.1
20	55.6	50.2	48.1	25.1	24.1	19.2	4.8	2.8
25	69.4	62.8	60.1	31.4	30.1	24.1	6.0	3.5
30	83.3	75.3	72.2	37.7	36.1	28.9	7.2	4.2
37.5	104.2	94.1	90.2	47.1	45.1	36.1	9.0	5.2

Continued

TRANSFORMERS

Table 19.2 Three-Phase Transformers *(continued)*

kVA Ratings	Full Load Current—Amperes							
	208 V	230 V	240 V	460 V	480 V	600 V	2400 V	4160 V
45	125.0	113.0	108.3	56.5	54.1	43.3	10.8	6.2
50	138.9	125.5	120.3	62.8	60.1	48.1	12.0	6.9
60	166.7	150.6	144.3	75.3	72.2	57.7	14.4	8.3
75	208.3	188.3	180.4	94.1	90.2	72.2	18.0	10.4
100	277.8	251.0	240.6	125.5	120.3	96.2	24.1	13.9
112.5	312.5	282.4	270.6	141.2	135.3	108.3	27.1	15.6
150	416.7	376.5	360.8	188.3	180.4	144.3	36.1	20.8
200	555.6	502.0	481.1	251.0	240.6	192.5	48.1	27.8
225	625.0	564.8	541.3	282.4	270.6	216.5	54.1	31.2
300	833.3	753.1	721.7	376.5	360.8	288.7	72.2	41.6
400	1111.1	1004.1	962.3	502.0	481.1	384.9	96.2	55.5
500	1388.9	1255.1	1202.8	627.6	601.4	481.1	120.3	69.4
750	2083.3	1882.7	1804.2	941.3	902.1	721.7	180.4	104.1
1000	2777.8	2510.2	2405.6	1255.1	1202.8	962.3	240.6	138.8

kVA = (FLA × Line Voltage × 1.732) / 1000
Amperes = (kVA × 1000) / Line Voltage × 1.732

COMMON TRANSFORMER CONNECTIONS

Single-Phase Transformers

Figure 19.1 shows a single-phase transformer connection with subtractive polarity. Figure 19.2 shows a single-phase transformer connection with additive polarity. Figures 19.3 and 19.4 are diagrams of transformer connections with secondary windings connected in parallel and connected in series, respectively.

Three-Phase Transformers

The symbols that represent three-phase transformer connections are shown in Figure 19.5. Figure 19.6 shows a common three-phase delta-connected transformer. Figure 19.7 shows a three-phase wye-connected transformer. Three single-phase transformers connected to form a delta transformer bank and to form a wye transformer bank are shown in Figures 19.8 and 19.9, respectively.

Figures 19.10 through 19.13 show further single-phase transformers connected to form three-phase transformer banks. The figures show both the primary and secondary connections. The configurations include delta to delta, delta to wye, wye to wye, and wye to delta-connected transformers.

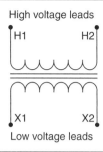

Figure 19.1 Subtractive polarity.

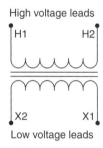

Figure 19.2 Additive polarity.

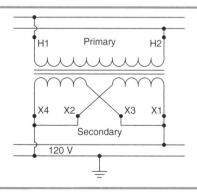

Figure 19.3 Secondary windings connected in parallel.

302

TRANSFORMERS

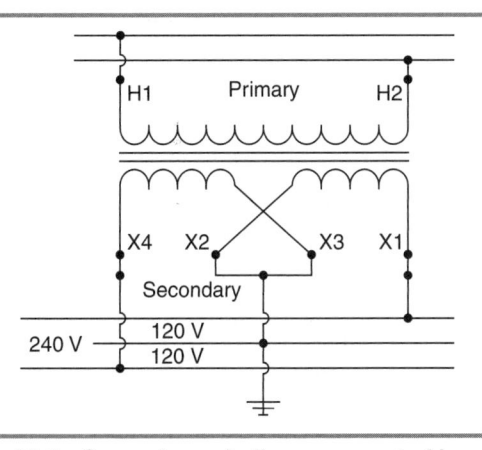

Figure 19.4   Secondary windings connected in series.

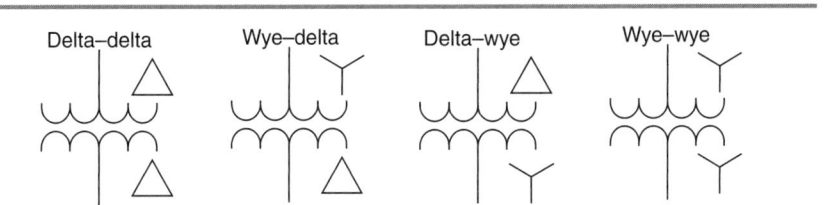

Figure 19.5   Symbols representing three-phase transformer connections.

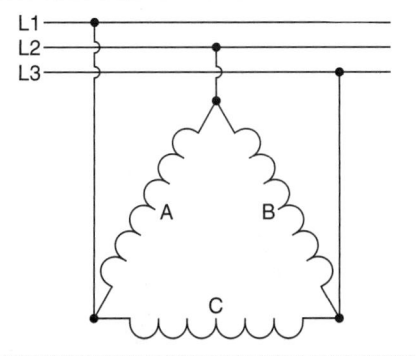

Figure 19.6   Three-phase delta-connected transformer.

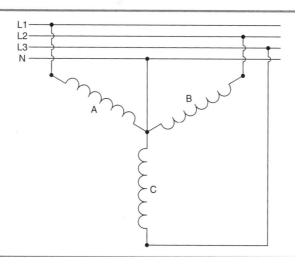

Figure 19.7   Three-phase wye-connected transformer.

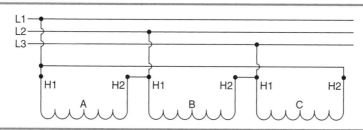

Figure 19.8   Three single-phase transformers connected to form a delta transformer bank.

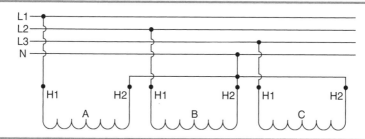

Figure 19.9   Three single-phase transformers connected to form a wye transformer bank.

304

TRANSFORMERS

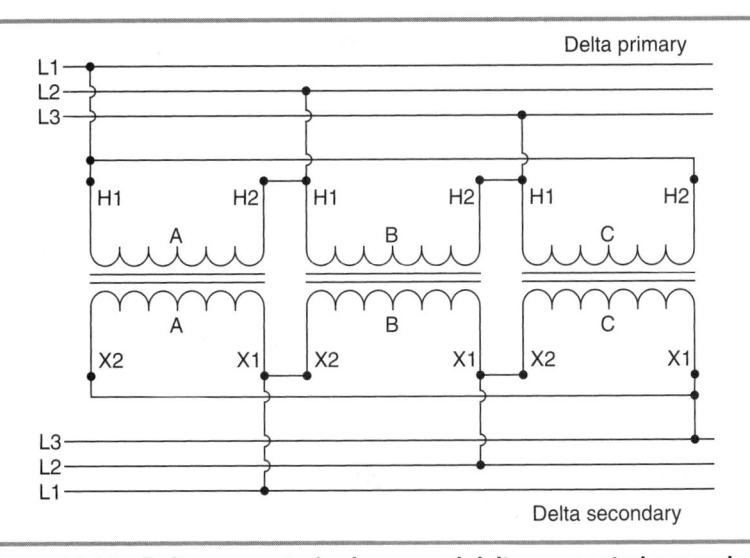

Figure 19.10   Delta connected primary and delta connected secondary.

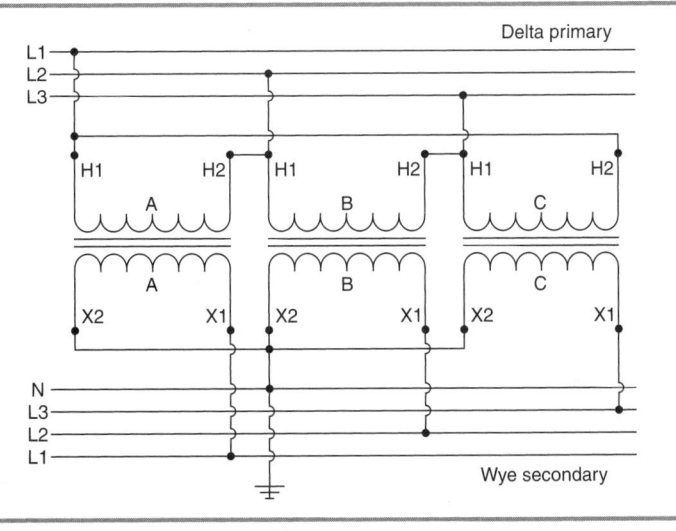

Figure 19.11   Delta connected primary and wye connected secondary.

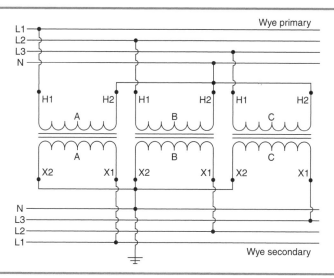

Figure 19.12 Wye connected primary and wye connected secondary.

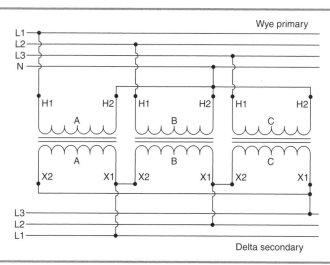

Figure 19.13 Wye connected primary and delta connected secondary.

TRANSFORMERS

TRANSFORMERS

TRANSFORMER FORMULAS USING VOLTAGE, CURRENT, AND TURNS

The relationship between the primary voltage and the secondary voltage is the same as the relationship between the number of turns in the primary and the number of turns in the secondary. The following formula can be transposed to fit the need of the problem.

$$\frac{T_P}{T_S} = \frac{E_P}{E_S} = \frac{I_S}{I_P}$$

Where:

T_P = number of turns in the primary

T_S = number of turns in the secondary

E_P = voltage in the primary

E_S = voltage in the secondary

I_P = current in the primary

I_S = current in the secondary

FORMULA VARIATIONS

Example 1

To find the number of turns in the primary when the voltages and secondary turns are known, use the following formula to solve for T_P:

$$T_P = \frac{E_P \times T_S}{E_S}$$

Solution   Given that the primary voltage is 120 volts, the secondary voltage is 24 volts, and the transformer has 50 turns in the secondary, what is the number of turns in the primary?

$$T_P = \frac{120 \times 50}{24} = 250 \text{ turns}$$

Example 2

To find the number of turns in the secondary when the voltages and primary turns are known:

$$T_S = \frac{E_S \times T_P}{E_P}$$

Solution The primary voltage is 240 volts, the secondary voltage is 120 volts, and the transformer has 200 turns in the primary. What is the number of turns in the secondary?

$$T_S = \frac{120 \times 200}{240} = 100 \text{ turns}$$

Example 3

To find the number of turns in the primary when the currents and secondary turns are known:

$$T_P = \frac{I_S \times T_S}{I_P}$$

Solution The primary current is 1 ampere, the secondary current is 5 amperes, and the transformer has 40 turns in the secondary. What is the number of turns in the primary?

$$T_P = \frac{5 \times 40}{1} = 200 \text{ turns}$$

Example 4

To find the number of turns in the secondary when the currents and primary turns are known:

$$T_S = \frac{I_P \times T_P}{I_S}$$

Solution The primary current is 8 amperes, the secondary current is 2 amperes, and the transformer has 200 turns in the primary. What is the number of turns in the secondary?

$$T_S = \frac{8 \times 200}{2} = 800 \text{ turns}$$

Example 5

To find the primary voltage when the turns and secondary voltage are known:

$$E_P = \frac{T_P \times E_S}{T_S}$$

Solution A transformer has 1000 turns in the primary, 100 turns in the secondary, and 24 volts in the secondary. What is the primary voltage?

$$E_P = \frac{1000 \times 24}{100} = 240 \text{ volts}$$

TRANSFORMERS

TRANSFORMERS

Example 6

To find the secondary voltage when the turns and primary voltage are known:

$$E_P = \frac{T_S \times E_P}{T_P}$$

Solution A transformer has 50 turns in the primary, 250 turns in the secondary, and 120 volts in the primary. What is the secondary voltage?

$$E_S = \frac{250 \times 120}{50} = 600 \text{ volts}$$

Example 7

To find the primary voltage when the currents and secondary voltage are known:

$$E_P = \frac{I_S \times E_S}{I_P}$$

Solution The primary current is 2 amperes, the secondary current is 8 amperes, and the secondary voltage is 120 volts. What is the primary voltage?

$$E_P = \frac{8 \times 120}{2} = 480 \text{ volts}$$

Example 8

To find the secondary voltage when the currents and primary voltage are known:

$$E_S = \frac{I_P \times E_P}{I_S}$$

Solution The primary current is 10 amperes, the secondary current is 50 amperes, and the primary voltage is 120 volts. What is the secondary voltage?

$$E_S = \frac{10 \times 120}{50} = 24 \text{ volts}$$

Example 9

To find the primary current when the turns and secondary current are known:

$$I_P = \frac{T_S \times I_S}{T_P}$$

Solution A transformer has 1200 turns in the primary, 300 turns in the secondary, and 96 amperes in the secondary. What is the primary current?

$$I_P = \frac{300 \times 96}{1200} = 24 \text{ amperes}$$

Example 10

To find the secondary current when the turns and primary current are known:

$$I_S = \frac{T_P \times I_P}{T_S}$$

Solution A transformer has 200 turns in the primary, 1000 turns in the secondary, and 125 amperes in the primary. What is the secondary current?

$$I_S = \frac{200 \times 125}{1000} = 25 \text{ amperes}$$

Example 11

To find the primary current when the voltages and secondary current are known:

$$I_P = \frac{E_S \times I_S}{E_P}$$

Solution The secondary current is 40 amperes, the primary voltage is 120 volts, and the secondary voltage is 480 volts. What is the primary current?

$$I_P = \frac{480 \times 40}{120} = 160 \text{ amperes}$$

Example 12

To find the secondary current when the voltages and primary current are known:

$$I_S = \frac{E_P \times I_P}{E_S}$$

Solution The primary current is 30 amperes, the primary voltage is 460 volts, and the secondary voltage is 230 volts. What is the secondary current?

$$I_S = \frac{460 \times 30}{230} = 60 \text{ ampere}$$

DELTA AND WYE TRANSFORMER FORMULAS

Center-Tap Grounded, Three-Phase, Four-Wire Delta Formulas

In the center-tap grounded, three-phase, four-wire delta diagram shown in Figure 19.14, the phase voltage is equal to the line voltage and is expressed with the formula:

$$E_P = E_L$$

310

TRANSFORMERS

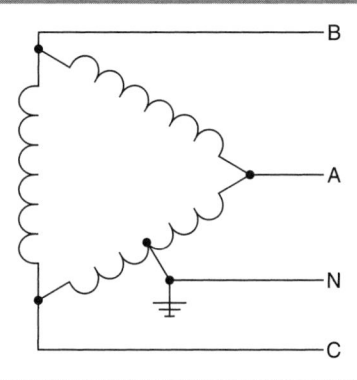

Figure 19.14   Center-tap grounded, three-phase, four-wire delta transformer.

Phase current is less than line current.

Phase current is equal to line current divided by 1.732.

$I_P = I_L \div 1.732$

Line current is more than phase current.

Line current is equal to phase current multiplied by 1.732.

$I_L = I_P \times 1.732$

The following four formulas work with both delta and wye transformers.

$VA = E_P \times I_P \times 3$

$VA = E_L \times I_L \times 1.732$

$I_P = \dfrac{VA}{E_P \times 3}$

$I_L = \dfrac{VA}{E_L \times 1.732}$

Where:

E_P = Phase Voltage

E_L = Line Voltage

I_P = Phase Current

I_L = Line Current

Note: To find the high-leg (wild-leg or stinger) voltage to ground, divide the phase voltage by 2 and then multiply by 1.732.

Example

The line voltage on a center-tap grounded, three-phase, four-wire delta-connected transformer is 230 volts. What is the voltage from the high leg to ground?

Solution Since line voltage is the same as phase voltage in this type of transformer, the phase voltage is 230 volts. Half of the phase voltage is 115 volts. Multiply 115 by 1.732 to find the high-leg voltage to ground ($115 \times 1.732 = 199.18$). The voltage from the high leg to ground is 199 volts.

Three-Phase, Four-Wire Wye Formulas

In the three-phase, four-wire wye-connector diagram shown in Figure 19.15, the phase voltage is less than the line voltage. Phase voltage is equal to line voltage divided by 1.732.

$E_P = E_L \div 1.732$

Line voltage is more than phase voltage.

Line voltage is equal to phase voltage multiplied by 1.732.

$E_L = E_P \times 1.732$

Phase current is equal to line current.

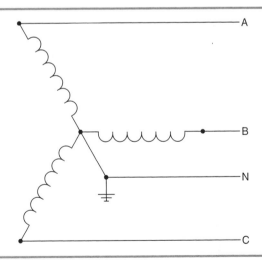

Figure 19.15 Three-phase, four-wire wye transformer.

312

TRANSFORMERS

$I_P = I_L$

The following four formulas work with both wye and delta transformers.

$VA = E_P \times I_P \times 3$

$VA = E_L \times I_L \times 1.732$

$I_P = \dfrac{VA}{E_P \times 3}$

$I_L = \dfrac{VA}{E_L \times 1.732}$

Where:

E_P = Phase Voltage

E_L = Line Voltage

I_P = Phase Current

I_L = Line Current

Example

Find the neutral current in a three-phase, four-wire wye system when the currents in line 1, line 2, and line 3 are known:

$$I_N = \sqrt{I_A^2 + I_B^2 + I_C^2 - (I_A \times I_B) - (I_B \times I_C) - (I_C \times I_A)}$$

Solution   The current in line 1 (phase A) is 50 amperes, the current in line 2 (phase B) is 60 amperes, and the current in line 3 (phase C) is 100 amperes in a three-phase, four-wire wye transformer. What is the neutral current?

$$I_N = \sqrt{50^2 + 60^2 + 100^2 - (50 \times 60) - (60 \times 100) - (100 \times 50)}$$

$$I_N = \sqrt{2500 + 3600 + 10,000 - 3000 - 6000 - 5000}$$

$$I_N = \sqrt{16,100 - 14,000} = \sqrt{2100} = 48 \text{ amperes}$$

BUCK-BOOST TRANSFORMERS

An economical means to correct a higher or lower voltage is with a buck-boost transformer. As the name indicates, *boost* increases voltage and *buck* decreases voltage. Depending on how it is connected, the same transformer can either buck or boost voltage. Depending on the need, either winding can serve as primary or secondary (see Figure 19.16). Buck-boost transformers usually have dual windings in both the primary and the secondary. Three common transformer ratings include 120/240 to 12/24 volts, 120/240 to 16/32 volts, and 240/480 to 24/48 volts. Always refer to the manufacturer's instructions and connection diagrams for proper installation.

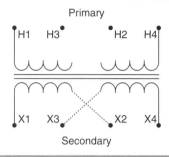

Figure 19.16   Typical diagram of transformer terminations.

See Table 19.3 if an increase in voltage is needed and Table 19.4 if a decrease in voltage is needed. The first two columns include transformer ratings for both primary and secondary voltage. The third column contains the input voltage on the line side of the transformer. This is the voltage before installing the transformer. The fourth column contains the output voltage on the load side of the transformer. This is the desired voltage after installing the buck-boost transformer. After determining the transformer rating, and input and output voltages, look in the last column for the connection diagram. Figures 19.17 through 19.23 contain diagrams for boosting voltage. Figures 19.24 through 19.27 contain diagrams for bucking voltage.

Table 19.3   Boosting

Transformer Rating				
Primary Voltage	Secondary Voltage	Input Voltage (Line)	Output Voltage (Load)	Connection Diagram
120/240	16/32	84	115	F
120/240	16/32	91	115	B1
120/240	12/24	96	115	B1
120/240	16/32	100	115	E
120/240	16/32	102	115	A1
120/240	12/24	105	115	A1
120/240	16/32	88	120	F
120/240	16/32	95	120	B1

Continued

TRANSFORMERS

Table 19.3 Boosting *(continued)*

Transformer Rating				
Primary Voltage	Secondary Voltage	Input Voltage (Line)	Output Voltage (Load)	Connection Diagram
120/240	12/24	96	120	F
120/240	12/24	100	120	B1
120/240	16/32	104	120	E
120/240	16/32	106	120	A1
120/240	12/24	109	120	A1
240/480	24/48	173	208	B1
120/240	16/32	184	208	D1
120/240	12/24	189	208	D1
240/480	24/48	189	208	A1
120/240	16/32	195	208	C1
120/240	12/24	198	208	C1
120/240	12/24	208	229	D1
240/480	24/48	192	230	B1
120/240	16/32	199	230	G
120/240	16/32	203	230	D1
120/240	12/24	207	230	G
120/240	12/24	209	230	D1
240/480	24/48	209	230	A1
120/240	16/32	216	230	C1
120/240	12/24	219	230	C1
120/240	16/32	220	235	C1
120/240	16/32	208	236	D1
240/480	24/48	200	240	B1
120/240	16/32	208	240	G
120/240	16/32	212	240	D1
120/240	12/24	216	240	G
120/240	12/24	218	240	D1
240/480	24/48	218	240	A1
120/240	16/32	225	240	C1
120/240	12/24	229	240	C1
120/240	12/24	220	242	D1
120/240	16/32	240	272	D1
240/480	24/48	230	277	B1
240/480	24/48	345	380	D1
240/480	24/48	362	380	C1
240/480	24/48	364	400	D1
240/480	24/48	381	400	C1
240/480	24/48	377	415	D1
240/480	24/48	395	415	C1
240/480	24/48	418	460	D1
240/480	24/48	438	460	C1
240/480	24/48	436	480	D1
240/480	24/48	457	480	C1
240/480	24/48	460	483	C1

Table 19.4 Bucking

Transformer Rating				
Primary Voltage	Secondary Voltage	Input Voltage (Line)	Output Voltage (Load)	Connection Diagram
120/240	12/24	127	115	A2
120/240	16/32	130	115	A2
120/240	12/24	138	115	B2
120/240	16/32	146	115	B2
120/240	12/24	132	120	A2
120/240	16/32	136	120	A2
120/240	12/24	144	120	B2
120/240	16/32	152	120	B2
120/240	12/24	218	208	C2
120/240	16/32	222	208	C2
120/240	12/24	229	208	D2
240/480	24/48	229	208	A2
120/240	16/32	236	208	D2
240/480	24/48	250	208	B2
120/240	12/24	242	230	C2
120/240	16/32	245	230	C2
120/240	12/24	253	230	D2
240/480	24/48	253	230	A2
120/240	16/32	261	230	D2
240/480	24/48	276	230	B2
120/240	12/24	252	240	C2
120/240	16/32	256	240	C2
120/240	12/24	264	240	D2
240/480	24/48	264	240	A2
120/240	16/32	272	240	D2
240/480	24/48	288	240	B2
240/480	24/48	399	380	C2
240/480	24/48	418	380	D2
240/480	24/48	420	400	C2
240/480	24/48	440	400	D2
240/480	24/48	436	415	C2
240/480	24/48	457	415	D2
240/480	24/48	483	460	C2
240/480	24/48	506	460	D2
240/480	24/48	504	480	C2
240/480	24/48	528	480	D2

Boosting Connection Diagrams

Figures 19.17 through 19.23 contain connection diagrams for boosting voltage. These diagrams are referenced in the last column of Table 19.3. Look in Table 19.3 to determine the type of transformer needed for the application. Follow the row across to the last column and see which diagram is listed. Find the diagram and connect the transformer accordingly.

316

TRANSFORMERS

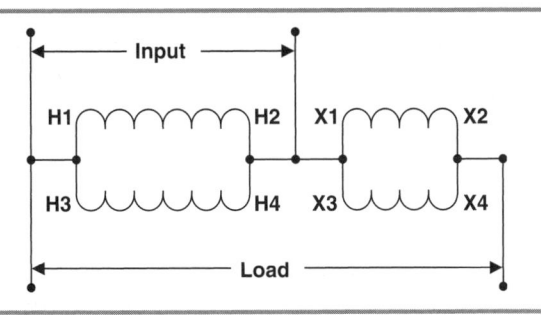

Figure 19.17   Diagram A1.

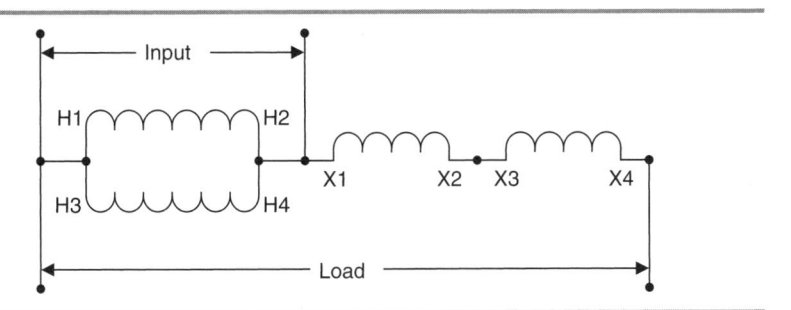

Figure 19.18   Diagram B1.

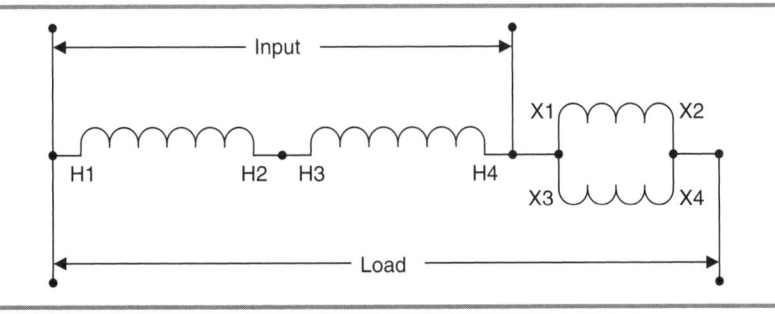

Figure 19.19   Diagram C1.

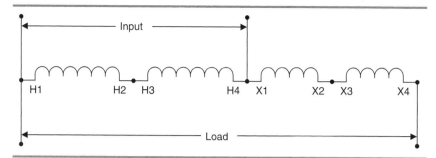

Figure 19.20   Diagram D1.

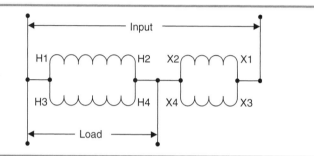

Figure 19.21   Diagram E.

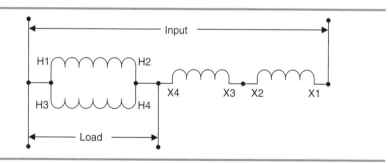

Figure 19.22   Diagram F.

318

TRANSFORMERS

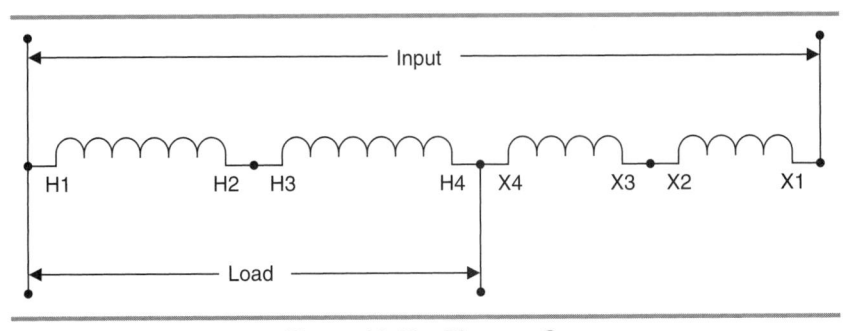

Figure 19.23 Diagram G.

Bucking Connection Diagrams

Figures 19.24 through 19.27 contain connection diagrams for bucking voltage. These diagrams are referenced in the last column of Table 19.4. Look in Table 19.4 to determine the type of transformer needed for the application. Follow the row across to the last column and see which diagram is listed. Find the diagram and connect the transformer accordingly.

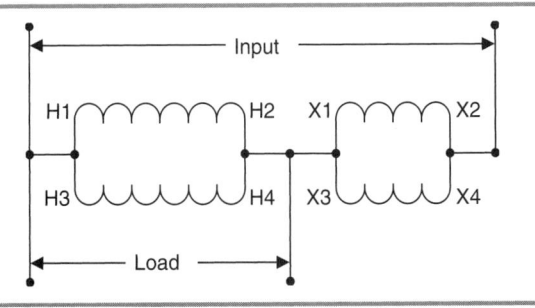

Figure 19.24 Diagram A2.

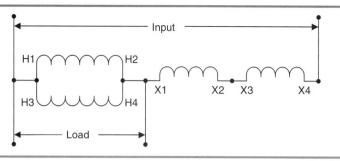

Figure 19.25   Diagram B2.

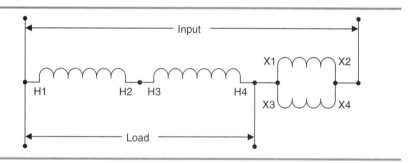

Figure 19.26   Diagram C2.

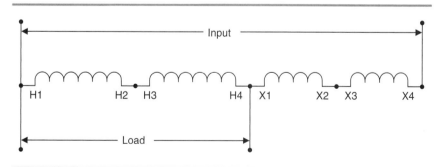

Figure 19.27   Diagram D2.

TRANSFORMERS

CHAPTER 20
GENERAL REFERENCE

Chapter 20 provides general-use information that is not found elsewhere in this pocket guide, including guidance on tightening torques, drill and hole saw sizes, pulley calculations, fire extinguishers, knots, crane hand signals, and phone and data jack wiring diagrams.

TIGHTENING TORQUES

For the testing of wire connectors for which the manufacturer has not assigned another value appropriate for the design, Tables 20.1 through 20.4 provide data on the tightening torques that Underwriters Laboratories uses. These tables should be used for guidance only if no tightening information on the specific wire connector is available. They should not be used to replace the manufacturer's instructions, which should always be followed.

Table 20.1 Tightening Torques for Screws* in Pound-Inches

Wire Size (AWG or kcmil)	Slotted Head No. 10 and Larger		Hexagonal Head– External Drive Socket Wrench	
	Slot Width to 3/64 in. or Slot Length to 1/4 in.**	Slot Width Over 3/64 in. or Slot Length Over 1/4 in.**	Split-Bolt Connectors	Other Connectors
30–10	20	35	80	75
8	25	40	80	75
6	35	45	165	110
4	35	45	165	110
3	35	50	275	150
2	40	50	275	150
1	—	50	275	150
1/0	—	50	385	180
2/0	—	50	385	180
3/0	—	50	500	250
4/0	—	50	500	250
250	—	50	650	325
300	—	50	650	325
350	—	50	650	325
400	—	50	825	325

Continued

GENERAL REFERENCE

Table 20.1 Tightening Torques for Screws* in Pound–Inches *(continued)*

Wire Size (AWG or kcmil)	Slotted Head No. 10 and Larger		Hexagonal Head– External Drive Socket Wrench	
	Slot Width to 3/64 in. or Slot Length to 1/4 in.**	Slot Width Over 3/64 in. or Slot Length Over 1/4 in.**	Split-Bolt Connectors	Other Connectors
500	—	50	825	375
600	—	50	1000	375
700	—	50	1000	375
750	—	50	1000	375
800	—	50	1100	500
900	—	50	1100	500
1000	—	50	1100	500
1250	—	—	1100	600
1500	—	—	1100	600
1750	—	—	1100	600
2000	—	—	1100	600

*Clamping screws with multiple tightening means. For example, for a slotted hexagonal head screw, use the torque value associated with the tool used in the installation. UL uses both values when testing.
**For values of slot width or length other than those specified, select the largest torque value associated with conductor size.

Table 20.2 Torques in Pound–Inches for Slotted Head Screws* Smaller Than No. 10, for Use with 8 AWG and Smaller Conductors

Screw-Slot Length (in.)**	Screw-Slot Width Less than 3/64 in.	Screw-Slot Width 3/64 in. and Larger
To 5/32	7	9
5/32	7	12
3/16	7	12
7/32	7	12
1/4	9	12
9/32	—	15
Above 9/32	—	20

*Clamping screws with multiple tightening means. For example, for a slotted hexagonal head screw, use the torque value associated with the tool used in the installation. UL uses both values when testing.
**For slot lengths of intermediate values, select torques pertaining to next-shorter slot length.

Table 20.3 Torques for Recessed Allen Head Screws

Socket Size Across Flats (in.)	Torque (lb–in.)
1/8	45
5/32	100
3/16	120
7/32	150
1/4	200
5/16	275
3/8	375
1/2	500
9/16	600

Table 20.4 Lug-Bolting Torques for Connection of Wire Connectors to Busbars

Bolt Diameter	Tightening Torque (lb–ft)
No. 8 or smaller	1.5
No. 10	2.0
1/4 in. or less	6
5/16 in.	11
3/8 in.	19
7/16 in.	30
1/2 in.	40
9/16 in. or larger	55

DRILL AND HOLE SAW SIZES

Table 20.5 shows drill sizes for machine screws. Different sizes are shown for different applications. Two columns provide drill sizes for holes that will be tapped for machine screws. The drill size depends upon the material that will be tapped. Two columns show drill sizes for holes that will not be tapped. One size is shown for drilling holes just large enough for the screw. The other size will provide a loose fit for the screw. Table 20.6 provides the same type of information as Table 20.5 except for metric sizes. Table 20.7 provides hole saw sizes for electrical conduits.

GENERAL REFERENCE

Table 20.5 Tap Drill Sizes

Machine Screw Size			Tap Drills				Clearance Hole Drills			
			Non Ferrous and Plastics 75% Thread		Stainless Steel, Steels and Iron 50% Thread		All Materials			
							Close Fit		Free Fit	
No. or Dia.	Decimal (inch)	Threads per Inch	Drill Size	Decimal Equiv.	Drill Size	Decimal Equiv.	Drill Size	Decimal Equiv.	Drill Size	Decimal Equiv.
0	0.06	80	3–64	0.0469	55	0.052	52	0.0635	50	0.07
1	0.073	64	53	0.0595	1/16	0.0625	48	0.076	46	0.081
		72	53	0.0595	52	0.0635				
2	0.086	56	50	0.07	49	0.073	43	0.089	41	0.096
		64	50	0.07	48	0.076				
3	0.099	48	47	0.0785	44	0.086	37	0.104	35	0.11
		56	45	0.082	43	0.089				
4	0.112	40	43	0.089	41	0.096	32	0.116	30	0.136
		48	42	0.0935	40	0.098				
5	0.125	40	38	0.1015	7/64	0.1094	30	0.1285	29	0.136
		44	37	0.104	35	0.11				
6	0.138	32	36	0.1065	32	0.116	27	0.144	25	0.1495
		40	33	0.113	31	0.12				
8	0.164	32	29	0.136	27	0.144	18	0.1695	16	0.177
		36	29	0.136	26	0.147				

Continued

Table 20.5 Tap Drill Sizes (continued)

Machine Screw Size	No. or Dia. Decimal (inch)	Threads per Inch	Tap Drills Non Ferrous and Plastics 75% Thread		Tap Drills Stainless Steel, Steels and Iron 50% Thread		Clearance Hole Drills All Materials Close Fit		Clearance Hole Drills All Materials Free Fit	
			Drill Size	Decimal Equiv.	Drill Size	Decimal Equiv.	Drill Size	Decimal Equiv.	Drill Size	Decimal Equiv.
10	0.19	24	25	0.1495	20	0.161	9	0.196	7	0.201
		32	21	0.159	18	0.1695				
12	0.216	24	16	0.177	12	0.189	2	0.221	1	0.228
		28	14	0.182	10	0.1935				
		32	13	0.185	9	0.196				
1/4	0.25	20	7	0.201	7/32	0.2188	F	0.257	H	0.266
		28	3	0.213	1	0.228				
		32	7/32	0.2188	1	0.228				
5/16	0.3125	18	F	0.257	J	0.277	P	0.323	Q	0.332
		24	I	0.272	9/32	0.2812				
		32	9/32	0.2812	L	0.29				
3/8	0.375	16	5/16	0.3125	Q	0.332	W	0.386	X	0.397
		24	Q	0.332	S	0.348				
		32	11/32	0.3438	T	0.358				
7/16	0.4375	14	U	0.368	25/64	0.3906	29/64	0.4531	15/32	0.4687
		20	25/64	0.3906	13/32	0.4062				
		28	Y	0.404	Z	0.413				

Continued

GENERAL REFERENCE

Table 20.5 Tap Drill Sizes *(continued)*

Machine Screw Size		Threads per Inch	Tap Drills				Clearance Hole Drills			
			Non Ferrous and Plastics 75% Thread		Stainless Steel, Steels and Iron 50% Thread		All Materials			
							Close Fit		Free Fit	
No. or Dia.	Decimal (inch)		Drill Size	Decimal Equiv.	Drill Size	Decimal Equiv.	Drill Size	Decimal Equiv.	Drill Size	Decimal Equiv.
1/2	0.5	13	27/64	0.4219	29/64	0.4531	33/64	0.5156	17/32	0.5312
		20	29/64	0.4531	15/32	0.4688				
		28	15/32	0.4688	15/32	0.4688				
9/16	0.5625	12	31/64	0.4844	33/64	0.5156	37/64	0.5781	19/32	0.5938
		18	33/64	0.5156	17/32	0.5312				
		24	33/64	0.5156	17/32	0.5312				
5/8	0.625	11	17/32	0.5312	9/16	0.5625	41/64	0.6406	21/32	0.6562
		18	37/64	0.5781	19/32	0.5938				
		24	37/64	0.5781	19/32	0.5938				
11/16	0.6875	24	41/64	0.6406	21/32	0.6562	45/64	0.7031	23/32	0.6562
3/4	0.75	10	21/32	0.6562	11/16	0.6875	49/64	0.7656	25/32	0.7812
		16	11/16	0.6875	45/64	0.7031				
		20	45/64	0.7031	23/32	0.7188				
13/16	0.8125	20	49/64	0.7656	25/32	0.7812	53/64	0.8281	27/32	0.8438
7/8	0.875	9	49/64	0.7656	51/64	0.7969	57/64	0.8906	29/32	0.9062
		14	13/16	0.8125	53/64	0.8281				
		20	53/64	0.8281	27/32	0.8438				
15/16	0.9375	20	57/64	0.8906	29/32	0.9062	61/64	0.9531	31/32	0.9688

Continued

Table 20.5 Tap Drill Sizes (continued)

Machine Screw Size	No. or Dia.	Decimal (inch)	Threads per Inch	Tap Drills						Clearance Hole Drills			
				Non Ferrous and Plastics		Stainless Steel, Steels and Iron				All Materials			
				75% Thread		50% Thread				Close Fit		Free Fit	
				Drill Size	Decimal Equiv.	Drill Size	Decimal Equiv.	Drill Size	Decimal Equiv.	Drill Size	Decimal Equiv.	Drill Size	Decimal Equiv.
	1	1	8	7/8	0.875	59/64	0.9219	1-1/64	1.0156	1-1/32	1.0313		
			12	15/16	0.9375	61/64	0.9531						
			20	61/64	0.9531	31/32	0.9688						
1-1/16		1.0625	18	1	1	1-1/64	1.0156	1-5/64	1.0781	1-3/32	1.0938		
1-1/8		1.125	7	63/64	0.9844	1-1/32	1.0313	1-9/64	1.1406	1-5/32	1.1562		
			12	1-3/64	1.0469	1-5/64	1.0781						
			18	1-1/16	1.0625	1-5/64	1.0781						
1-3/16		1.1875	18	1-1/8	1.125	1-9/64	1.1406	1-13/64	1.2031	1-7/32	1.2188		
1-1/4		1.25	7	1-7/64	1.1094	1-5/32	1.1562	1-17/64	1.2656	1-9/32	1.2812		
			12	1-11/64	1.1719	1-13/64	1.2031						
			18	1-3/16	1.1875	1-13/64	1.2031						
1-5/16		1.3125	18	1-1/4	1.25	1-17/64	1.2656	1-21/64	1.3281	1-11/32	1.3438		
1-3/8		1.375	6	1-7/32	1.2187	1-17/64	1.2656	1-25/64	1.3906	1-13/32	1.4062		
			12	1-19/64	1.2969	1-21/64	1.3281						
			18	1-5/16	1.3125	1-21/64	1.3281						
1-7/16		1.4375	18	1-3/8	1.375	1-25/64	1.3906	1-29/64	1.4531	1-15/32	1.4688		
1-1/2		1.5	6	1-11/32	1.3437	1-25/64	1.3906	1-33/64	1.5156	1-17/32	1.5312		
			12	1-27/64	1.4219	1-7/16	1.4375						
			18	1-7/16	1.4375	1-29/64	1.4531						

Continued

Table 20.5 Tap Drill Sizes (continued)

Machine Screw Size		Threads per Inch	Tap Drills						Clearance Hole Drills			
			Non Ferrous and Plastics		Stainless Steel, Steels and Iron				All Materials			
			75% Thread		50% Thread							
							Close Fit				Free Fit	
No. or Dia.	Decimal (inch)		Drill Size	Decimal Equiv.	Drill Size	Decimal Equiv.	Drill Size	Decimal Equiv.	Drill Size	Decimal Equiv.	Drill Size	Decimal Equiv.
1-9/16	1.5625	18	1-1/2	1.5	1-33/64	1.5156	1-37/64	1.5781			1-19/32	1.5938
1-5/8	1.625	18	1-9/16	1.5625	1-37/64	1.5781	1-41/64	1.6406			1-21/32	1.6562
1-11/16	1.6875	18	1-5/8	1.625	1-41/64	1.6406	1-45/64	1.7031			1-23/32	1.7188
1-3/4	1.75	5	1-11/16	1.6875	1-5/8	1.625	1-49/64	1.7659			1-25/32	1.7812

Table 20.6 Metric Tap Drill Sizes

Screw Size			Tapping Drills		Clearance Drills			
					Close Fit		Free Fit	
Nom. Dia. (mm)	Pitch (mm)	Series	Drill Size Approx. 75% Thread	Dec. Equiv.	Drill Size	Dec. Equiv.	Drill Size	Dec. Equiv.
1	0.2	Fine	0.8	0.0315	1.05	0.0413	1.2	0.0472
	0.25	Coarse	0.75	0.0295				
1.1	0.2	Fine	0.9	0.0354	1.15	0.0453	1.3	0.0512
	0.25	Coarse	0.85	0.0335				
1.2	0.2	Fine	1	0.0394	1.3	0.0512	1.5	0.059
	0.25	Coarse	0.95	0.0374				
1.4	0.2	Fine	1.2	0.0472	1.5	0.0591	1.7	0.0669
	0.3	Coarse	1.1	0.0433				
1.6	0.2	Fine	1.4	0.0551	1.7	0.0669	2	0.0787
	0.35	Coarse	1.25	0.0492				
1.8	0.2	Fine	1.6	0.063	1.9	0.0748	2.2	0.0866
	0.35	Coarse	1.45	0.0571				
2	0.25	Fine	1.75	0.0689	2.2	0.0866	2.6	0.1024
	0.4	Coarse	1.6	0.063				
2.2	0.25	Fine	1.95	0.0768	2.4	0.0945	2.8	0.1102
	0.45	Coarse	1.75	0.0689				
2.5	0.35	Fine	2.15	0.0846	2.7	0.1063	3.1	0.122
	0.45	Coarse	2.05	0.0807				
3	0.35	Fine	2.65	0.1043	3.2	0.126	3.6	0.1417
	0.5	Coarse	2.5	0.0984				
3.5	0.35	Fine	3.15	0.124	3.7	0.1457	4.2	0.1653
	0.6	Coarse	2.9	0.1142				
4	0.5	Fine	3.5	0.1378	4.3	0.1693	4.8	0.189
	0.7	Coarse	3.3	0.1299				
4.5	0.5	Fine	4	0.1575	4.8	0.189	5.3	0.2087
	0.75	Coarse	3.7	0.1457				
5	0.5	Fine	4.5	0.1772	5.3	0.2087	5.8	0.2283
	0.8	Coarse	4.2	0.1654				
6	0.75	Fine	5.25	0.2067	6.4	0.252	7	0.2756
	1	Coarse	5	0.1969				
7	0.75	Fine	6.25	0.2461	7.4	0.2913	8	0.315
	1	Coarse	6	0.2362				

Continued

GENERAL REFERENCE

Table 20.6 Metric Tap Drill Sizes

Screw Size			Tapping Drills		Clearance Drills			
					Close Fit		Free Fit	
Nom. Dia. (mm)	Pitch (mm)	Series	Drill Size Approx. 75% Thread	Dec. Equiv.	Drill Size	Dec. Equiv.	Drill Size	Dec. Equiv.
8	0.75	Fine	7.25	0.2854	8.4	0.3307	10	0.3937
	1	Fine	7.5	0.2953				
	1.25	Coarse	6.8	0.2677				
9	0.75	Fine	8.25	0.3248	9.5	0.374	10.5	0.4134
	1	Fine	8	0.315				
	1.25	Coarse	7.8	0.3071				
10	1.25	Fine	8.8	0.3465	10.5	0.4134	12	0.4724
	0.75	Fine	9.25	0.3642				
	1	Fine	9	0.3543				
	1.5	Coarse	8.5	0.3346				
11	0.75	Fine	10.25	0.4035	12	0.4724	13	0.5118
	1	Fine	10	0.3937				
	1.5	Coarse	9.5	0.374				
12	1	Fine	11	0.4331	13	0.5118	15	0.5905
	1.25	Fine	10.75	0.4232				
	1.5	Fine	10.5	0.4134				
	1.75	Coarse	10.2	0.4016				
14	1	Fine	13	0.5118	15	0.5905	17	0.6693
	1.25	Fine	12.8	0.5039				
	1.5	Fine	12.5	0.4921				
	2	Coarse	12	0.4724				
15	1	Fine	14	0.5512	16	0.6299	18	0.7087
	1.5	Fine	13.5	0.5315				
16	1	Fine	15	0.5906	17	0.6693	19	0.748
	1.5	Fine	14.5	0.5709				
	2	Coarse	14	0.5512				
17	1	Fine	16	0.6299	18	0.7087	20	0.7874
	1.5	Fine	15.5	0.6103				
18	1	Fine	17	0.6693	19	0.748	21	0.8268
	1.5	Fine	16.5	0.6496				
	2	Fine	16	0.6299				
	2.5	Coarse	15.5	0.6102				

Continued

Table 20.6 Metric Tap Drill Sizes

Screw Size			Tapping Drills		Clearance Drills			
					Close Fit		Free Fit	
Nom. Dia. (mm)	Pitch (mm)	Series	Drill Size Approx. 75% Thread	Dec. Equiv.	Drill Size	Dec. Equiv.	Drill Size	Dec. Equiv.
20	1	Fine	19	0.748	21	0.8268	24	0.9449
	1.5	Fine	18.5	0.7283				
	2	Fine	18	0.7087				
	2.5	Coarse	17.5	0.689				
22	1	Fine	21	0.8268	23	0.9055	26	1.0236
	1.5	Fine	20.5	0.8071				
	2	Fine	20	0.7874				
	2.5	Coarse	19.5	0.7677				
24	1	Fine	23	0.9055	25	0.9842	28	1.1024
	1.5	Fine	22.5	0.8858				
	2	Fine	22	0.8661				
	3	Coarse	21	0.8268				
25	1	Fine	24	0.9449	26	1.0236	30	1.1811
	1.5	Fine	23.5	0.9252				
	2	Fine	23	0.9055				
27	1	Fine	26	1.0236	28	1.1024	32	1.2598
	1.5	Fine	25.5	1.0039				
	2	Fine	25	0.9843				
	3	Coarse	24	0.9449				

GENERAL REFERENCE

GENERAL REFERENCE

Table 20.7   Hole Saw Sizes for Conduits

Conduit	Actual Hole Size		
Size	Inch	Decimal	mm
1/2	7/8	0.875	22.2
3/4	1-1/8	1.125	28.6
1	1-3/8	1.375	35.0
1-1/4	1-3/4	1.75	44.4
1-1/2	2	2	50.8
2	2-1/2	2.5	63.5
2-1/2	3	3	76.2
3	3-5/8	3.625	92.1
3-1/2	4-1/8	4.125	104.8
4	4-1/2	4.5	114.3

PULLEY CALCULATIONS

Calculating Pulley Diameters

To find the diameter of the driven pulley:

$$\text{Driven Pulley Diameter} = \frac{\text{Driver Pulley Diameter} \times \text{Driver Pulley RPM}}{\text{Driven Pulley RPM}}$$

To find the diameter of the driver pulley:

$$\text{Driver Pulley Diameter} = \frac{\text{Driven Pulley Diameter} \times \text{Driven Pulley RPM}}{\text{Driver Pulley RPM}}$$

To find the revolutions per minute of the driven pulley:

$$\text{Driven Pulley RPM} = \frac{\text{Driver Pulley Diameter} \times \text{Driver Pulley RPM}}{\text{Driven Pulley Diameter}}$$

To find the revolutions per minute of the driver pulley:

$$\text{Driver Pulley RPM} = \frac{\text{Driven Pulley Diameter} \times \text{Driven Pulley RPM}}{\text{Driver Pulley Diameter}}$$

Calculating Belt Length

The following variables are used in the belt length formula.

L = belt length

C = pulley centerline distance

D = outside diameter of large pulley
d = outside diameter of small pulley

To find the belt length when the pulley diameters are the same size:

$L = (D \times 3.1416) + (C \times 2)$

To find the belt length when the pulley diameters are not the same size:

$L = (D \times 1.57) + (d \times 1.57) + (C \times 2) + \dfrac{(D-d)^2}{(C \times 4)}$

PORTABLE FIRE EXTINGUISHERS

The different classes of portable fire extinguishers are shown in Table 20.8.

Table 20.8 Classes of Fires

Class	Type of Fire	Examples of Fuels
A	Common combustibles	Wood, paper, cloth, rubber, household rubbish, some plastics
B	Flammable liquids	Oils, greases, tar, lacquers, flammable gasses, oil-based paints, some plastics
C	Energized electrical equipment	TV's, computers, building wiring, fuse boxes, circuit breakers, conveyor belt motors, transformers, generators, power tools, lamps and lighting fixtures, radios, stage lighting or sound equipment, appliances
D	Combustible metals	Magnesium, titanium, zirconium, sodium, lithium, potassium
K	Combustible cooking media	Vegetable or animal fats and oils

The traditional labeling system for extinguishers is shown in Figure 20.1 and the pictograph labeling system is shown in Figure 20.2.

GENERAL REFERENCE

Ordinary

A

Combustibles

Extinguishers suitable for Class A fires should be identified by a triangle containing the letter "A." If colored, the triangle is colored green.*

Flammable

B

Liquids

Extinguishers suitable for Class B fires should be identified by a square containing the letter "B." If colored, the square is colored red.*

Electrical

C

Equipment

Extinguishers suitable for Class C fires should be identified by a circle containing the letter "C." If colored, the circle is colored blue.*

Combustible

D

Metals

Extinguishers suitable for fires involving metals should be identified by a five-pointed star containing the letter "D." If colored, the star is colored yellow.*

*Recommended colors, per PMS (Pantone Matching System) include the following:

GREEN — Basic Green
RED — 192 Red
BLUE — Process Blue
YELLOW — Basic Yellow

Figure 20.1   Traditional labeling system for extinguishers.
Source: NFPA 10, *Standard for Portable Fire Extinguishers,* 2002 edition, Figure B.1.1.

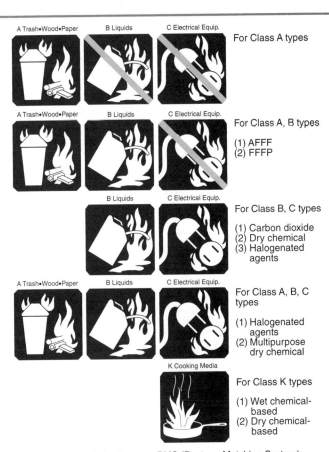

Figure 20.2 Pictograph labeling system for extinguishers.
Source: NFPA 10, *Standard for Portable Fire Extinguishers,* 2002 edition, Figure B.2.2.

GENERAL REFERENCE

GENERAL REFERENCE

KNOTS

Figures 20.3 through 20.22 show various helpful knots.

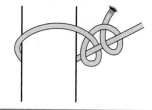

Figure 20.3 Figure eight knot.

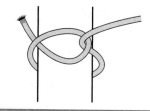

Figure 20.4 Half hitch.

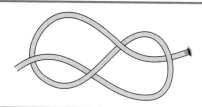

Figure 20.5 Two half hitches.

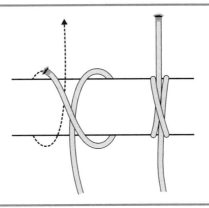

Figure 20.6   Clove hitch.

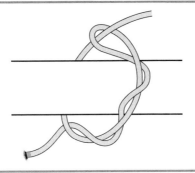

Figure 20.7   Timber hitch.

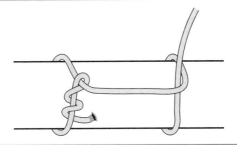

Figure 20.8   Timber hitch with half hitch.

GENERAL REFERENCE

GENERAL REFERENCE

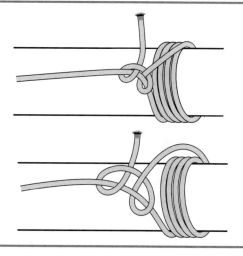

Figure 20.9 Pipe hitch.

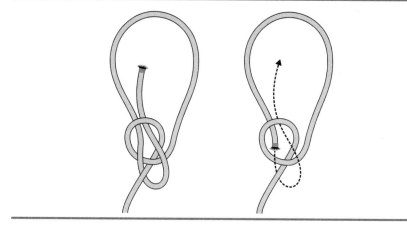

Figure 20.10 Bowline.

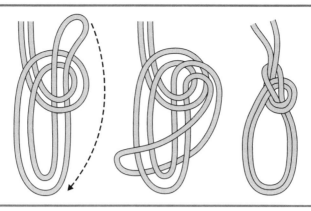

Figure 20.11 Bowline on a bight.

Figure 20.12 Square knot.

Figure 20.13 Surgeon's knot.

GENERAL REFERENCE

GENERAL REFERENCE

Figure 20.14 Fisherman's knot.

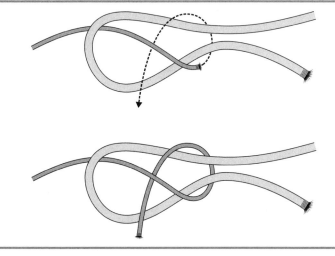

Figure 20.15 Sheet bend.

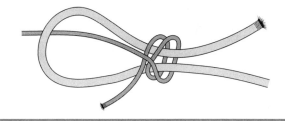

Figure 20.16 Double sheet bend.

Figure 20.17   Single blackwell.

Figure 20.18   Double blackwell.

342

GENERAL REFERENCE

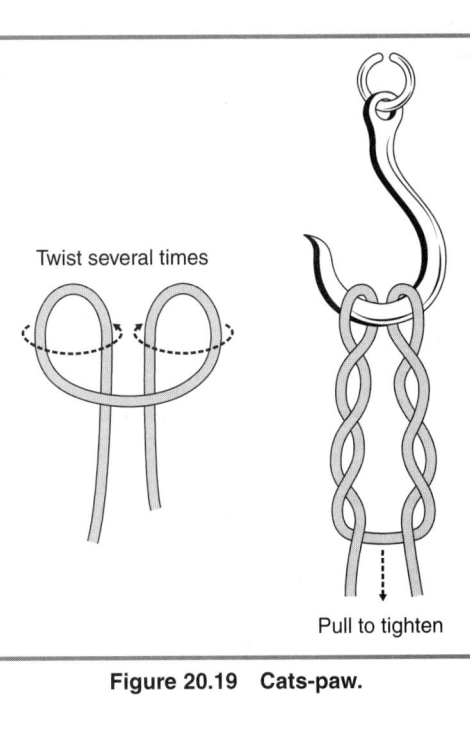

Twist several times

Pull to tighten

Figure 20.19   Cats-paw.

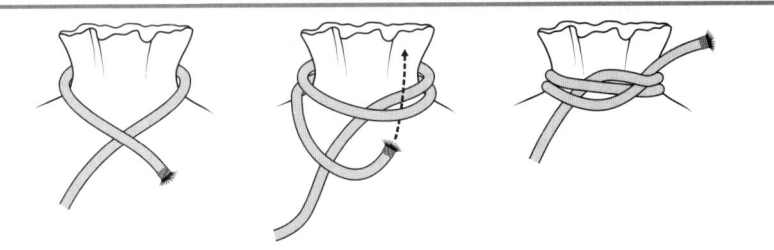

Figure 20.20   Miller's (constrictor) knot.

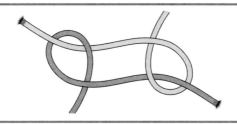

Figure 20.21   Strap knot.

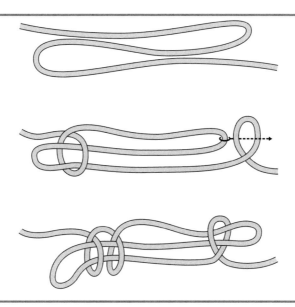

Figure 20.22   Sheepshank.

344

GENERAL REFERENCE

STANDARD CRANE HAND SIGNALS

Figures 20.23 through 20.40 show standard crane hand signals, for use if electronic communication with the crane operator is not available.

Figure 20.23   Stop: Arm extended, palm down, move forearm and hand back and forth in a horizontal chopping motion.

Figure 20.24 Emergency stop: Same as stop, but with both forearms rapidly chopping in a horizontal motion.

Figure 20.25 Use main hoist: Tap fist on head; then use regular signals.

GENERAL REFERENCE

346

GENERAL REFERENCE

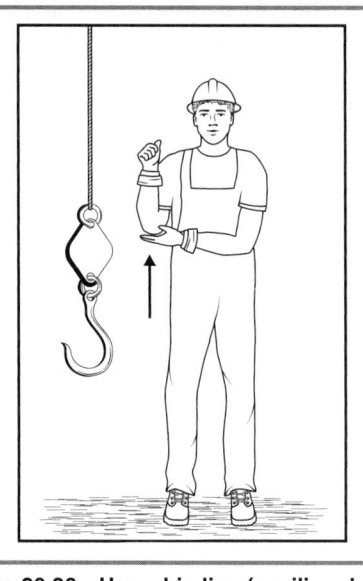

Figure 20.26 Use whip line (auxiliary hoist):
Tap elbow with open palm; then use regular signals.

Figure 20.27 Raise load (hoist): With forearm vertical
and forefinger pointing upward, move hand in small horizontal circles.

Figure 20.28 Lower load: With arm extended downward and forefinger pointing downward, move hand in small horizontal circles.

Figure 20.29 Raise boom: Arm extended, fingers closed, thumb pointing upward.

GENERAL REFERENCE

348

GENERAL REFERENCE

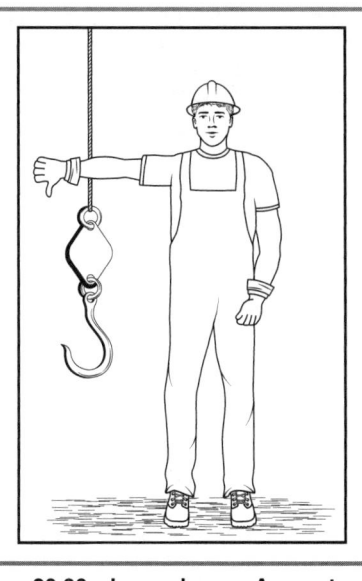

Figure 20.30 Lower boom: Arm extended,
fingers closed, thumb pointing downward.

Figure 20.31 Extend boom (telescoping booms):
Both fists in front of body with thumbs pointing outward.

Figure 20.32   Retract boom (telescoping booms):
Both fists in front of body with thumbs pointing toward each other.

Figure 20.33   Swing boom: Arm extended,
point with finger in the direction boom should swing.

GENERAL REFERENCE

GENERAL REFERENCE

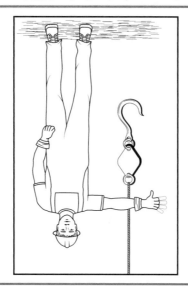

Figure 20.34 Raise the boom and lower the load: With arm extended and thumb pointing up, flex fingers in and out as long as load movement is desired.

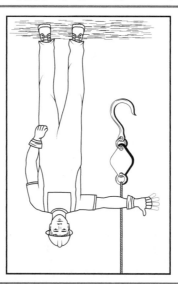

Figure 20.35 Lower the boom and raise the load: With arm extended and thumb pointing down, flex fingers in and out as long as load movement is desired.

Figure 20.36 Travel: Arm extended forward, hand open and slightly raised, make pushing motion in direction of travel.

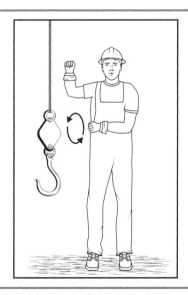

Figure 20.37 Travel (one track): Lock the track on side indicated by raised fist. Travel opposite track in direction indicated by circular motion of other fist, rotated vertically in front of body. (For crawler cranes only.)

GENERAL REFERENCE

GENERAL REFERENCE

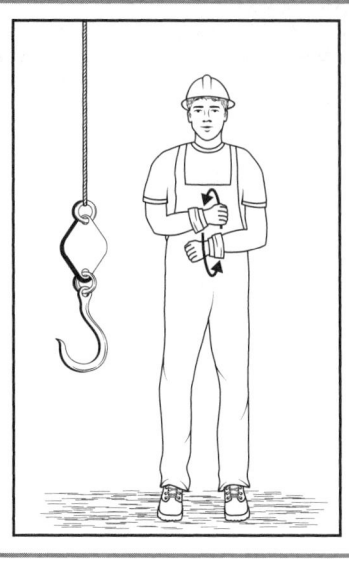

Figure 20.38 Travel (both tracks): Use both fists in front of body, making circular motion about each other, indicating direction of travel; forward or backward. (For crawler cranes only.)

Figure 20.39 Move slowly: Use one hand to give any motion signal and place other hand motionless in front of hand giving the motion signal. (Hoist slowly shown as example.)

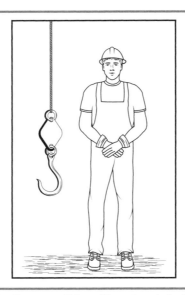

Figure 20.40 Dog everything: Clasp hands in front of body.

PHONE AND DATA JACK WIRING DIAGRAMS

The configuration for a 4-position/4-wire jack is shown in Table 20.9 and Figure 20.41.

Table 20.9 USOC (Universal Service Order Code) 2-Pair Configuration

Pin Number	Designation	Band-Striped Wire	Solid-Color Wire	Pair
1	Tip	White/Orange	Black	2
2	Ring	Blue/White	Red	1
3	Tip	White/Blue	Green	1
4	Ring	Orange/White	Yellow	2

GENERAL REFERENCE

353

354

GENERAL REFERENCE

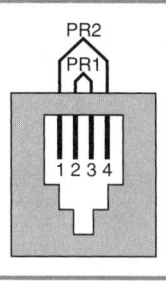

Figure 20.41   Front view of female connector in a 4-position/4-wire jack.

Table 20.10 and Figure 20.42 illustrate a 6-position/4-wire jack.

Table 20.10   USOC 2-Pair Configuration

Pin Number	Designation	Band-Striped Wire	Solid-Color Wire	Pair
2	Tip	White/Orange	Black	2
3	Ring	Blue/White	Red	1
4	Tip	White/Blue	Green	1
5	Ring	Orange/White	Yellow	2

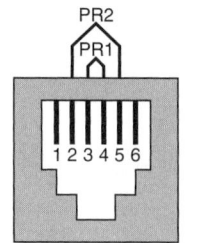

Figure 20.42   Front view of female connector in a 6-position/4-wire jack.

Table 20.11 and Figure 20.43 illustrate a 6-position/6-wire jack.

Table 20.11 USOC 3-Pair Configuration

Pin Number	Designation	Band-Striped Wire	Solid-Color Wire	Pair
1	Tip	White/Green	White	3
2	Tip	White/Orange	Black	2
3	Ring	Blue/White	Red	1
4	Tip	White/Blue	Green	1
5	Ring	Orange/White	Yellow	2
6	Ring	Green/White	Blue	3

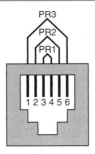

Figure 20.43 Front view of female connector in a 6-position/6-wire jack.

Tables 20.12 through 20.14 and Figures 20.44 through 20.46 illustrate three types of 8-position/8-wire jack.

Table 20.12 USOC 4-Pair Configuration

Pin Number	Designation	Band-Striped Wire	Pair
1	Tip	White/Brown	4
2	Tip	White/Green	3
3	Tip	White/Orange	2
4	Ring	Blue/White	1
5	Tip	White/Blue	1
6	Ring	Orange/White	2
7	Ring	Green/White	3
8	Ring	Brown/White	4

GENERAL REFERENCE

356

GENERAL REFERENCE

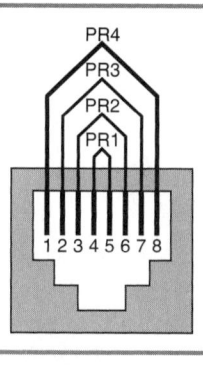

Figure 20.44   Front view of female connector in a USOC 8-position/8-wire jack.

Table 20.13   RJ-45 TIA/EIA 568A 4-Pair Configuration

Pin Number	Designation	Band-Striped Wire	Pair
1	Tip		3
2	Ring	Green/White	3
3	Tip	White/Orange	2
4	Ring	Blue/White	1
5	Tip	White/Blue	1
6	Ring	Orange/White	2
7	Tip	White/Brown	4
8	Ring	Brown/White	4

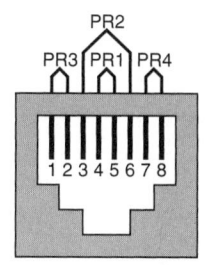

Figure 20.45   Front view of female connector in a 568A 8-position/8-wire jack.

Table 20.14 RJ-45 TIA/EIA 568B 4-Pair Configuration

Pin Number	Designation	Band-Striped Wire	Pair
1	Tip	White/Orange	2
2	Ring	Orange/White	2
3	Tip	White/Green	3
4	Ring	Blue/White	1
5	Tip	White/Blue	1
6	Ring	Green/White	3
7	Tip	White/Brown	4
8	Ring	Brown/White	4

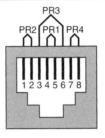

Figure 20.46 Front view of female connector in a 568B 8-position/8-wire jack.

GENERAL REFERENCE

CHAPTER 21

WEBSITE ADDRESSES

Chapter 21 contains Internet contact information for organizations with interests related to electrical installations.

WEBSITE ADDRESSES

Company	Internet Address
3M Electrical	www.3m.com
Accubid Systems	www.accubid.com
Accu-Tech Corporation	www.accu-tech.com
Ace Hardware	www.acehardware.com
Acterna	www.acterna.com
ADC Telecommunications	www.adc.com
ADI Communications	www.adilink.com
Advance Transformer Co.	www.advancetransformer.com
AEMC Instruments	www.aemc.com
AFC Cable Systems, Inc.	www.afcweb.com
Alcan Cable	www.cable.alcan.com
Alcoa Fujikura Ltd.	www.alcoa.com
Alflex Corporation	www.alflex.com
Allied Moulded Products, Inc.	www.alliedmoulded.com
American Express	www.americanexpress.com
American National Standards Institute	www.ansi.org
American Power Conversion	www.apcc.com
American Saw & Mfg. Co.	www.lenoxsaw.com
American Subcontractors Association	www.asaonline.com
Amprobe Instruments	www.amprobe.com
Arlington Industries	www.aifittings.com
Arrow Fastener Co.	www.arrowfastener.com
Associated Builders & Contractors	www.abc.org

Continued

WEBSITE ADDRESSES

Company	Internet Address
Associated Specialty Contractors	www.assoc-spec-con.org
Automatic Switch Co.	www.asco.com
AVO International	www.avointl.com
Band-It-Idex Inc.	www.band-it-idex.com
Belden Wire & Cable Co.	www.belden.com
BISCI	www.bisci.org
B-Line Systems	www.b-line.com
Bobcat	www.bobcat.com
Bosch Power Tools	www.boschtools.com
Brady USA	www.whbrady.com
Bridgeport Fittings, Inc.	www.bptfittings.com
Bud Industries, Inc.	www.budind.com
Cable Management Solutions	www.icc.com
Cablofil, Inc.	www.cablofil.com
Cantex	www.cantexinc.com
Cementex	www.cementexusa.com
Chalfant Cable Tray	www.chalfantcabletray.com
Charles R. Miller	www.charlesRmiller.com
CommScope	www.commscope.com
ConEst Software Systems	www.conest.com
Controlled Power Co.	www.controlledpwr.com
Cooper Lighting Corp.	www.cooperlighting.com
Cooper Wiring Devices	www.cooperwiringdevices.com
Copper Development Association	www.copper.org
Corning, Inc.	www.corningcablesystems.com
Cutler-Hammer/Eaton Corp.	www.cutler-hammer.eaton.com
Dabmar Lighting	www.dabmar.com
Daybrite/Capri/Omega	www.dcolighting.com
Delmar Learning	www.delmarlearning.com
Ditch Witch	www.ditchwitch.com
Dranetz BMI	www.dranetz-bmi.com
E/T Software	www.etsoftware.biz

Table continues below.

Company	Internet Address
Edwards Co.	www.edwards-signals.com
Electrical Contractor Magazine	www.ecmag.com
Electric-Find.com	www.electric-find.com
ElectricSmarts	www.electricsmarts.com
Encore Wire & Cable	www.encorewire.com
Environmental Protection Agency	www.epa.gov
Erico, Inc.	www.erico.com
Estimation Inc.	www.estimation.com
Fluke Corp.	www.fluke.com
Fluke Networks	www.flukenetworks.com
Fotec Inc.	www.fotec.com
Fulham Co., Inc.	www.fulham.com
Gamewell Corp.	www.gamewell.com
GB Electrical	www.gardnerbender.com
GE Lighting System Inc.	www.gelighting.com
General Electric Supply	www.ge.com
Genesis Cable Systems	www.genesiscable.com
Genie Industries	www.genielift.com
Grainger.com	www.grainger.com
Graybar	www.graybar.com
Greenlee/Textron	www.greenlee.textron.com
GS Metal	www.flextray.com
Hadco	www.hadcolighting.com
Halex Co.	www.halexco.com
Harger Lightning & Grounding	www.harger.com
Harris Electric Supply	www.harriselectricsupply.com
Hilti Inc.	www.us.hilti.com
Hioki USA Corp.	www.hiokiusa.com
Hoffman Commercial	www.hoffmanonline.com
Home Depot	www.homedepot.com
Hubbell Inc.	www.hubbell.com
i2 Technologies	www.i2tradeservice.com
ICC	www.icc.com

Continued

WEBSITE ADDRESSES

Company	Internet Address
IDEAL Industries	www.idealindustries.com
IGE-XAO	www.see-technical.com
ILSCO	www.ilsco.com
Independent Electrical Contractors	www.ieci.org
Intermatic	www.intermatic.com
International Association of Electrical Inspectors (IAEI)	www.iaei.org
King Safety Products	www.kingsafety.com
Klein Tools, Inc.	www.kleintools.com
Leviton Manufacturing Co.	www.leviton.com
Lightolier	www.lightolier.com
Littelfuse	www.littelfuse.com
Lowe's	www.lowes.com
Lutron Electronics Co, Inc.	www.lutron.com
M&S Systems	www.mssystems.com.
Makita Tools	www.makita.com
McCormick Systems, Inc.	www.mccormicksys.com
Milwaukee Electric Tool	www.milwaukeetools.com
National Electrical Contractors Association (NECA)	www.necanet.org
National Electrical Manufacturers Association (NEMA)	www.nema.org
National Fire Protection Association (NFPA)	www.nfpa.org
NSI Industries	www.nsipolaris.com
Occupational Safety & Health Administration (OSHA)	www.osha.gov
Optical Cable Corp.	www.occfiber.com
Ortronics	www.ortronics.com
Osram Sylvania	www.sylvania.com
Panduit Corp.	www.panduit.com
Pass & Seymour/Legrand	www.passandseymour.com
Penn-Union Corp.	www.penn-union.com
Porter-Cable Corp.	www.porter-cable.com

Table continues below.

Company	Internet Address
Progress Lighting	www.progresslighting.com
Regal Manufacturing	www.regalfittings.com
Reliance Controls Corp.	www.reliancecontrols.com
Ruud Lighting, Inc.	www.ruudlighting.com
Sea Gull Lighting Products	www.seagulllighting.com
Siemens Energy & Automation	www.sea.siemens.com
Siemon Co.	www.siemon.com
Simplex Grinnell	www.simplexgrinnell.com
Southwire Co.	www.southwire.com
Square D Co.	www.squared.com
Sunstar Lighting	www.sunstarlighting.com
Suttle	www.suttleonline.com
Technical Consumer Products, Inc.	www.tcpi.com
Tempo-Textron	www.tempo.textron.com
Thomas & Betts	www.tnb.com
TradePower	www.tradepower.com
Triplett Corporation	www.triplett.com
Underwriters Laboratories (UL)	www.ul.com
Unicom Electric	www.unicomlink.com
United Rentals Inc.	www.unitedrentals.com
Universal Lighting	www.universalballast.com
US House of Representatives	www.house.gov
US Senate	www.senate.gov
Vision InfoSoft Corp.	www.visioninfosoft.com
Vista Professional Outdoor Lighting	www.vistapro.com
W.A.C. Lighting	www.waclighting.com
WH Salisbury & Co.	www.whsalisbury.com
White House	www.whitehouse.gov
Wiremold Co.	www.wiremold.com
Zircon Corp.	www.zircon.com

CHAPTER 22

FIRST AID EMERGENCY PROCEDURES

Chapter 22 provides first aid emergency procedures for electricians in the field to easily reference. The material in Chapter 22 is courtesy of Coyne First Aid, Inc.

COYNE FIRST AID, INC. Phone 215.723.0926
P.O. Box 390 Fax 215.723.0929
Sellersville, PA 18960 www.coynefirstaid.com

NOTE: This reference section is intended only as a source of immediate reference in a medical emergency or for periodic review. By itself it does not constitute complete or comprehensive training.

Whenever you approach the scene of an accident or someone who has just been injured or taken ill, always assess personal danger. Before proceeding, determine the cause of the problem and make certain that there are no unsafe conditions that might require immediate attention.

WHEN PROVIDING FIRST AID ASSISTANCE, AVOID COMING IN DIRECT CONTACT WITH ANOTHER'S BLOOD OR OTHER POTENTIALLY DANGEROUS BODILY FLUIDS. PRACTICE UNIVERSAL PRECAUTIONS.

In any medical emergency proper care of the patient follows *Priorities for Proper Care:*

1. Open Airway
2. Restore Breathing and Circulation
3. Stop Bleeding
4. Treat for Traumatic Shock

CAUTION... DO NOT MOVE THE INJURED PERSON UNLESS:

- There is danger of further injury.
- You must do so to effectively provide lifesaving care.

FIRST AID EMERGENCY PROCEDURES

RESCUE BREATHING—ADULT

Signs/Symptoms:

- No visible signs of normal breathing. Failure of the chest or upper abdomen to move.
- No detectable movement of air from mouth or nose.

Emergency Treatment:

1. Tap shoulders and shout "Are you OK?"
2. If no response CALL EMS
3. Without moving patient Look, Listen and Feel for breathing for about 5 seconds.
   - If breathing normally, put in Recovery Position if no indications of head, neck or back injury.
   - If not breathing normally, place the patient on back while supporting head and neck.
4. Open airway by tilting head back & lifting chin. Reassess breathing for about 5 seconds.
   - If patient is breathing normally put in Recovery Position.
   - If not breathing normally, give two slow full breaths.
5. Look for signs of circulation (example: coughing, movement, normal breathing)
   - If there are signs of circulation but no normal breathing, begin Rescue Breathing. Give 1 breath every 5 seconds, about 12 breaths per minute.
   - If no signs of circulation, begin CPR.
6. Reassess breathing and circulation after first minute, and every few minutes thereafter.

ONE RESCUER CPR—ADULT

Signs/Symptoms:

No visible signs of normal breathing . . . failure of chest or upper abdomen to move.

- No visible signs of circulation.
- Emergency Treatment:

1. Tap shoulders and shout "Are you OK"? If no response CALL EMS.
2. Without moving patient Look, Listen and Feel for breathing for about 5 seconds.
   - If breathing normally, put in Recovery Position if no indications of head, neck or back injury.
   - If not breathing normally, place the patient on back while supporting head and neck.

3. Open airway by tilting head back & lifting chin. Reassess Breathing.
   - If patient is breathing normally, put in Recovery Position.
   - If not breathing normally, give two slow full breaths.
4. Look for signs of circulation. Example; coughing, movement, normal breathing.
5. If no signs of circulation begin CPR.
   - Give 15 chest compressions (push straight down on chest 1-1/2 to 2 in.) and 2 slow breaths about 2 seconds each.
6. After 4 cycles of breaths & compressions, reassess for signs of circulation.
   - If no signs of circulation, resume CPR beginning with 15 compressions.
   - If there are signs of circulation, but no breathing, begin Rescue Breathing.
   - If there is breathing and signs of circulation, place in Recovery Position.

CPR—ARRIVAL OF SECOND RESCUER

Essentially, two rescuers will alternate in performing one-rescuer CPR. If the 2nd rescuer arrives when CPR is in progress, he should first be certain that the EMS has been called, then assist in CPR by maintaining open airway, and making certain that ventilations and compressions being given are adequate.

Once begun, continue CPR until:

1. Circulation and breathing have been restored.
2. Resuscitation efforts are transferred to a properly qualified individual, who continues basic life support efforts.
3. Exhaustion makes further resuscitation efforts impossible, or places other lives in danger.

OBSTRUCTED AIRWAY—ADULT

Assessment:

1. Ask: Are you choking?
   - If person can cough forcefully and/or speak, encourage him/her to cough and DO NOT INTERFERE. Watch closely, if problem continues, notify EMS.
   - If person is coughing weakly, making wheezing, crowing noises, TREAT AS COMPLETE OBSTRUCTION.

Emergency Treatment:

1. Begin abdominal thrusts:
   - Stand behind choking person; wrap your arms around him/her.
   - Make a fist with one hand & place thumb side just above navel & well below xiphoid.

FIRST AID EMERGENCY PROCEDURES

- Grasp fist with the other hand & press into the abdomen with quick, upward thrusts.
- Repeat thrusts until object is expelled.
- Wounds/Bleeding

EXTERNAL BLEEDING

Minor Bleeding

Emergency Treatment for Small Cuts and Abrasions:
- Wash the wound with clean running water
- Apply an antiseptic
- Cover with a sterile dressing to keep the wound clean.

Severe Bleeding

Emergency Treatment for Major Bleeding:

Practice Universal Precautions to avoid direct unprotected contact with patient's blood.

3 Techniques to stop severe bleeding:

1. *Direct Pressure on the Wound*—press palm of hand over the entire area of the wound using a thick gauze compress, cleanest cloth available or gloved hand or fingers. If blood soaks through add additional layers while continuing to apply direct pressure.
2. *Elevation*—If there is no evidence of a fracture, a severely bleeding open wound of the head, neck, arm or leg, should be raised above the level of the patient's heart. Direct pressure should be maintained on the wound.
3. *Pressure on the Supplying Artery* (a last resort for severe bleeding only)—If direct pressure and elevation have not stopped severe bleeding, continue both while applying pressure on the supplying artery.

Note:

- Keep the wound clean to avoid infection.
- Treat for traumatic shock.
- Notify the EMS for all serious injuries.

INTERNAL BLEEDING

Signs/Symptoms

- Eyes dull, pupils enlarged
- Cold, clammy pale skin

- Weak, rapid pulse
- Vomiting or coughing blood or passage of blood in urine or by rectum
- Uncontrolled restlessness and excessive thirst
- Patient may be faint, drowsy, confused or unconscious.
- Body areas that are tender, swollen, bruised or hard.

Emergency Treatment:

- Maintain open airway. Perform Rescue Breathing if required.
- Notify the EMS
- Treat for traumatic shock due to blood loss, keep patient lying down.
- Do NOT give anything by mouth.

TRAUMATIC SHOCK

The condition that exists when there is inadequate blood pressure and blood flow to supply the organs of the body. Shock usually occurs when there is a massive loss of blood from either injury or internal bleeding.

Signs/Symptoms:

- A change in mental state, such as confusion.
- Nail beds will be pale or bluish.
- Skin may be cool and sweaty.
- Pulse may be weak and rapid.
- There may be nausea or vomiting.
- Person may be dizzy or lightheaded.

Emergency Treatment:

- Be certain patient is breathing and major external bleeding has been stopped. Reassure the patient.
- Call the EMS
- If there are possible neck or back injuries, DO NOT MOVE the patient unless there is danger of further injury or it is necessary to administer lifesaving first aid.
- If the injury permits, keep the patient lying down and covered lightly to prevent the loss of body heat. If the patient is lying on a cold and/or damp surface, place a pad or blanket under him/her. DO NOT ADD EXTRA HEAT.
- If no head, neck, back, hip or leg injuries are suspected, elevate feet *at least* 12 to 15 inches.

FIRST AID EMERGENCY PROCEDURES

BURNS

First Degree Burns Signs/Symptoms:

Skin is red, hypersensitive and painful. There are no blisters and usually no swelling.

Emergency Treatment:

- Cool the burn using JEL BURN DRESSING for 10–15 minutes, or by rinsing under cool water.
- DO NOT USE ICE as this constricts blood flow and may cause a hypothermic injury
- Topical moisturizers are helpful.

Second Degree Burns Signs/Symptoms:

Blisters will form on the skin. If the blisters break the skin underneath is red, painful and moist. There will be swelling on the surrounding areas.

Emergency Treatment:

- Extinguish the fire; cool injury by rinsing with cool water or using JEL BURN DRESSING for 15–20 minutes.
- Wrap blistered areas in dry dressings or place blankets over the patient.
- Keep patient warm and transfer to the nearest burn center.

Third Degree Burns Signs/Symptoms:

A third degree burn will destroy both the outer and inner layers of the skin. The outer layer (epidermis) is burned away and the underlying layer (dermis) is white and charred. The wound is unfeeling and dry. There will be swelling of the surrounding tissue.

Emergency Treatment:

- Remove from the source of the burn.
- Cool the injury by rinsing with cool water or by using JEL BURN DRESSING for 15–20 minutes or with JEL BURN BLANKETS. Wrap the affected areas with dry dressings or cover with blankets.
- Call the EMS.
- If there are facial burns or nasal/facial hair singeing the patient is at risk for loss of airway due to swelling.
- Keep patient warm and transfer to the nearest burn center.

ELECTRICAL BURNS/INJURY

Do not touch the person who has suffered the electrical shock and/or burns until he is clear of the source of the electricity. Many electrical injuries are associated with falls and contractures of large muscles which may fracture bones. Protect the patient's spine with a collar and back board.

Emergency Treatment:

- Call the EMS. If indicated, and you are the only rescuer present, provide one minute of CPR before doing so. Phone fast.
- Check for breathing and signs of circulation. Provide Rescue Breathing and CPR as needed. Do not move the casualty unless absolutely necessary. Ventricular arrhythmias are also common and should be treated by an A.E.D. as soon as possible.
- Cool any external burns with JEL BURN DRESSING. Then transfer to the burn center.
- Continue to monitor breathing and circulation.

CHEMICAL BURNS

Before working with any chemical and/or treating a patient for harmful chemical exposure, be familiar with the appropriate Material Safety Data Sheet (MSDS). Most chemicals can be irrigated with water, but some do react unfavorably. Always be sure to protect yourself with gowns, gloves, mask, etc.

Emergency Treatment:

- Brush off all loose chemicals; remove patient's clothing.
- Wash chemical off the skin with soap and water as quickly as possible. Continue for 30 minutes or until chemical has been removed.
- If there are burns, cool with JEL BURN DRESSING, then transfer to burn center.
- If chemical is in the eye, remove contact lenses, flush water into open eye from the inner corner to the outer corner until all chemicals are removed. Do not let patient rub his eye, bandage with clean dressing.
- Bring patient to emergency room where eye specialists are available.

372

FIRST AID EMERGENCY PROCEDURES

POISONS

Signs/Symptoms:

Someone who has been poisoned may exhibit one or more of the following:

- Unusual stains or burns, around the lips or mouth
- Unusual breath odor
- Headache, nausea or vomiting
- Coma, convulsions, delirium, shock, agitation

Emergency Treatment:

Inhaled Poison

- Get patient to fresh air immediately but do not place yourself in risk of inhaling the poison to do so.
- Notify EMS.
- Maintain open airway. Monitor pulse and breathing. Provide Rescue Breathing/ CPR as needed.

Swallowed Poison

- Search scene for clues (pills, drug paraphernalia, and witnesses.) Determine what was ingested, including name of product, ingredients, amount and time taken. Always bring container to the hospital.
- Telephone poison control center or EMS.
- Follow instructions given; DO NOT GIVE ANYTHING BY MOUTH UNTIL TOLD TO DO SO.

HEAT EXHAUSTION

Signs/Symptoms:

- Patient may complain of feeling faint, fatigued, a headache or exhibit confusion or agitation.
- Patient's body temperature is elevated or nearly normal.
- Skin is moist and warm.

Emergency Treatment:

- Move patient to a cool, shady place.
- Give liquids and have patient rest.

HEAT STROKE

Signs/Symptoms:

- A change in patient's mental functions. (May appear very fatigued, lethargic, confused, delirious or become unconscious.)
- Body temperature is very hot. Skin is hot, dry and flushed. Ability to sweat may be lost.
- Pulse is usually rapid and strong, becoming weaker. Breathing will be rapid.
- Nausea, vomiting, and diarrhea may be present.

Emergency Treatment:

- Move patient to cool shady place. Disrobe patient as much as possible.
- Apply ice packs to wrists, ankles, neck, armpits and groin areas to cool body temperature. Wet the entire body with cool water. If conscious, give cool water to drink (2 ounces every 5 seconds as tolerated.)
- Notify EMS. Continue efforts to cool the patient.

FROSTBITE

Signs/Symptoms:

- Skin appears glossy, chalky white or grayish yellow.
- Affected area becomes cold and numb. May sting and burn
- Patient's mental functions may change (for example: appear fatigued, confused or lose consciousness as hypothermia develops)

Emergency Treatment:

- Cover the frostbitten area and move the patient to a warm place. Do not attempt to thaw the area if it will be refrozen. Remove restrictive or wet clothes.
- Warm the affected area quickly by immersion in warm (not hot) water. If warm water immersion is not possible, wrap the affected area gently in warm material such as sheets and woolen blankets.
- If conscious, give the patient warm non-alcoholic drink.
- Get medical attention.
- DO NOT rub the affected part. DO NOT apply a hot water bottle or heat lamp. Keep away from close proximity to high heat.

FIRST AID EMERGENCY PROCEDURES

HYPOTHERMIA

Signs/Symptoms:

- Abnormally low body temperature, below 93 degrees F. Signs and symptoms may vary according to the degree of hypothermia as described below:
- 93 degrees; confusion, amnesia.
- 91-1/2 degrees; loss of muscular coordination, apathy.
- 89-1/2 degrees; stuporous.
- 88 degrees; shivering ceases.
- 82-1/2 degrees; ventricular fibrillation.
- 80-1/2 degrees; reflexes/voluntary motion ceases.

Emergency Treatment:

- Notify EMS.
- Open airway, check breathing and signs of circulation. If necessary, begin Rescue Breathing and/or CPR.
- Move person to a warm place. Remove all wet clothing and cover him in blankets and/or dry clothing.
- If conscious, give him warm, non-alcoholic fluids.
- DO NOT Immerse in a warm water bath.

NOTE: Following initial training, Coyne First Aid recommends an annual refresher in CPR and retraining in First Aid every three years. This is in accordance with OSHA Guideline CPL 2-2.53